Manual of
Physico-Chemical Analysis of Aquatic Sediments

Edited by

Alena Mudroch
José M. Azcue
Paul Mudroch

LEWIS PUBLISHERS

Boca Raton   New York   London   Tokyo

Acquiring Editor:	Neil Levine
Project Editor:	Carole Sweatman
Marketing Manager:	Greg Daurelle
Direct Marketing Manager:	Arline Massey
Typesetter:	Pamela Morrell
Cover design:	Denise Craig
PrePress:	Kevin Luong
Manufacturing:	Sheri Schwartz

Library of Congress Cataloging-in-Publication Data

Manual of physico-chemical analysis of aquatic sediments / edited by
   Alena Mudroch, Jose Azcue, Paul Mudroch.
       p.   cm.
   Includes bibliographical references and index.
   ISBN 1-56670-155-4
   1. River sediments--Analysis. 2. Lake sediments--Analysis.
   3. Marine sediments--Analysis. I. Mudroch, Alena. II. Azcue, José M.
   III. Mudroch, Paul.
   GB1399.6.M36 1996
   551.3′04—dc20

                                       96-27690
                                            CIP

No claim to original U.S. Government works
International Standard Book Number 1-56670-155-4
Library of Congress Card Number 96-27690
Printed in the United States of America 1 2 3 4 5 6 7 8 9 0
Printed on acid-free paper

Preface

After decades of intensive industrial development accompanied by the production of thousands of different chemicals, the world has become concerned with the preservation of existing natural resources and contamination of air, soil, and water. Sustainable industrial development and recycling of used materials appear to be the first steps in attempting to preserve the earth's natural resources. However, during the past, contaminants have entered the environment through intensive industrial activities and remain in the earth's ecosystem. The duration of their presence in the ecosystem depends on their persistence and the ability of currently developing techniques for remediation of contaminated compartments of the ecosystem.

The quality of aquatic ecosystems is of great interest to the entire world. Water, a part of the aquatic ecosystem, is one of the most important life-supporting media on the earth. The effects of contaminants in stream, lake, and ocean waters on human health and aquatic life have been studied for many years. Knowledge of the fate and pathways of different contaminants in aquatic ecosystems has been used in the development of remedial techniques for cleaning the world's streams and lakes. Many studies have shown that aquatic sediments are a sink as well as a source of contaminants. Sediment-associated contaminants may be released under specific conditions into overlying waters and pose a hazard to aquatic life and human health. Therefore, characterization of sediments and the identification of quality and quantity of contaminants in the sediments are very important in the assessment of sediment quality, in predictions of release of the sediment contaminants into water, and in the development of techniques for remediation of aquatic ecosystems.

The objective of this book is to provide, for those involved in the evaluation of sediment quality, information on analytical techniques used in the determination of physico-chemical properties of aquatic sediments. A second book is planned for the near future, containing information on recently developed methods for sediment bioassessment.

Alena Mudroch
José M. Azcue
Paul Mudroch

Acknowledgments

We wish to thank all the authors for their efforts in the preparation of the chapters and for their excellent cooperation during editing of the manuscripts. Further, we would like to express our thanks to the many people who assisted in the preparation of this book. In particular, we are grateful to Ms. Angela K. Lee, National Water Research Institute (NWRI), Burlington, Ontario, for all the (re)typing and formatting of the manuscript, which she undertook with great skill, patience, and dedication. We would like to give special recognition and greatly appreciate the editorial assistance of Mrs. Dianne Crabtree, also of NWRI. The excellent contribution of Graphic Arts Unit members of NWRI are also greatly acknowledged.

Alena Mudroch
José M. Azcue
Paul Mudroch

Editors

Alena Mudroch, M.Sc., is a research scientist with the National Water Research Institute, Environment Canada, Burlington, Ontario. Mrs. Mudroch graduated with a diploma from the Chemistry Department, State College, Prague, Czech Republic, and obtained her M.Sc. degree in 1974 from the Department of Geology, McMaster University, Hamilton, Ontario.

Currently, Mrs. Mudroch is a member of the Canadian Association on Water Pollution Research and Control and of the International Association for Great Lakes Research (IAGLR), where she served as President (1989–90). She holds membership in several national and international committees and is the Chief of the Sediment Remediation Project, Aquatic Ecosystem Restoration Branch, National Water Research Institute. Mrs. Mudroch has published over 100 scientific papers and reports and has presented over 50 papers at national and international conferences and workshops. She is co-editor and co-author of the books *Handbook of Techniques for Aquatic Sediments Sampling*, published in 1991 by CRC Press, Inc., Boca Raton, FL, *Handbook of Techniques for Aquatic Sediments Sampling*, second edition, published in 1994 by Lewis Publishers/CRC Press, Inc., Boca Raton, FL, and *Manual of Aquatic Sediment Sampling*, published in 1995 by Lewis Publishers/CRC Press, Inc., Boca Raton, FL. Her current major research interests include the characterization of aquatic sediments, defining the role of the pathways, fate, and effects of sediment-associated contaminants in aquatic ecosystems, and remediation of contaminated sediments.

José M. Azcue, Ph.D., is a research scientist with the National Water Research Institute, Environment Canada, Burlington, Ontario. In 1981, Dr. Azcue graduated from the University of Basque Country in Bilbao, Spain, with a B.Sc. degree in biology and ecology. In 1983, he received a Technical Agronomist diploma from the Ministry of Agronomy, Spain. He obtained his M.Sc. degree in 1987, in biophysics from the Federal University of Rio de Janeiro and in 1992, his Ph.D. degree in geochemistry from the University of Waterloo, having conducted his research on the mobility of arsenic in abandoned mine tailings.

During 1990, he lectured for the FURJ, where he co-organized an international course entitled "Sampling of Aquatic Environments". Among other publications, Dr. Azcue is co-author of the books *Metales en Sistemas Biologicos*, published in 1992 by the University of Barcelona, Spain, and *Manual of Aquatic Sediment Sampling*, published in 1995 by Lewis Publishers/CRC Press, Inc., Boca Raton, FL. His current major research interests include geochemical cycling and analytical determination of contaminants in the environment, industrial and residential waste treatment, technology transfer, and environmental education.

Paul Mudroch, B.Sc., is a physical scientist with the Federal Programs Division, Environmental Protection Service, Environment Canada, Nepean, Ontario. Mr. Mudroch obtained his B.Sc. degree in geology and environmental studies in 1978 from the Department of Geology, University of Pennsylvania, Philadelphia.

His work experience includes assessment of contamination of the aquatic environment at abandoned mining sites and evaluation of the effects of oil and gas exploration on sediment quality in the Beaufort Sea. For 4 years he supervised analysis of different environmental materials in the laboratory of Indian Affairs and Northern Development, Yellowknife, Northwest Territories. His current activities include assessment of dredging projects at the federal facilities in Ontario, particularly the methods of sediment sampling and evaluation of sediment quality.

Contributors

Haig Agemian is presently Chief of Environmental Chemistry at the National Laboratory for Environmental Testing of Environment Canada. He has over 20 years experience in various areas of environmental analytical chemistry, including method development, laboratory automation and information management systems, quality assurance and laboratory accreditation. He received degrees in physics and chemistry from Carleton University in 1970 and 1972.

Pierre Brassard is assistant professor of geology and chemistry at McMaster University in Hamilton, Ontario. He studies the exchange of metals and contaminants with aquatic particles and how the release and adsorption of metals are affected by the formation of flocs. In collaboration with the National Water Research Institute, he also examines the mechanisms that promote oxidation of acid mine wastes in order to find efficient methods for their neutralization and safe disposal.

Dennis R. Gere is a senior applications chemist in the supercritical fluid extraction laboratory at the Hewlett-Packard Little Falls Site in Wilmington, DE. Dr. Gere has been with HP for 22 years in several different capacities and locations. He has been involved in HPLC, SFC, and SFE development, marketing, and customer applications. He is also active in presenting training courses in HPLC, SFE, and other analytical technologies. Dr. Gere is a native of Iowa, and holds a B.S. degree from Iowa State University. He also has a M.S. and a Ph.D. from Kansas State University. Previously, he served 3 years in the U.S.A.F. He also worked for Varian Aerograph and Suprex Corporation. He learned the basic principles of science while working as a lab technician for 4 years in a sewage treatment plant in Ames, Iowa.

José Marcus Godoy obtained a B.Sc. in chemistry from the Catholic University of Rio de Janeiro, a M.Sc. degree in nuclear engineering from the Federal University of Rio de Janeiro and a Ph.D. in radiochemistry from the Technical University of Munich. Dr. Godoy is a senior researcher at the Radiological Protection and Dosimetry Institute in Rio de Janeiro and professor at the Catholic University of Rio de Janeiro. For the last 20 years he has been involved with environmental radio-chemistry and the application of nuclear techniques to environmental studies.

Gwendy E.M. Hall was educated at Exeter University and Imperial College, London. She joined the Geological Survey of Canada in 1973 as a research chemist in applied geochemistry. In 1978 she became head of the Geochemistry Laboratories, whose mandate gradually changed from one of providing routine analytical services to geochemists within the GSC to that of research into new methods and applications. This laboratory focuses on the techniques of ICP mass spectrometry and anodic stripping voltammetry. Realizing the potential of ICP-MS, Dr. Hall promoted government funding of the Canadian firm, Sciex, in the early 1980s, leading to the introduction of the first commercial instrument (the Elan) in 1983. Thus, her laboratory was responsible for much of the early application of ICP-MS to the analysis of geological materials.

Recent research has centered on the development of methods to enhance the capability of ICP-MS for the rapid and accurate analysis of fresh water for more than 50 elements to levels below their natural abundances. She has applied these methods in hydrogeochemical surveys, involving the collection of thousands of waters, in support of both exploration and environmental studies. Another area of activity recently has been in the development and application of ICP-MS to the analysis of selective leaches, thus providing information (at ppb levels in sediments and soils) crucial to interpretation of element sources and origin. Gwendy has published 130 scientific papers (including 5 book chapters), sits on the editorial boards of *Chemical Geology* and the *Journal of Geochemical Exploration* and is past president of the Association of Exploration Geochemists (AEG). She now heads the "Metals in the Environment" program at the GSC and continues analytical research in both exploration and environmental geochemistry.

Hing-Biu (Bill) Lee is a research scientist in the Priority Substances Project, Aquatic Ecosystem Protection Branch, National Water Research Institute, Environment Canada, located at Canada Centre for Inland Waters in Burlington, Ontario. Dr. Lee was born in Hong Kong and has been working at CCIW since 1978. He holds a B.Sc. degree from the Chinese University of Hong Kong and a Ph.D. degree from McMaster University, both in chemistry. In his earlier years at CCIW, he worked as a quality assurance chemist and was responsible for the development of certified reference materials for toxic organics in sediments, as well as the implementation of several national and international interlaboratory QA/QC programs for organics. His current research interests involve the development of analytical methods for priority substances in water and sediments. Dr. Lee has over 50 journal publications, 30 scientific reports, and several review articles in these areas.

Patricia J. Lindsay is the supervisor of the Sedimentology Laboratory of the Terrain Sciences Division, Geological Survey of Canada, Ottawa, Ontario, where she has 20 years experience in physical property testing. She has directed a number of technique developments that have been adopted by private industries specializing in sample preparation for geochemical analysis and heavy mineral separations. She graduated as an Environmental Engineering Technician from Fanshawe College, London, Ontario, in 1975.

Ken F. Maley completed a B.Sc. at Trent University in environmental science and biology in 1987. His work at Lakefield Research over the past 8 years has led to specialization in geological, metallurgical, and on-stream process applications of XRF. Current areas of investigation involve automation of sample preparation and data handling.

Jeanne B. Percival is a research scientist in the Mineral Resources Division of the Geological Survey of Canada (GSC), Ottawa, Ontario. Her studies in geology (B.Sc. Concordia University) and geography (M.Sc. Queen's University) provided opportunity to do environmental research over the past 20 years. After teaching chemistry in high school, she joined the radioactive waste disposal program of Atomic Energy of Canada Limited in Ottawa at the GSC. Following her work in pore structure and rock properties, she began doctoral research on the Cigar Lake uranium deposit at Carleton University. Dr. Percival continues her research in environmental geochemistry and clay mineralogy. She is currently involved in projects related to abandoned mine sites (Athabasca basin and Cobalt, Ontario), Venice lagoonal sediments, and various sea floor studies.

Contents

Chapter 1 Introduction 1
Alena Mudroch, José M. Azcue, and Paul Mudroch
References 6

**Chapter 2 Measurement of Physical Properties of
Sediments** 7
Jeanne B. Percival and Patricia J. Lindsay
2.1 Introduction 7
2.2 Sampling Considerations 9
2.3 Bulk or Index Properties 10
    2.3.1 Introduction 10
    2.3.2 Bulk Density 11
        2.3.2.1 Gravimetric Methods 12
        2.3.2.2 Radiation Method 14
    2.3.3 Water Content 15
        2.3.3.1 Direct Methods 15
        2.3.3.2 Indirect Methods 16
        2.3.3.3 Time Domain Reflectometry 16
        2.3.3.4 Radiation Method 18
    2.3.4 Porosity 18
        2.3.4.1 Indirect Methods 18
    2.3.5 Recommendations 19
2.4 Particle Size Distribution 20
    2.4.1 Introduction 20
    2.4.2 Particle Size Analysis 21
        2.4.2.1 Sample Preparation 21
        2.4.2.2 Classical Methods 23
            2.4.2.2.1 Sieve Analysis 23
            2.4.2.2.2 Pipette Method 25
        2.4.2.3 Instrumental Methods 26
            2.4.2.3.1 Electrical Sensing Zone
                        (Coulter Counter®) 26
            2.4.2.3.2 X-Ray Sedimentation
                        (Sedigraph®) 26

2.4.2.3.3   Laser-Time of Transition
          Theory (Brinkman PSA®)............29
2.4.2.3.4   Laser Diffraction
          Spectroscopy 30
2.4.2.3.5   Image Analysis 31
  2.4.3   Recommendations...31
2.5   Specific Surface Area 34
  2.5.1   Introduction..34
  2.5.2   Methods for Determination of Specific
        Surface Area 35
      2.5.2.1   Ethylene Glycol Monoethyl Ether
            Method 36
      2.5.2.2   Methylene Blue Method..............................37
  2.5.3   Recommendations...37
Acknowledgments 38
References..38

Chapter 3   **Measurement of Eh and pH in Aquatic**
         Sediments..47
         Pierre Brassard
3.1   Introduction...47
3.2   Measurement of Eh..48
  3.2.1   General Considerations..48
  3.2.2   Methods Used for Field Measurements of Eh 51
3.3   Measurement of pH...55
  3.3.1   General Considerations..55
  3.3.2   The Suspension Effect 56
3.4   General Procedure for Measurements of Eh and pH in
      Bottom Sediments 59
  3.4.1   Equipment and Solutions Used in the
        Measurements...59
  3.4.2   Preparation of the Equipment Before the
        Field Trip...60
  3.4.3   Measurements in the Field 61
3.5   Reference Electrode...62
3.6   Conclusions 64
References..65

Chapter 4   **Rapid and Cost-Effective Analysis for Aquatic Sediment**
         Samples by X-Ray Fluorescence Spectrometry.............69
         Ken F. Maley
4.1   Introduction...69
4.2   Theory of X-Ray Fluorescence Spectrometry 70
  4.2.1   Excitation 70
  4.2.2   Lines and Notation 71

4.2.3   Absorption..71
4.2.4   Scatter..73
4.3   Practical Application for Analysis of Environmental
      Samples..73
      4.3.1   Detection and Counting...73
      4.3.2   Sample Preparation..75
      4.3.3   Calibration..77
4.4   Discussion..80
      4.4.1   Advantages and Limitations......................................80
Acknowledgments..81
References..81

**Chapter 5   Determination of Trace Elements in
             Sediments**...85
             Gwendy E.M. Hall
5.1   Introduction...85
      5.1.1   Standard Reference Materials...................................89
5.2   Decomposition Techniques...91
      5.2.1   Acid Digestion...92
             5.2.1.1   Acid Properties...92
             5.2.1.2   Open Systems...94
             5.2.1.3   Closed Systems...96
      5.2.2   Fusion..100
5.3   Atomic Absorption Spectrometry...................................102
      5.3.1   Principles and Instrumentation of Atomic
             Absorption Spectrometry.......................................102
      5.3.2   Applications of Atomic Absorption
             Spectrometry..103
             5.3.2.1   Flame Atomic Absorption
                       Spectrometry...103
             5.3.2.2   Quartz Tube Atomic Absorption
                       Spectrometry...104
             5.3.2.3   Graphite Furnace Atomic Absorption
                       Spectrometry...107
             5.3.2.4   Slurry Atomic Absorption
                       Spectrometry...111
5.4   Inductively Coupled Plasma Atomic Emission
      Spectrometry for Sediment Analysis...............................112
      5.4.1   Principles and Instrumentation of ICP-AES...........112
      5.4.2   Optimization in ICP-AES..115
      5.4.3   Alternatives to Conventional Nebulization............117
      5.4.4   Applications of ICP-AES..118
5.5   Inductively Coupled Plasma Mass Spectrometry.............124
      5.5.1   Principles and Instrumentation of ICP-MS.............125
      5.5.2   Alternatives to Conventional Nebulization............127

      5.5.3   Applications of ICP-MS 131
References 136

Chapter 6   Neutron Activation Analysis 147
      José Marcus Godoy
  6.1   Introduction 147
  6.2   Theory 147
  6.3   Methods in Instrumental Neutron Activation
        Analysis 154
      6.3.1   Relative Method 154
      6.3.2   Absolute Method 154
  6.4   Material and Methods in Activation Analysis 156
  6.5   Analytical Sensitivity, Detection Limits, and
        Accuracy 163
  6.6   Application to Sediment Samples 167
  6.7   Final Remarks 168
References 168

**Chapter 7   Determination of Nutrients in Aquatic
             Sediments** 175
      Haig Agemian
  7.1   Introduction 175
  7.2   Chemical Forms of Nutrients in Aquatic Sediments:
        Literature Review 178
      7.2.1   Nutrients in Suspended Particulate Matter
            (i.e., Suspended Sediments) 178
      7.2.2   Chemical Forms of Carbon, Nitrogen, and
            Phosphorus in Sediments 179
            7.2.2.1   Carbon Species in Sediments 179
            7.2.2.2   Nitrogen Species in Sediments 180
            7.2.2.3   Phosphorus Species in Sediments 180
  7.3   Methods Used in Determination of Carbon, Nitrogen,
        and Phosphorus in Sediments: Literature Review 181
      7.3.1   Determination of Carbon in Sediments 181
      7.3.2   Determination of Nitrogen in Sediments 182
      7.3.3   Determination of Phosphorus in Sediments 184
  7.4   Considerations for the Selection of Methods for
        Determination of Nutrients in Sediments 188
      7.4.1   Carbon 189
      7.4.2   Nitrogen 190
      7.4.3   Phosphorus 191
  7.5   Description of Selected Methods for Determination of
        Nutrients in Sediments 193
      7.5.1   Determination of Carbon and Nitrogen by Dry
            Combustion 195

7.5.1.1   Principle 195
7.5.1.2   Apparatus 197
7.5.1.3   Reagents 197
7.5.1.4   Procedure 197
        7.5.1.4.1   Sample Preparation.................... 197
        7.5.1.4.2   Instrumental Analysis 198
        7.5.1.4.3   Analysis of Suspended Matter from
                    Filtered Water Samples 198
        7.5.1.4.4   Analysis of Sediments............... 198
7.5.1.5   Comments.. 198
7.5.2   Determination of Nitrogen by Kjeldahl
Digestion... 198
7.5.2.1   Principle 198
7.5.2.2   Apparatus 199
7.5.2.3   Reagents 199
        7.5.2.3.1   Reagents for Sediment
                    Digestion....................................... 199
        7.5.2.3.2   Reagents for Ammonia
                    Analysis 202
7.5.2.4   Procedure 202
        7.5.2.4.1   Sediment Digestion.................... 202
        7.5.2.4.2   Manual Analysis.......................... 202
        7.5.2.4.3   Automated Analysis 202
7.5.2.5   Comments.. 203
7.5.3   Determination of Exchangeable Ammonia, Nitrate,
Nitrite, and Urea by KCl (2 N) Extraction................ 203
7.5.3.1   Principle 203
7.5.3.2   Apparatus 203
7.5.3.3   Reagents 205
        7.5.3.3.1   Reagents for NH_4^+ 205
        7.5.3.3.2   Reagents for NO_3^- and NO_2^- 205
        7.5.3.3.3   Reagents for Urea Analysis....... 206
7.5.3.4   Procedure 207
        7.5.3.4.1   KCl (2 N) Extraction of
                    NH_4^+, NO_3^-, and NO_2^- 207
        7.5.3.4.2   Automated or Manual Analysis
                    of NH_4^+ 207
        7.5.3.4.3   Automated Analysis of NO_3^-
                    and NO_2^- 207
        7.5.3.4.4   Manual Analysis of NO_3^-
                    and NO_2^- 207
        7.5.3.4.5   KCl (2 N) Extraction and
                    Determination of Urea 207
7.5.3.5   Comments.. 208

7.5.4   Determination of Total, Inorganic, and Organic Phosphorus by Dry Combustion Followed by Digestion by HCl (1 N) 208

    7.5.4.1   Principle 208

    7.5.4.2   Apparatus 208

    7.5.4.3   Reagents 210

        7.5.4.3.1   Phosphate Standards 210

        7.5.4.3.2   Reagents Required for Method I (Figure 7-9) 210

        7.5.4.3.3   Reagents Required for Method II (Figure 7-9) 211

        7.5.4.3.4   Reagents Required for Manual Analysis 211

    7.5.4.4   Procedure 211

        7.5.4.4.1   Ignition of Sediment 211

        7.5.4.4.2   Extraction of Orthophosphate from Sediments 212

    7.5.4.5   Analysis 212

        7.5.4.5.1   Automated Analysis 212

        7.5.4.5.2   Manual Analysis 212

    7.5.4.6   Comments 212

7.5.5   Determination of Total Phosphorus by $H_2SO_4–K_2S_2O_8$ Digestion 213

    7.5.5.1   Principle 213

    7.5.5.2   Apparatus 213

    7.5.5.3   Reagents 213

    7.5.5.4   Procedure 214

        7.5.5.4.1   Bomb Digestion 214

        7.5.5.4.2   Hot Block Digestion 214

    7.5.5.5   Analysis 214

    7.5.5.6   Comments 215

7.5.6   Determination of Phosphorus in Chemical Fractions 215

    7.5.6.1   Principle 215

    7.5.6.2   Apparatus 217

    7.5.6.3   Reagents 217

    7.5.6.4   Procedure 217

        7.5.6.4.1   Fraction A 217

        7.5.6.4.2   Fraction B 217

        7.5.6.4.3   Fraction C 218

        7.5.6.4.4   Fraction D 218

        7.5.6.4.5   Fraction E 218

    7.5.6.5   Comments 218

7.6   Data Quality Control 219

References 220

Chapter 8 Supercritical Fluid Extraction of Organic Contaminants in Sediments... 229
        Hing-Biu Lee and Dennis R. Gere
8.1   Introduction.. 229
8.2   Principles of Supercritical Fluids................................... 231
       8.2.1   Supercritical Fluid Extraction Hardware and the Flowing Fluid Process 234
       8.2.2   Advantages of the Supercritical Fluid Extraction Method.. 240
       8.2.3   Practical Candidate Fluids for Supercritical Fluid Extraction.. 240
8.3   Applications of Supercritical Fluid Extraction in Environmental Analysis... 242
       8.3.1   Polychlorinated Biphenyls (PCBs) and Chlorobenzenes 242
                8.3.1.1   Recommended Extraction Procedure for PCBs in Solid Matrices 245
                8.3.1.2   Discussion.. 246
       8.3.2   Polynuclear Aromatic Hydrocarbons (PAHs) 246
                8.3.2.1   Recommended Procedure for the Extraction of PAHs in Solid Samples........................... 249
                8.3.2.2   Discussion.. 250
       8.3.3   Total Recoverable Protroleum Hydrocarbons (TPHs) 251
                8.3.3.1   Recommended Procedure for the Extraction of TPHs from Solid Samples 252
                8.3.3.2   Discussion.. 253
       8.3.4   Pesticides and Herbicides 253
                8.3.4.1   Recommended Procedure for the Extraction of Organochlorine Insecticides................... 256
                8.3.4.2   Discussion.. 257
       8.3.5   Phenols... 257
                8.3.5.1   Recommended Procedure for the Extraction and Acetylation of Chlorinated Phenolics.................................. 259
                8.3.5.2   Discussion.. 259
       8.3.6   Resin and Fatty Acids.. 260
       8.3.7   Polychlorinated Dibenzo-p-Dioxins and Furans.. 260
       8.3.8   Organotins 261
8.4   Auxilliary Techniques in Supercritical Fluid Extraction... 263
       8.4.1   Sulfur Removal... 263
       8.4.2   Lipid Removal 264

8.4.3 *In Situ* Derivatization 264
8.4.4 Supercritical Fluid Extraction of Organics in Water
       Samples via Solid Phase Extraction 265
References... 267

Index... 275

Abbreviations

AAS	atomic absorption spectrophotometry
AMD	acid mine drainage
APDC	ammonium pyrrolidine dithiocarbamate
ASV	anodic stripping voltametry
AVS	acid-volatile sulfide
BD	bicarbonate-dithionite
BNC	base neutralizing capacity
CDB	citrate-dithionite-bicarbonate
CEC	cation exchange capacity
CHN	carbon/hydrogen/nitrogen
CID	charge injection device
COD	chemical oxygen demand
CRM	certified reference materials
CVAAS	cold vapor atomic absorption spectrometry
DCB	dithionate/citrate/bicarbonate
DCP	direct current plasma
DDP	differential pulse polargraphy
DDW	double distilled water
DIN	direct injection nebulizer
DOC	dissolved organic carbon
DSDP	Deep Sea Drilling Projects
DTPA	diethylene-triamine pentaacetic acid
EDTA	ethylene-diamine tetraacetic acid
ENAA	epithermal neutron activation analysis
ETV	electrothermal vaporization
FAAS	flame atomic absorption spectrometry
FI	flow injection
GC	gas chromatography
GFAAS	graphite furnace atomic absorption spectrometry
GRAPE	gamma ray attenuation porosity evaluation
GSC	Geological Survey of Canada
HGAAS	hydride generation inductively coupled plasma
HPLC	high-pressure liquid chromatography
IAEA	International Atomic Energy Agency

IBI	Index of Biotic Integrity
ICP	inductively coupled plasma
ICP-AES	inductively coupled plasma atomic emission spectrometry
ICP-MS	inductively coupled plasma mass spectrometry
INAA	instrumental neutron activation analysis
IP	ionization potential
IR	infrared
ISO	International Standards Organization
MIBK	methyl isobutyl ketone
MPN	most probable number
MTAA	Modern Trends in Activation Analysis
NAA	neutron activation analysis
NBS	National Bureau of Standards
NIST	National Institute of Standards and Technology
NH_3	ammonia
NH_4^+	ammonium ion
NMM	nitrate minimal medium
NO_3^-	nitrate
NO_2^-	nitrite
NRCC	National Research Council of Canada
NWRI	National Water Research Institute
ODP	Ocean Drilling Program
PAHs	polynuclear aromatic hydrocarbons
PCBs	polychlorinated byphenyls
PCR	polymerase chain reaction
POC	particulate organic carbon
PON	particulate organic nitrogen
ppb	ng/g
ppm	µg/g
ppq	fg/g
ppt	pg/g
QA/QC	quality assurance/quality control
REE	rare earth element
RM	reference material
γRNA	ribosomal ribonucleic acid
rpm	revolutions per minute
RSD	relative standard deviation
SEM/SRM	scanning electron microscopy/X-ray microanalysis
SFC	supercritical fluid chromatography
SFE	supercritical fluid extraction
SPE	solid phase extraction
SQC	sediment quality criteria
SRB	sulfate reducing bacterium
SRM	standard reference material

TDR	time domain reflectometry
TDS	total dissolved salt
TKN	total Kjeldahl nitrogen
TON	total organic nitrogen
UV/VIS	ultraviolet/visible
XPS	X-ray photoelectron spectroscopy (10–30 Å)
XRF	X-ray fluorescence spectrometry
XRPD	X-ray powder diffraction spectrometry

List of tables

Chapter 2 **Measurement of Physical Properties of Sediments**........7
Jeanne B. Percival and Patricia J. Lindsay

Table 2-1 Types of Bulk Physical Properties 8

Table 2-2 Standard Classification for Particles into Wentworth Size Classes, Phi Scale, and Comparative Sieve Sizes..22

Table 2-3 Typical Results Obtained from Image Analysis..............32

Table 2-4 Summary of Methods for Optimal Particle-Size Detection 34

Chapter 4 **Rapid and Cost-Effective Analysis for Aquatic Sediment Samples by X-Ray Fluorescence Spectrometry**..............69
Ken F. Maley

Table 4-1 Partial K and L X-ray Spectra for Ba 71

Table 4-2 Examples of Elemental Analyses and Typical Lower Limits of Detection Reported by XRF................................76

Chapter 5 **Determination of Trace Elements in Sediments**...........85
Gwendy E.M. Hall

Table 5-1 Glossary of Acronyms Used in This Chapter...................86

Table 5-2 International Sediment Standard Reference Materials (SRM). Parentheses Indicate Not All the SRMs in That Series Are Certified for That Element................................90

Table 5-3 Comparison of GFAAS Results with Accepted Values for BCSS-1; Values in ppm...108

Table 5-4 Detection Limits Quoted by Commercial Laboratories for Trace Elements Determined by ICP-AES Following an *Aqua Regia* or HF-HClO$_4$-HNO$_3$-HCl Decomposition 120

Table 5-5 Analysis of the Reference Marine Sediment, PACS-1 by ICP-MS..132

Chapter 6   **Neutron Activation Analysis** 147
José Marcus Godoy

Table 6-1   Neutron Sources for Activation Analysis 149

Table 6-2   Principal Characteristics of the Radionuclides Used in the Instrumental Neutron Activation Analysis of Sediment Samples 157

Table 6-3   Instrumental Neutron Activation Analysis Detection Limits in Relation to the Elements Average Concentration in the Earth's Crust (values in µg/g, when not indicated) 166

Chapter 8   **Supercritical Fluid Extraction of Organic Contaminants in Sediments** 229
Hing-Biu Lee and Dennis R. Gere

Table 8-1   Candidates for Supercritical Fluid Extraction Fluids 241

Table 8-2   Examples of *In Situ* Derivatization of Contaminants in Soils and Aquatic Sediments Under Supercritical Fluid Extraction Conditions 266

List of figures

Chapter 1   Introduction..1
Alena Mudroch, José M. Azcue, and Paul Mudroch
Figure 1-1   Relationship among study objectives, sediment sampling, and analysis...5

Chapter 2   **Measurement of Physical Properties of Sediments**...7
Jeanne B. Percival and Patricia J. Lindsay
Figure 2-1   Examples of time domain reflectometry (TDR) transmission lines used as soil probes: (a) simulated coaxial (unbalanced) and (b) twin-lead (balanced pair).............17
Figure 2-2   Example of a nest of metal sieves, set on an automated shaker, used for dry sieving...24
Figure 2-3   (a) Configuration of the Coulter Counter® featuring, from right to left, the sample chamber, electrical sensing zone unit, and computer. (b) Close up view of the sample chamber showing electrolyte solution with suspended particles...27
Figure 2-4   Example of a Sedigraph® unit featuring, from right to left, the automated sampler, sedigraph analyzer, and computer..28
Figure 2-5   The Brinkman Particle Size Analyzer® showing, from right to left, the sample reservoir, peristaltic pump, video monitor and computer. During analysis, images of particles can be viewed on the video monitor to ensure particle disaggregation...29
Figure 2-6   (a) Backscattered electron image of one scan-field in an image analysis automated run. (b) Example of the screen capture of the detected features in the scan-field (field width is 40 µm)...33

Chapter 3   **Measurement of EH and pH in Aquatic Sediments**...47
Pierre Brassard

Figure 3-1   The sliding subsampler used by Bagander and
             Niemistö (1978). A slice of sediment is cut from the
             core by the sliding motion of two plastic sheets. The
             subsample is protected from air contamination. For
             measurement, a probe assembly is mounted on the
             top sheet as shown...53
Figure 3-2   The simple probe used by Hargis and Twilley (1994)
             to obtain multiple Eh readings with minimal
             disturbance 54
Figure 3-3   A microprofiler design for fine pH and pCO₂ work
             at the surface of oceanic sediments.....................................57
Figure 3-4   Experimental set-up used by Jenny et al. (1950) to
             study the suspension effect. The four inserts represent
             all possible combinations for inserting a reference and a
             pH electrode in and above the sediment............................58
Figure 3-5   Designs of reference electrode junction. A) porous plug; B)
             cone and sleeve junction; C) "J" junction which relies on
             the heavier density of saturated KCl..................................63

Chapter 4   Rapid and Cost-Effective Analysis for Aquatic Sediment
             Samples by X-Ray Fluorescence Spectrometry.............69
             Ken F. Maley
Figure 4-1   X-ray absorption curve for barium showing K, L, and M
             absorption edges...72
Figure 4-2   Wavelength scan of a blank sample using a rhodium tube
             operated at 60 kV with NaI scintillation and Ar/CH₄ flow
             detection..74
Figure 4-3   Philips PW1480. A sequential wavelength dispersive
             spectrometer fitted with a 72 position sample
             changer...75
Figure 4-4   Claisse Fluxy three burner fusion system. Each unit
             produces about 60 fused disks per 8-h shift 77

Chapter 5   Determination of Trace Elements in Sediments...........85
             Gwendy E.M. Hall
Figure 5-1   Approximate detection limits in pure solution associated
             with the techniques under discussion 88
Figure 5-2   A double-beam AA spectrometer 103
Figure 5-3   Apparatus for the CVAAS measurement of Hg 105
Figure 5-4   Determination of 16 elements by AAS............................110
Figure 5-5   Conventional simultaneous ICP atomic emission
             spectrometer...113
Figure 5-6   Hydride generation system used in the author's laboratory
             at the GSC..119

Figure 5-7   Scheme used by Chemex Laboratories of Vancouver for ultratrace analysis by ICP-AES.. 123
Figure 5-8   A conventional quadrupole ICP mass spectrometer..... 125

Chapter 6   Neutron Activation Analysis... 147
José M. Godoy
Figure 6-1   Neutron spectrum in a swimming pool reactor 150
Figure 6-2   Typical Neutron Activation Analysis scheme, particularly the instrumental neutron activation analysis on the left.. 151
Figure 6-3   Gamma-ray spectrum obtained after the irradiation of a sediment sample.. 158
Figure 6-4   Germanium detectors inside lead shielding, above, and a germanium detector with a sample holder utilized to fix the sample-detector distance, below................................ 159
Figure 6-5   PC-based counting room, above, and a monitor showing a spectrum in detail, below 160
Figure 6-6   Typical intrinsic germanium detector energy vs. efficiency curve 161
Figure 6-7   Sediment samples and gold and cobalt flux monitors ready for irradiation.. 162
Figure 6-8   Detection limits range for several elements in a 250 mg sediment sample.. 165

Chapter 7   Determination of Nutrients in Aquatic Sediments... 175
Haig Agemian
Figure 7-1   The Perkin-Elmer Model 2400 CHN elemental analyzer for the dry combustion analysis of carbon and nitrogen in aquatic sediments.. 194
Figure 7-2   Technicon (Bran and Luebbe) autoanalyzer Model AA II system for the automated segmented flow analysis of ammonia, nitrate, nitrite, or phosphorus 194
Figure 7-3   Technicon (Bran and Luebbe) autoanalyzer Model TRAACS 800 system for the automated segmented flow analysis of ammonia, nitrate, nitrite, or phosphorus.. 195
Figure 7-4   Dilution manifold for the automated dilution of sediment extracts for the analysis of nitrogen and phosphorus species.. 196
Figure 7-5   Automated flow manifold for the determination of ammonia by AA II Technicon (Bran and Luebbe) autoanalyzer.. 200
Figure 7-6   Commercially available Technicon hot block digestor for the acid digestion of sediments 201

Figure 7-7   Schematic of a custom built aluminum hot block for the acid digestion of sediments 201

Figure 7-8   Automated flow manifold for the determination of nitrate plus nitrite in sediment extracts by AA II Technicon (Bran and Luebbe) autoanalyzer 204

Figure 7-9   Automated flow manifold for the determination of orthophosphate in sediment extracts by AA II Technicon (Bran and Luebbe) autoanalyzer 209

Figure 7-10 Flow chart for the fractionation of phosphorus species in aquatic sediments 216

Figure 7-11 Application of a robot arm to the automated preparation of TKN samples for hot block digestion 220

Chapter 8   Supercritical Fluid Extraction of Organic Contaminants in Sediments 229
Hing-Biu Lee and Dennis R. Gere

Figure 8-1   Phase diagram of water 231

Figure 8-2   Generic supercritical fluid extraction hardware 235

chapter one

Introduction

Alena Mudroch, José M. Azcue, and Paul Mudroch

Materials on the bottom of lakes, streams, and oceans, known as bottom sediments, originate from soil erosion and precipitation from chemical and biological processes in the water. Depending on their origin, bottom sediments contain particles of different sizes, shapes, and mineralogical and chemical composition. Prior to settling on the bottom, particles are transported in the water and become sorted and deposited according to their textural properties in different areas of the lakes, streams, and oceans. Generally, coarse materials, such as sand and pebbles, settle in the nearshore zone of a lake and in fast flowing streams. Fine-grained particles, such as silt and clay, become deposited in deep waters in areas with restricted currents. The particles accumulate on the bottom at different rates, varying between tenths of millimeters to several centimeters per year.

Sediments have been studied for many years to characterize their nature and properties for different purposes. Many analytical methods used in the general characterization of bottom sediments in environmental studies have been derived from those used in soil and sediment analysis in mineral exploration. The understanding of the origin of bottom sediments, particularly the erosion, transport, and deposition of terrestrial soil particles into lakes and streams, led to the application of analytical methods for soil analysis in characterization of bottom sediments. In addition, some techniques used in the determination of physico-chemical properties of soil, rocks, and sediments used in mineral exploration were adapted in assessment of the quality of bottom sediments in environmental studies.

Historically, the investigation of freshwater sediments developed within the past few decades has become essential in many limnological studies. In the 1920s, phosphorus and nitrogen were recognized as the nutrients that usually limit primary production in oligotrophic lakes (Juday et al., 1927). The role of these nutrients in the eutrophication of

1-56670-155-4/97/$0.00+$.50
© 1997 by CRC Press, Inc.

the lakes was summarized by Vollenweider (1968). It was recognized that eutrophication, generally known as a natural process in the aging of lakes, was accelerated by the cultural activities of man. Agricultural and urban activities were indicated as most responsible for increased sediment and nutrient inputs into the lakes. Many studies indicated that the input of phosphorus and nitrogen has been retained in the lake and stream bottom sediments. In the 1960s and 1970s, research was initiated to determine the concentrations and chemical forms of nutrients, particularly phosphorus and nitrogen, in sediments with the objective of assessing the rate of release of the nutrients retained in the sediments into the overlying water.

In these early studies, analytical methods, developed in soil science for the determination of nutrients in soils, were used in the analysis of the sediments (for example, Frink, 1969; Shukla et al., 1971). Although soil science is relatively new compared to the older disciplines of mathematics, physics, chemistry, etc., methods for chemical analysis of soils are numerous and have been continuously improved and modified over many years. In addition to the methods for the general characterization of the physico-chemical properties of soils, techniques for quantitative determination of many elements and compounds have developed, particularly in studies of plant nutrition, essentiality and toxicity, and physiological substitutions. With the recognition that bottom sediments are generally derived from soils, the application of the methods used in soil analysis to the sediments was justified.

In addition to the application of analytical techniques used in soil science, methods used in geochemical prospecting were applied to the analysis of bottom sediments. Geochemical prospecting for mineral deposits has rapidly developed in the past 30 years. Recently developed geochemical and geophysical techniques have been used frequently in most mineral exploration projects. Geochemical prospecting for minerals, as defined by common usage, includes any method of mineral exploration based on systematic measurement of one or more chemical properties of a naturally occurring material (Rose et al., 1979). Consequently, many reliable analytical methods are available for quantitative determination of elements of interest in natural materials, including aquatic sediments, to support the effectiveness of the geochemical methods of exploration.

In early environmental studies of bottom sediments, the preparation of sediment samples was developed after their retrieval from the lake, stream, or ocean floor, according to the needs of analytical methods used in soil science and mineral exploration. Most of these analytical techniques require dry material. Therefore the bottom sediments, which contain up to 95% water, were oven-, air- or freeze-dried prior to analysis. Particle size distribution and quantitative determination of major and trace elements, including inorganic and organic carbon and nitrogen in the sediments, were the most common parameters used in early environmental studies of the bottom sediments, such as the geochemical survey of the North American Great Lakes (for example, Thomas et al., 1972, 1976). However, in recognition of different conditions in the sediments deposited at various areas in the lakes, few field measurements were carried out during sediment sampling in these early studies. The field measurements were mainly the determination of pH and redox potential carried out immediately after the retrieval of the sediments from the lake bottom.

The development of the understanding of different biogeochemical processes in bottom sediments and their role in aquatic ecosystems, particularly under anoxic conditions, led to the design of different analytical techniques. Techniques for sediment sampling, sample storage, and preparation relevant to the studies of the biogeochemical processes and pathways of sediment contaminants were revised and improved (Mudroch and MacKnight, 1994; Mudroch and Azcue, 1995). The major objective was to preserve the conditions in sampled sediments similar to those on the bottom of the streams and lakes. New techniques for *in situ* sampling of sediment pore water and methods for the determination of chemical species in the sediments with respect to its redox potential were developed (Hesslein, 1976; Tessier et al., 1979; Carignan et al., 1985). In addition, many methods for quantitative determination of different organic contaminants were tested and applied in studies of sediment quality.

Recently, contaminated sediments have become one of the most important environmental issues. The main objectives of the recent studies of bottom sediments were the assessment of the sediment quality and the assessment of the effects of sediment-associated contaminants on aquatic ecosystems. The contaminants in the sediments are often identified as "in-place toxics" that can pose a high risk to the environment; their existence is a serious and costly environmental issue whose management may require a special approach, including sampling and analysis of the sediments, interpretation of the results, establishment of the guidelines, and remedial action plans. Extensive survey, monitoring, and research activities (generally very expensive) are required to assess the extent and severity of sediment contamination, to evaluate the effects of contaminated sediments on aquatic ecosystems, and to prepare a plan for proper remedial action. These activities require sampling and physico-chemical analysis of all compartments of the aquatic system, including the bottom sediments. Bioassessment of sediment quality is becoming very important in the evaluation of the quality of bottom sediments. However, a detailed description of sampling techniques, analytical methods used in the determination of the physico-chemical properties of the sediments, and techniques used in sediment bioassessment has been neglected in many reports on sediment studies.

In addition, to compare the results of the studies on bottom sediments carried out at sites with different environmental conditions and involving contamination by different compounds, identical sampling and analytical techniques and bioassays must be used.

The first step in the collection of sediment samples should be the planning of the sediment sampling program. The program should include a list of analyses and tests that need to be carried out to obtain all necessary data for the characterization of the sediments and/or assessment of sediment quality. Further, the quantity of sediments to be collected needs to be calculated to obtain sufficient material for all analyses and tests, including the quality assurance/quality control program. The use of incorrectly collected and analyzed samples may lead to a waste of money and effort and to erroneous conclusions. The importance of and the relationship among the individual steps of sediment sampling and analysis in the assessment of sediment quality are outlined in Figure 1-1. The logistics used in the preparation of the sediment sampling program were described recently in conjunction with the sediment sampling techniques (Mudroch and MacKnight, 1994; Mudroch and Azcue, 1995). In some cases, it will not be necessary to carry out all the analyses and tests outlined in Figure 1-1. The selection of the type of physico-chemical analysis and sediment bioassessment will depend on the objectives for the characterization of the sediments and/or the assessment of sediment quality.

The objective of this book is to provide sufficient information on analytical methods to those involved in the characterization of sediments for different purposes, and in studies of sediment/water interaction processes. Further, the book provides information on recently developed methods for sediment bioassessment which contribute to the evaluation of the health of aquatic ecosystems. The methods described were designed for freshwater ecosystems. However, many of them may be used or modified for investigations of the marine environment. The information is divided into the following chapters. The measurements of physical properties of sediments, such as water content, bulk density, particle size distribution, etc., are discussed in Chapter 2. A review of methods used in the determination of the redox potential and pH in sediments is given in Chapter 3. The application of X-ray fluorescence spectrometry in the determination of major and trace elements in sediments is described in Chapter 4. Analytical methods for determination of trace elements in sediments using atomic absorption spectrometry (AAS), inductively coupled plasma atomic emission spectrometry (ICP-AES), and inductively coupled plasma mass spectrometry (ICP-MS) are described in Chapter 5. Determination of trace elements by neutron activation is discussed in Chapter 6. Chapter 7 contains descriptions of analytical methods for determination of nutrients, particularly nitrogen, phosphorus, and carbon, in sediments.

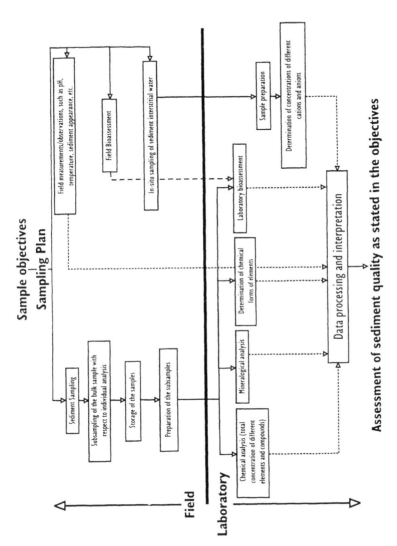

Figure 1-1 Relationship among study objectives, sediment sampling, and analysis; — ⟶ indicates transfer of subsamples for additional testing; - -⟶ indicates flow of data from completed analysis.

Recently developed methods for extraction of organic contaminants in sediments are reviewed in Chapter 8.

References

Carignan, R., Rapin, F., and Tessier, A., Sediment pore water sampling for metal analysis: a comparison of techniques, *Geochim. Cosmochim. Acta*, 49, 2493, 1985.

Frink, C.R., Fractionation of phosphorus in lake sediments, *Soil Sci. Soc. Am. Proc.*, 33, 326, 1969.

Hesslein, R.H., An *in situ* sampler for close interval pore water studies, *Limnol. Oceanogr.*, 21, 912, 1976.

Juday, C., Birge, E.A., Kemmerer, G.I., and Robinson, R.J., Phosphorus content of lake waters of northeastern Wisconsin, *Trans. Wis. Acad. Sci. Arts Lett.*, 23, 233, 1927.

Mudroch, A. and Azcue, J.M., *Manual of Aquatic Sediment Sampling*, CRC/Lewis, Boca Raton, FL, 1995, 219.

Mudroch, A. and MacKnight, S.D., Eds., *Handbook of Techniques for Aquatic Sediment Sampling*, CRC/Lewis, Boca Raton, FL, 1994, 236.

Rose, A.W., Hawkes, H.E., and Webb, J.S., *Geochemistry in Mineral Exploration*, 2nd ed., Academic Press, London, 1979, 657.

Shukla, S.S., Syers, J.K., Williams, J.D.H., Armstrong, D.E., and Harris, R.F., Sorption of inorganic phosphate by lake sediments, *Soil Sci. Soc. Am. Proc.*, 35, 344, 1971.

Tessier, A., Campbell, P.G.C., and Bisson, M., Sequential extraction procedure for the speciation of particulate trace metals, *Anal. Chem.*, 51, 844, 1979.

Thomas, R.L., Jaquet, J.-M., and Kemp, A.L.W., Surficial sediments of Lake Erie, *J. Fish. Res. Board Can.*, 33, 385, 1976.

Thomas, R.L., Kemp, A.L.W., and Lewis, C.F.M., Distribution, composition and characteristics of the surficial sediments of Lake Ontario, *J. Sed. Petrol.*, 42, 66, 1972.

Vollenweider, R.A., Scientific Fundamentals of the Eutrophication of Lakes and Flowing Waters, with Particular Reference to Nitrogen and Phosphorus as Factors in Eutrophication, Organization for Economic Cooperation and Development, Directorate for Scientific Affairs, Paris, France, 1968, 160.

chapter two

Measurement of physical properties of sediments

Jeanne B. Percival and Patricia J. Lindsay

2.1 Introduction

Sediments are complex deposits of inorganic particles, organic matter, and adsorbed and dissolved constituents. With increased interest in monitoring and measurement of aquatic effects in the receiving environment due to industrial and domestic activities, it is imperative to understand the nature and character of the sediment before embarking on detailed studies.

Sediments can be characterized in terms of several readily measured physical properties that reflect their provenance and depositional environment. Depositional and postdepositional processes (e.g., compaction, diagenesis, authigenic mineral formation), as well as mineral composition, ultimately influence the texture, bulk properties, and chemical characteristics of sediments. Harrison et al. (1964) distinguished four categories of bulk physical properties (Table 2-1). The categories include (1) primary properties: those dependent upon provenance and depositional environment; (2) secondary properties: those affected by postdepositional changes; (3) intermediate properties: those dependent upon (1) and (2); and (4) related physical properties. Additional parameters, such as compressional and shear wave velocities, or permeability and consolidation characteristics, commonly determined during the Legs of the Ocean Drilling Program (ODP) and its predecessor Deep Sea Drilling Projects (DSDP) organized by the Texas A&M University, would be classified as secondary to intermediate properties. Although the list of physical properties is extensive, it is important to note that only some are relevant to environmentally related sediment studies. For example, the geotechnical properties, such as Atterberg limits, cohesion, shear strength, and sensitivity, are useful in engineer-

1-56670-155-4/97/$0.00+$.50
© 1997 by CRC Press, Inc.

7

Table 2-1 Types of Bulk Physical Properties

Physical properties	Influence
Primary properties	
Atterberg limits	Origin of sediments
Liquid limit	Depositional conditions
Plastic limit	
Plasticity index	
Secondary properties	
Liquidity index	Post-depositional changes
Preconsolidation stress	Environmental effects, stress
(Consolidation)	
Intermediate properties	
Shear strength	Original sediment
Bulk density	Environmental changes
Bulk chemical composition	
Color	
Sensitivity	
(Shear wave velocity)	
(Compressional wave velocity)	
(Permeability)	
Related physical properties	
Specific gravity	
Grain size distribution	
Sorting	
Clay mineralogy	
Water content	
Void ratio/porosity	
Degree of saturation	
Gas content	
Temperature	
(Specific surface area)	

Note: Additional properties given in parentheses. Properties examined in
           this chapter are italicized.

Modified from Harrison, W., Lynch, M.P., and Altschaeffl, A.G., *J. Sed.
Petrol.*, 34, 727, 1964.

ing studies (Wroth and Wood, 1978; Zavoral et al., 1989) and detailed
sedimentological studies (for example, any ODP and DSDP Initial
Reports). It is important to determine the physical properties that com-
plement and, in part, control chemical characteristics in comprehensive
descriptions of sediments in aquatic environments.

The list of physical properties that can be measured in sediments
is extensive. Some properties provide basic information that is useful
in trace contaminant studies; others are particular to engineering or
geological studies. It is important in any sediment study to measure

bulk or index properties as well as particle size as a means of characterizing the sediments.

Some methods not listed in Table 2-1 may have potential benefits in aquatic sediment studies in the future. Electrical properties (i.e., resistivity, formation factor) are traditionally measured in rocks to characterize the interconnectivity of pores with respect to mobility of hydrocarbons. Electrical measurements, over a wide frequency range, are currently being tested at the Geological Survey of Canada for the purpose of identifying different types of clay minerals in unconsolidated sediments. Electrical measurements may also provide useful data on the anisotropic characteristics of sediments. Anisotropy is related to changes in texture, structure, or mineral composition. Magnetic susceptibility measurements of core samples may be used to correlate synchronous levels at many coring sites (Engstrom and Wright, 1984). The physical nature of sediments in part controls the chemical behavior. Therefore, knowledge of the physical characteristics is fundamental to any aquatic sediment study.

2.2   *Sampling considerations*

Methods to collect suspended or bottom sediments were outlined in detail in Mudroch and MacKnight (1994a) and Mudroch and Azcue (1995). Once the samples have been collected, the sediments should be inspected visually before processing. Information with respect to color, texture, structure, and odor should be recorded. Color can be quantified by using an international system, such as Munsell® Soil Color Charts. Descriptions of the samples should include measurements of the thickness of units, consistency of the material, observed structures, such as graded bedding or cross-bedding, estimated particle size, and presence of organic matter, shells, or other coarse fragments (Mudroch and Azcue, 1995). The samples should also be photographed. In some studies, X-radiography may be used to examine the detailed internal structure of sediment cores (Mudroch and MacKnight, 1994b). This is a nondestructive technique that is commonly used in stratigraphic studies.

Once the samples have been described, they must be preserved for analysis. The samples or subsamples are placed in air-tight containers or tightly sealed bags and stored at 4°C in a humidity-controlled room. Storage at these temperatures should prevent microbial degradation or oxidation (Mudroch and Azcue, 1995). Although the sediment will keep for several months without any changes to the physical properties, the samples should be analyzed as soon as possible. For detailed information regarding the preservation, processing, and storage of sediment samples, see Mudroch and Bourbonniere (1994) and Mudroch and Azcue (1995).

2.3   Bulk or index properties

2.3.1   Introduction

Bulk density, water content, and porosity are fundamental physical properties of sediment. Measurement of these properties is rapid, inexpensive, and highly reproducible (Whalley, 1981).

Bulk density measurements allow a conversion of water percentages by weight to water content by volume and are used for calculating porosity and void ratios when particle density is known (Blake and Hartge, 1986a,b). Dadey et al. (1992) stated that the determination of dry-bulk density of a sedimentary deposit is a necessary component to estimate sediment accumulation rates. They defined a mass accumulation rate as the product of a linear sedimentation rate and dry-bulk density, and suggested that this determination may be more useful than sedimentation rates alone in mass balance calculations.

Jones and Bowser (1978) identified four areas where accurate measurements of sedimentation rates are required: (1) to determine the temporal variation and mass accumulation rates of minerals, organic matter, nutrients, and trace metals in sediments; (2) to calculate the regenerative fluxes from sediments, requiring knowledge of diffusive fluxes of elements from the sediments and mass accumulation rates; (3) to determine the relative importance of river inputs, coastal erosion, and endogenic (material derived from within the water column) precipitation in aquatic sediments; and (4) to study temporal variations in lacustrine sedimentation and the relationship with climate and anthropogenically induced changes in lakes. Studies of lake sediments in the North American Great Lakes suggested that knowledge of sedimentation rates is the key to understanding inputs of anthropogenic metals, nutrients, and organic contaminants to the lakes (Kemp and Thomas, 1976; Durham and Oliver, 1983; Christensen and Goetz, 1987; Mudroch, 1993).

Bulk density measurements have also been used to calculate effective overburden stress-depth profiles, assuming hydrostatic conditions, for deep sea sediments in the northwest Pacific (The Marine Geotechnical Consortium, 1985; Schultheiss, 1985). From these profiles and other geotechnical measurements, the state of consolidation of the sediments was determined.

Bulk physical properties are interdependent with the consolidation state of sediments. In normally consolidated sediments, bulk density increases with depth, whereas water content and porosity decrease. In a geotechnical study of the Lau basin, Lavoie et al. (1994) determined that sediments were probably underconsolidated based on measurements of bulk properties measurements. They found that in certain zones of high fluid circulation, changes with depth were negligible for bulk density, water content, porosity, and compressional wave velocity. Lavoie et al. (1994) suggested that the fluid lost during the consolidation process was immediately replaced, resulting in a lack of gradients for the physical properties.

Water content forms the basis of the geotechnical index properties, the Atterberg Limits (Whalley, 1981). Estimates of water content of sediments are necessary for expressing contaminant concentration on a dry weight basis (Bengtsson, 1979). Determination of water content was used in other applications, for example, to calculate a "critical limit" between areas of erosion and transportation in lakes (Hakanson, 1982). A water content of about 50% could be used to delineate this zone. Kemp and Thomas (1976) observed that color and firmness of lake sediments were reflected in their water contents and redox potential. Water contents decreased from more than 90% by weight at the sediment–water interface to about 50 to 60% at 1 m depth, indicating increased compaction of the sediments. In addition, they observed that the water contents also covaried with particle size distribution, with the coarse-grained sediments having the lower water contents.

Porosity of sediments can be derived from the bulk density and water content of wet sediments (Boyce, 1976). The relationship between porosity and other physical properties, such as bulk density, has been studied extensively in the petroleum industry because of their importance to oil production and well logging (Boyce, 1973). Wetzel (1990) determined that for fine-grained, slowly deposited, marine sediments, porosity was closely related to the liquid limit. He suggested that the porosity of the surface sediment (i.e., upper 2 cm) reflected the sediment composition and the ionic composition of the pore water. A correlation between porosity and bulk density with consolidation for marine sediments was established by Hamilton (1960). Dadey and Klaus (1992) plotted porosity against compressional wave velocity to distinguish between the extent of consolidation and diagenesis/cementation in volcaniclastic sediments in the Izu-Bonin arc region, Japan. They pointed out that increasing consolidation increases velocity through a decrease in porosity, and cementation increases velocity through increased rigidity of the sediment with little change in porosity.

2.3.2   Bulk density

Bulk density is the mass of material divided by its volume, including the volume of its pore spaces. In contrast, particle density or specific gravity exclude the volume of the pore spaces. Mineralogy, void ratio, particle size distribution, and particle shape all influence the density

(Whalley, 1981). A variety of methods exist to determine the wet- or dry-bulk density of sediments. They can be divided into destructive and nondestructive methods. Destructive or gravimetric methods include core, clod, excavation, and displacement techniques, used in soil science, and a Syringe Technique (variation of the core method) utilized in DSDP projects. The nondestructive or radiation methods use both the scattering or transmission of gamma rays *in situ* or in the laboratory (Boyce, 1973, 1976; Lonardi, 1977; Blake and Hartge, 1986a).

2.3.2.1   Gravimetric methods

There are several ways to determine bulk density gravimetrically. Some of these methods are applicable to soft sediments only; others are adapted for stony or rocky sediments. The core method is used to determine dry-bulk density in soil materials and is well described in the literature, such as Sheldrick (1984), Blake and Hartge (1986a), Culley (1993), and American Society for Testing and Materials (1994a) (Standard Method D2937-83). In the ASTM method, a cylindrical (single or double sleeve) metal sampler is pressed or hammered into the material to a desired depth. The sampler is removed carefully to preserve a known volume of material. The sample is then dried at 105°C for 1 to 3 d (depending upon volume of material) and then weighed. Bulk density is calculated according to Blake (1965):

$$\rho_{db} = M_{ods}/V_f \qquad (1)$$

where, ρ_{db} = dry-bulk density (g/cm^3), M_{ods} = mass of oven-dried sample (g), and V_f = field volume (cm^3).

This method can be modified to characterize samples collected from box cores or grab samples. Blake (1965) noted that if soils are overly wet, however, friction along the sides of the sampler and vibration from the hammer may cause viscous flow and compression of the sample. This phenomenon is common in coring sediments using open-barrel and piston cores (Ross and Riedel, 1967; Mudroch and Mac-Knight, 1994b; Mudroch and Azcue, 1995). Conversely, in dry materials, hammering may shatter the sample (Blake, 1965).

A variation of the core method to determine wet-bulk density of marine sediments is the Syringe Technique described by Boyce (1973, 1976). This standard method is used for soft sediments only. The technique allows for determination of porosity, wet-bulk density, and water content on the same sample, a desirable feature when sample material may be limited. Sampling can be carried out in the central part of a split core or in a grab sample where there is minimal disturbance. The end of a 1-cm^3 syringe is cut off, squared, and sharpened so the leading edge is flush with the inside diameter of the syringe. The cylinder and end of the plunger are placed flush with the surface of the sediment;

the plunger is held stationary while the cylinder is slowly pushed into the sediment, similar to the action of a piston core. The volume of the sediment is determined before the sediment is weighed. The sediment is then extruded and weighed wet. The wet-bulk density is calculated by:

$$\rho_{wb} = M_{ws}/V_{ws} \tag{2}$$

where, ρ_{wb} = wet-bulk density (g/cm³), M_{ws} = mass of wet sediment (g), and V_{ws} = volume of wet sediment (cm³). The sample is then oven-dried at 105°C for 24 h, cooled in a desiccator for 2 h, and reweighed to facilitate calculation of water content and porosity. Boyce (1976) noted that a minimum volume of 0.3 cm³ is optimal with this method.

For stiff, hard or stony sediments, the clod (Blake, 1965) or rock chunk (Boyce, 1973, 1976) method is preferable. The volume of the clod (soil term for an artificially formed aggregate of soil particles) or rock chunk is determined by making use of Archimedes' principle. The clod must be firm and stable during processing. The clod is air dried and placed in the wire basket of a suitable balance; its weight in air is then recorded after predetermining the tare weights of the container in air and water. The clod and container are dipped in melted paraffin (60°C) or in Saran resin (Blake and Hartge, 1986a), and the excess is allowed to drain. The clod, paraffin, and container are weighed in air, and then suspended in water and reweighed (Blake, 1965; Blake and Hartge, 1986b). The dry-bulk density is calculated by the equation

$$\rho_{db} = \rho_w \times M_{ods}/[M_{sa} - M_{spw} + M_{pa} - (M_{pa} \times \rho_w/\rho_p)] \tag{3}$$

where, ρ_{db} = dry-bulk density (g/cm³), ρ_w = density of water at temperature of determination (g/cm³), M_{sa} = net mass of sample in air (g), M_{ods} = oven-dried mass of sample (g), M_{spw} = net mass of sample plus paraffin in water (g), M_{pa} = mass of paraffin in air (g), and ρ_p = density of paraffin (~0.9 g/cm³). In this method densities may be higher than those for the core method, as the pore spaces are not considered, and the sample volume is the air-dry volume rather than field wet.

Another method that can be used to determine bulk density of gravelly materials is the excavation method as detailed in Blake (1965) and Blake and Hartge (1986a). A quantity of material is exacavated and oven-dried at 105°C; then the sample is weighed. The volume of the sample can be determined by either the sand funnel method or the rubber balloon method. In the sand funnel method, the volume of the sample is determined by filling the excavation with sand of known volume per unit mass. In the rubber balloon method, the volume is determined by placing a balloon into the hole and filling it with water until the hole is just full. The volume of the sample is equivalent to the volume of displaced fluid (Blake and Hartge, 1986a). The sand funnel method is similar to the standard method described in ASTM (1994a) as procedure D1556-90. This is a multitask procedure that requires extreme care. Using large volumes of material reduces the errors, but there is a lack of discrimination for thin horizons.

2.3.2.2   Radiation method

Single source gamma ray attenuation techniques were used extensively in DSDP projects to estimate the wet-bulk density of sediments (Boyce, 1976). The method was also described in ASTM (1994a) as procedure D2922-91. A dual source method has been used to measure both bulk density and volume of water in soils (Soane, 1967; Phogat et al., 1991). The advantages of the gamma ray attenuation method over the gravimetric method are that measurements can be continuous or repeated at the same point and are nondestructive and rapid. The high cost of the equipment and the potential hazard of working with gamma rays are some of the disadvantages.

An unsplit core, in its liner, was passed through the GRAPE (Gamma Ray Attenuation Porosity Evaluator) apparatus where single source (^{133}Ba) gamma rays are absorbed or scattered by the sediment (Boyce, 1973, 1976). The rate of attenuation of gamma rays by sediments or rocks is related to the density of the material (Boyce, 1973). Individual minerals have different attenuation rates, which are distinct from that of water. The attenuation coefficient for water is 10% higher than that of quartz. The densities determined by the GRAPE system must be corrected to give a true wet-bulk density, especially if the attenuation coefficients of minerals differ from that of quartz by more than ±3% (Boyce, 1976).

The GRAPE system measures the intensity of a 3-mm wide gamma ray that passes through the core continuously at a rate of 3.17 mm for every 2-sec counting interval. For each 2-sec counting period, Lonardi (1977) indicated an accuracy of ±10% for bulk density determinations. Increasing this to a 2-min counting time reduces the error to ±2%. The measurements are calibrated against reference standards of aluminum cylinders and empty plastic liners of comparable size to the core barrel. It is also assumed that the sediment fills the core tube completely. Hence, this method works most efficiently with soft sediments. If stiff sediments or rocks in the core have smaller diameters than the liners, then corrections can be applied to calculate the true wet-bulk density of the material (Boyce, 1976). It is interesting to note that Schultheiss (1985) reported that GRAPE-determined wet-bulk density measurements correlated very poorly with shore-based laboratory values of bulk density.

2.3.3 Water content

Water (or moisture) content is the ratio of the weight of water to the weight of the sediment in a given volume of sediment, expressed as a percentage (Whalley, 1981; Gardner, 1986). Boyce (1973, 1976) distinguished between a wet-water content as the ratio of the weight of water to the weight of wet-saturated sediment and a dry-water content as the ratio of the weight of water to the weight of dry sediment. Water content is useful when contaminant concentration is to be reported on a dry weight basis, even if determined on wet samples (Environment Canada, 1993). If water content and bulk density are both known, then porosity of a sample can be calculated. Also, a weight-based water content can be converted to a volume-based water content if the bulk density is known.

Water content can be assessed by direct or indirect methods. Direct methods include determining the amount of water removed by evaporation, leaching, or chemical reaction (Gardner, 1986). Indirect methods include measurement of a physico-chemical property of the material using a variety of instruments and probes.

2.3.3.1 Direct methods

Although there are numerous methods to determine water content directly, gravimetry with oven-drying (conventional or microwave) is the most common standard method used. One problem, according to Gardner (1986), is the definition of the "dry" state. Sediments, like soils, contain colloidal and noncolloidal inorganic particles, organic matter, water, dissolved substances, and other volatiles. Drying, usually by heat at a temperature of 105°C, will not only drive off the free and absorbed water but may also oxidize and decompose organic matter (Gardner, 1986). Loss of compounds more than or as volatile as water can also interfere with the determination (Environment Canada, 1993).

Both conventional and microwave ovens can be used to dry samples. Gardner (1986) suggested that it may be difficult to achieve accuracy and reproducibility in microwave ovens relative to conventional ovens because of potential oxidation of organic matter. Ryley (1969) indicated little difference between oven types. One advantage of the microwave oven is that the sample remains cool to the touch. Standard methods for conventional and microwave ovens are given in ASTM (1994a) as methods D2216-92 and D4643-93, respectively.

To determine the water content, the sediment is placed in a preweighed weighing bottle or metal can with a tight fitting lid. The wet sediment and container are weighed. The container is placed in the oven with lid off and dried at 105°C to a constant weight. This may take from 1 to 3 d, depending upon the volume of sample, number of samples in the oven, and type of oven. The sample is removed, and the lid is placed on the container, which is then transferred to a desiccator to cool for about 2 h. The sample and container are reweighed, and water content is calculated according to the equation (Boyce, 1976)

$$\Theta_w = M_w/M_{ws} \times 100 \tag{4}$$

$$= M_{ws} - M_{ds}/M_{ws} \times 100 \tag{5}$$

similarly,

$$\Theta_d = M_{ws} - M_{ds}/M_{ds} \times 100 \tag{6}$$

where, Θ_w = wet-water content (%), Θ_d = dry-water content (%), M_w = mass of water (g), M_{ws} = mass of wet sediment (g), and M_{ds} = mass of dry sediment (g).

To express dry-water content on a volume basis

$$\Theta_v = \rho_{db}/\rho_w \times \Theta_d \tag{7}$$

where, Θ_v = volumetric water content (%), ρ_{db} = dry-bulk density of sediment (g/cm^3), and ρ_w = density of water (g/cm^3).

2.3.3.2   Indirect methods

Indirect measurements involve quantification of some property of the material that is affected by or dependent upon water content. These methods are nondestructive but usually require some type of calibration. They are only briefly described because their application in aquatic sediment studies is unproven.

2.3.3.3   Time domain reflectometry

Time Domain Reflectometry (TDR) is an *in situ* electrical testing technique that has been adapted to measure the water content of soils. Time domain reflectometry is also known as cable radar because it was initially used to detect faults in cables and transmission lines (Topp et al., 1994). The technique measures the time required for a high frequency voltage pulse to travel down to and reflect back from the edge of a wave guide or probe inserted in the soil (Topp, 1992, 1993). The velocity of the pulse is dependent on the dielectric constant of the materials. As the dielectric constant of water at 80 is so much higher than that of soil at 2 to 7 or air at 1 (Topp, 1993), the velocity is essentially a function of the soil water content.

The TDR instrument includes a voltage pulser, a receiver, a precise timing system, and a display screen. The timing system synchronizes the pulser, receiver, and data display, whereas the display screen shows the time and voltage magnitude as a reflection coefficient (Topp, 1992,

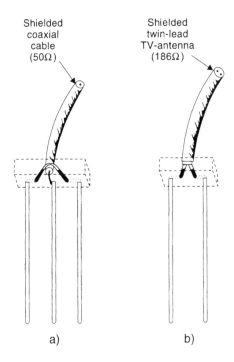

Figure 2-1 Examples of time domain reflectometry (TDR) transmission lines used as soil probes: (a) simulated coaxial (unbalanced) and (b) twin-lead (balanced pair). (After Topp, G.C., in *Soil Moisture Modeling and Monitoring for Regional Planning*, Eley, F.J., Granger, R., and Martin, L., Eds., National Hydrological Research Institute, Saskatoon, Symposium No. 9, 1992, 155. With permission.)

1993). The probe, as shown in Figure 2-1, contains 2 or 3 parallel metal rods that are not more than 0.1 m apart and greater than 0.1 m long (Topp, 1992). To determine the water content, the transition time of the voltage pulse over a known length is measured. Typically, this transit time is about 2 ns for 0.3 m probe length (Dalton, 1992).

The TDR method has been used in a variety of soil applications. Kachanoski et al. (1992) and Elrick et al. (1992) demonstrated that TDR can be used to measure the mass flux of solute in soil tracer studies under steady-state leaching conditions of constant surface volume flux density of water. Hayhoe et al. (1983) used TDR to detect the presence of frost and the movement of freezing and thawing fronts. Dalton (1992) showed that TDR could be used to measure soil water content and bulk soil electrical conductivity (salinity) simultaneously. Dowding and Pierce (1994) suggested that TDR could be used to detect bridge scour and to monitor pier and abutment movements during floods. Topp (personal communication, 1995) indicated that TDR has the potential to measure water contents of sediments. A method to deploy the probe into soft sediment would need to be developed. The variety of applications demonstrated above for soils has the potential for use in aquatic sediment studies in bogs, shallow lakes, streams, and tailings ponds.

2.3.3.4   Radiation method

The gamma ray attenuation technique to determine bulk density can also be used to determine water content, as mentioned above. If the particles in the sediment and bulk density remain constant, then the changes in attenuation can be attributed to variable water content. As bulk density usually changes with depth in a sediment core, it is necessary to use two different gamma ray energies to measure bulk density and water content simultaneously (Gardner, 1986; Topp, 1993). This method has been used in soil studies (Soane, 1967) and has been combined with CAT (computerized axial tomography) scans to produce 3-dimensional imaging (Phogat et al., 1991). Details of this method with respect to soils were described by Gardner (1986) and Topp (1993).

2.3.4   Porosity

Porosity is usually determined indirectly from other physical parameters such as wet-bulk density and wet-water contents, from dry-bulk density and particle density, and from the void ratio of sediments. Porosity can also be determined directly by using techniques, such as mercury porosimetry (Rootare, 1970), helium porosimetry, or water immersion (American Petroleum Institute, 1960). These techniques also provide information about pore size distribution which is useful in assessing permeability and adsorption, characteristics of soils and rocks. Mercury porosimetry has also been shown to be useful in determining particle size distribution and surface area (Rootare, 1970). Porosity of unconsolidated sediment ranges from about 25 to 80% (Freeze and Cherry, 1979).

2.3.4.1   Indirect methods

Boyce (1976) showed that porosity can be calculated from wet-bulk density and wet-water content determinations. Combining equations (2) and (5) gives

$$\phi = 100/\rho_w \times (M_{ws} - M_{ds})/V_{ws} \tag{8}$$

$$= \rho_{wb} \times \Theta_w \tag{9}$$

where, ϕ = porosity (%), ρ_w = density of water (1.0 g/cm^3), M_{ws} = mass wet sediment (g), M_{ds} = mass dry sediment (g), V_{ws} = volume of wet sediment (cm^3), ρ_{wb} = wet-bulk density (g/cm^3), and Θ_w = wet-water content (%).

Porosity can also be determined using dry-bulk density and particle density. Usually a particle density of 2.65 g/cm³ is assumed, the value of quartz (Carter and Ball, 1993). The calculation is

$$\phi = 1 - (\rho_{db}/\rho_p) \tag{10}$$

where ρ_{db} = dry-bulk density (g/cm³) and ρ_p = particle density (g/cm³). If greater accuracy is required, then particle densities can be determined independently using the pycnometer method as described in Blake and Hartge (1986b).

Whalley (1981) stated that porosity could be determined from the void ratio of materials, such that

$$\phi = V_v/V \tag{11}$$

$$e = V_v/V_s \tag{12}$$

$$\phi = e/(1 + e) \tag{13}$$

where, V_v = volume of voids, V_s = volume of solid, V = total volume, and e = void ratio. The total volume of the sample equals the volume of solid and voids; thus, porosity can be calculated by measuring any two of these properties. Total volume can be measured directly with either a pycnometer or by mercury porosimetry (Whalley, 1981).

2.3.5   *Recommendations*

The methods for physical analyses of bulk properties described above are intended as an introduction to the reader. These methods have been described in detail in the literature. Method manuals are generally available either through publications by the various societies (e.g., Canadian Soil Science Society; American Petroleum Institute) or are supplied with specialized equipment. In carrying out these physical analyses, it has been common to use gravimetric rather than instrumental methods as the standard.

Instrumental methods can be a major capital investment. However, in many cases, the benefits of an efficient and precise instrument may outweigh the expenditures. Some of these instrumental methods are becoming more available through commercial analysis of samples.

Gravimetric methods are simple and require only a precision balance and oven. Gardner (1986) commented on the fact that although these gravimetric methods are considered standard, there is nothing standard about the ovens used. The balance and oven should be calibrated regularly. To ensure quality control in a study, about 10 to 20% of the samples should be analyzed in duplicate.

Many of the instrumental methods used to measure bulk properties have been developed for soil or engineering studies. This does not preclude their use in sediment studies. Creative ways to adapt the instrument or sample may be required. For example, the application of TDR to detect solute movement may be transferable to tailing ponds or shallow lakes.

2.4   Particle size distribution

2.4.1   Introduction

Quantitative analysis of the size distribution of particles is important in sediment transport studies, stratigraphic correlations, mapping, assessing modern and past geological environments, geotechnical studies, and surface reactions (Lewis, 1984; McCave and Syvitski, 1991). Particle size distribution can, particularly in glaciated terrains, reflect mineralogical partitioning and thus is an indicator of variation in physical and chemical properties of sediments (DiLabio, 1995; Shilts, 1995). In addition, spatial and temporal variability of trace element concentrations in sediments has been attributed, in part, to particle size (Jones and Bowser, 1978; Jenne et al., 1980; Förstner and Wittmann, 1981; de Groot et al., 1982; Horowitz and Elrick, 1987; Horowitz et al., 1989; Barbanti and Bothner, 1993). A strong correlation between concentrations of contaminants and fine-grained particles was observed in many studies (Ackermann, 1980; Förstner, 1982; Mudroch, 1984; Horowitz and Elrick, 1988). Further, it was found that organic carbon, known to bind different contaminants, is inversely related to the proportions of sand-sized and directly proportional to the silt- and clay-sized sediment particles (Damiani and Thomas, 1974). Barbanti and Bothner (1993) suggested that texture (i.e., grain size) may control trace metal contaminant concentration in marine sediments. It is well known that fine-grained sediments (e.g., <0.063 mm) are more "reactive", through their higher surface area and adsorption capacity, than coarse-grained sediments (Förstner and Salomons, 1980; Salomons and Förstner, 1984). The clay-sized fraction not only preferentially adsorbs contaminants due to its high cation exchange capacity, but it is the first to yield anthropogenically derived or naturally occurring trace elements as a result of sediment disturbance (Shilts, 1995). The Sedimentology Laboratory of the Geological Survey of Canada routinely concentrates the clay-sized particles of sediments by centrifugation for trace element geochemical studies (Lindsay and Shilts, 1995).

Fine-grained coatings, such as Mn- and Fe-oxides, carbonates, and organics, on otherwise chemically inert, coarser particles, such as quartz and feldspar, which dominate the fine sand and silt fractions, may enhance the ability of particulates to mobilize contaminants. Shilts

(1995), in studies of tills, observed that different grain size fractions have a geochemical signature that reflects the dominant minerals in that fraction. The dilemma is to select an optimum size fraction for analysis. This fraction should not be biased by sample-to-sample variation in the abundance of inert rock-forming minerals (DiLabio, 1995). A standardized procedure with regard to particle size is critical for the assessment and monitoring of contaminants in aquatic systems. Grain size correction methods were developed (Förstner, 1990) to minimize the influence of a chemically inert fraction (i.e., coarse grains) of the sediment. Förstner and Salomons (1980) recommended to use the <0.063 mm particle size fraction in environmental studies of aquatic sediments. Their reasons include observations that (1) contaminants are concentrated in the clay-silt fraction, (2) this fraction is nearly equivalent to material carried in suspension, and (3) sieving does not alter the contaminant concentrations.

2.4.2   *Particle size analysis*

Particle size distribution is the percentage, by mass, volume, or number, of particles in a range of specific sizes. The methods used to determine particle size distribution are classified as "classical" (or manual), and "instrumental". The instrumental methods are generally faster and as reproducible as the classical methods but also require extensive sample preparation.

McCave and Syvitski (1991) suggested that the accuracy of particle size analysis is dependent upon the definition of the size being determined: projected area size by image analysis, intermediate diameter size by sieve, volume size by electronic sensor counter, or quartz equivalent spherical sedimentation diameter by pipette method. Selection of a particular technique is not as important as standardization of procedure, and all samples to be compared are analyzed in precisely the same manner (Lewis, 1984). The use of standard reference material for particle size analysis also requires confirmation of a given methodology as being close to true accuracy (McCave and Syvitski, 1991). Thus precision, not accuracy, becomes an important consideration when choosing a method that will reflect the nature of an investigation. The standard classification of particles into Wentworth size classes, phi scale, and comparative sieve sizes is summarized in Table 2-2.

2.4.2.1   *Sample preparation*

Sample preparation is a critical aspect of particle size analysis to avoid interferences from flocculation, oxidation, and organic material. Commonly used pretreatment of samples may be necessary to enhance dispersion of aggregates, as some sediments may contain organic matter, salts, iron oxides, or carbonate coatings that bind particles together

Table 2-2 Standard Classification for Particles into Wentworth
Size Classes, Phi Scale, and Comparative Sieve Sizes

Wentworth[1] size class	Millimeters	Phi (ϕ)[2]	U.S. standard sieve mesh #
Gravel			
Boulder	4096	−12	
	1024	−10	
Cobble	256	−8	Use wire squares
	64	−6	
Pebble	32	−5	
	16	−4	
Granule	4	−2	5
	3.36	−1.75	6
	2.83	−1.5	7
	2.38	−1.25	8
Sand			
Very coarse sand	2.00	−1.0	10
	1.68	−0.75	12
	1.41	−0.5	14
	1.19	−0.25	16
	1.00	0.0	18
Coarse sand	0.84	0.25	20
	0.71	0.5	25
	0.59	0.75	30
	0.50	1.0	35
Medium sand	0.42	1.25	40
	0.35	1.5	45
	0.30	1.75	50
	0.25	2.0	60
Fine sand	0.210	2.25	70
	0.177	2.5	80
	0.149	2.75	100
	0.125	3.0	120
Very fine sand	0.105	3.25	140
	0.088	3.5	170
	0.074	3.75	200
Mud			
Coarse silt	0.0625	4.0	230
	0.053	4.25	270
	0.044	4.5	325
	0.037	4.75	
	0.031	5.0	
Medium silt	0.0156	6.0	
Fine silt	0.0078	7.0	
Very fine silt	0.0039	8.0	
	0.0020	9.0	

Table 2-2 Standard Classification for Particles into Wentworth Size Classes, Phi Scale, and Comparative Sieve Sizes *(continued)*

Wentworth[1] size class	Millimeters	Phi $(\phi)^2$	U.S. standard sieve mesh #
Clay	0.00098	10.0	
	0.00049	11.0	
	0.00024	12.0	
	0.00012	13.0	
	0.00006	14.0	

[1] From Wentworth, C.K., *J. Geol.*, 30, 377, 1922.

[2] From Krumbein, W.C. and Pettijohn, F.J., in *Manual of Sedimentary Petrography*, Mather, K.F., Ed., Appleton-Century-Crofts, New York, 1938, 91.

(Day, 1965; Gee and Bauder, 1986). It is important that methods be standardized because mechanical or ultrasonic pretreatment to disperse aggregates may fragment particles, and chemical pretreatments can result in destruction and dissolution of some minerals (Gee and Bauder, 1986).

Drying the sediments before particle size analysis should be avoided because it is difficult, if not impossible, to redisperse the finest silt- and clay-sized particles (Harrison et al., 1964). If drying is necessary, however, then freeze-drying is recommended as the most nondestructive method, especially for fine-grained silty clay sediments. Barbanti and Bothner (1993) suggested that freeze-drying increased aggregation between particles and resulted in higher metal concentrations in the coarse fractions due to the aggregation of finer metal-enriched particles.

Removal of organic material is often achieved by the addition of hydrogen peroxide (H_2O_2), but its effectiveness is reduced in the presence of manganese oxide (MnO_2) (Gee and Bauder, 1986). Coatings of various iron oxides, such as hematite and goethite, may act as cementing agents in sediments. Dispersion of aggregates can be enhanced with the use of a bicarbonate-buffered, sodium dithionite-citrate solution for iron oxide removal (Mehra and Jackson, 1960; Jackson, 1979; Gee and Bauder, 1986). This method was also found to be the least destructive of iron silicate clays as indicated by minimal loss of cation exchange capacity (Gee and Bauder, 1986). Carbonate removal is achieved by acidification using, for example, 1 M HCl and can be accelerated by heating.

2.4.2.2 Classical methods

2.4.2.2.1 Sieve analysis. Dry sieving is a convenient procedure for segregating particles coarser than silt-sized (0.063 mm). The largest sphere that can pass through a given sieve has a diameter equal to the mesh, and by stacking sieves with successively smaller openings, the material may be separated into any number of grades. Sieves used for sampling normally have square openings or mesh sizes, so the resultant particle diameter will not necessarily agree with other techniques (Saheurs et al., 1993). The probability of a particle passing a given sieve in a given time of shaking depends upon the nature of the particle, its orientation, and the properties of the sieve (Day, 1965). A particle whose shape permits its passage only in a certain orientation (e.g., elongated grains, such as zircon) would require prolonged shaking. Mechanical sieving, using a vibrating motion, with a stack of sieves in a sieve-shaker, produces thorough sieving and reproducibility.

The sieves are arranged on top of a catchment pan in a sequence decreasing in size from top to bottom (Figure 2-2). A dried, disaggre-gated, preweighed sample is transferred into the screen, using a brush to complete the transfer. A lid is placed on top of the nest of sieves and firmly fastened in the automated sieve shaker. The sieves are shaken for a standard period of time (e.g., 10 min). The material remaining on

Figure 2-2 Example of a nest of metal sieves, set on an automated shaker, used for dry sieving.

each sieve is weighed, and cumulative weight percentages are calculated.

$$\text{Grain Size (\%)} = M_s/M_t \times 100 \tag{14}$$

where, M_s = mass of sediment retained on sieve (g) and M_t = total mass of starting material (g).

Wet sieving can be used to clean fine-grained particles clinging to the surface of previously dry-sieved material or to recover the fine fractions (usually <0.063 mm) for analysis. Samples are soaked for a period of time (minimum 2 h) to aid the process of disaggregation (Mudroch and Bourbonniere, 1994). The slurry is transferred to the top sieve and gently swirled or agitated until the particles finer than the mesh size being used are allowed to pass through the openings into the receiving vessel, often a plastic bucket.

2.4.2.2.2 *Pipette method.* The pipette method of analysis is a sedimentation procedure that utilizes pipette sampling of a dispersed sample in solution, at controlled depths and times, based on Stokes' Law (Day, 1965). It is assumed that particles settling independently with no hindered settling effects are spherical with a density of quartz, there is no flocculation, and the temperature is constant (McCave and Syvitski, 1991). In the analysis, a very thin layer is pipetted from a perfectly homogeneous suspension (5% w/v sodium hexametaphosphate $[NaPO_3]_6$); however, the measurement itself disturbs the physical system (Hayley and Joyce, 1984). Once the most common method of analysis of fine-grained particles, the pipette method has been surpassed recently with automated systems that are faster and more precise. For that reason limited attention is devoted to description of the pipette procedure.

A preweighed sample (e.g., 20 to 50 g) of material, usually finer than 0.063 mm, is dispersed in a 5% $(NaPO_3)_6$ solution and brought to a constant volume in a graduated cylinder. The suspension is agitated to ensure complete mixing. A pipette is then carefully inserted to a specified depth (e.g., 10 cm) and aliquots (e.g., 25 mL) of suspension are withdrawn at predetermined times (see Jackson, 1979; Lewis, 1984; etc.). The aliquots are placed in a suitable preweighed container and placed in an oven set at 105°C for about 24 h. The weight of the dried material is then determined and corrected for weight of the dissolved salts (i.e., $[NaPO_3]_6$). A cumulative curve may then be drawn of the weights of the successively slower settling fractions (<0.020 mm, <0.005 mm, <0.002 mm).

The pipette method continues to be used as a standard method of analysis for fine-grained sediments and as a basis for comparison with results from other techniques. However, Syvitski et al. (1991b) pointed out that in an interinstrumental, interlaboratory study, the pipette method showed the most scattered results compared to all other instrumentation methods tested. There is much detail in the literature referenced below regarding theories and methodology for this procedure (Krumbein and Pettijohn, 1938; Day, 1965; Jackson, 1979).

The standard test method for particle size analysis of soils as outlined in ASTM (1994a, method D422-63) utilizes the hydrometer, another classical method based on density settling, for the determination of the silt- and clay-sized fractions. This method, however, is not recommended because it is not considered accurate (McCave and Syvitski, 1991).

2.4.2.3 Instrumental methods

2.4.2.3.1 Electrical sensing zone (Coulter Counter®). The electrical sensing zone principle is dependent on particles being suspended in an electrolyte and passing between electrodes. The volume of electrolyte displaced by the particles is proportional to the volume of the particles, and corresponds to the change in electrical resistivity measured by the electrodes at the aperture (McCave and Syvitski, 1991; Haley and Joyce, 1984). The change in resistivity due to a particle of a given diameter is

$$\Delta R = [(8r_f d)/(3\ D^4)][1 + 4/5(d/D)^2 + 24/35(d/D)4^4 \ ...] \qquad (15)$$

where r_f = resistivity of the fluid (ohms), D = aperture diameter (mm), and d = particle diameter (mm).

An example of a Coulter Counter® unit is shown in Figure 2-3. The technique is generally valid for particle sizes in a 20:1 ratio corresponding to between 2 and 40% of the effective aperture diameter. Particles finer than 2% of the aperture diameter are not detected and must be analyzed using a smaller aperture, whereas those larger than 40% of the aperture size must be removed to prevent blockage of the aperture. Multiple apertures are required if the sediment samples have a broad size distribution and the data can be blended.

When an electrolyte, containing suspended material and carrying a constant current, flows through an aperture, particles are sensed as they displace an equivalent volume of solution. The number of changes of resistivity per unit time is directly proportional to the number of particles, and the magnitude of the pulse change is proportional to the volume. Voltage values are recorded and added one by one for cumulative data, and adjusted electronically for data reporting (Hayley and Joyce, 1984).

2.4.2.3.2 X-ray sedimentation (Sedigraph®). The Sedigraph® (Figure 2-4) employs a sedimentation technique that measures the

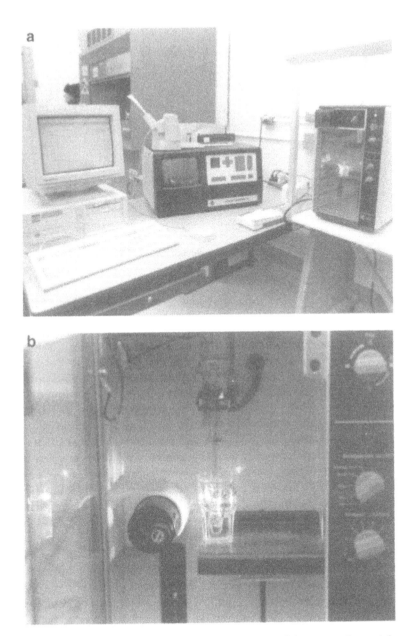

Figure 2-3   (a) Configuration of the Coulter Counter® featuring, from right to left, the sample chamber, electrical sensing zone unit, and computer. (b) Close-up view of the sample chamber showing electrolyte solution with suspended particles.

Figure 2-4 Example of a Sedigraph® unit featuring, from right to left, the automated sampler, sedigraph analyzer, and computer.

velocity of a particle falling through a viscous medium (e.g., 5% w/v [NaPO$_3$]$_6$), due to gravitational force, that can be related to the size of the particle based on Stokes' Law. As with other sedimentation techniques (e.g., pipette method), the density of quartz is assumed. The sedimentation velocity can be determined by measuring the concentration of particles remaining in suspension as a function of time (McCave and Syvitski 1991; Walker and Reynolds, 1991). A dispersed sample is circulated through a measurement cell using a peristaltic pump. When the analysis begins, the pump is turned off, allowing the particles to settle. The cell is irradiated perpendicularly by a finely collimated X-ray beam. The degree of attenuation of incident radiation by the suspension, as compared to the suspending solution alone, is related to the concentration of the sediment by weight. The transmittance of the suspension, which increases with time due to the sedimentation of the particles, is electronically transformed into concentration values and indicated linearly as cumulative mass percent on the Y-axis of an X-Y recorder (Walker and Reynolds, 1991; Hayley and Joyce, 1984). The sample cell is continuously lowered relative to the X-ray beam so that the effective sedimentation depth decreases with time. This cell movement is coincidental with the equivalent spherical diameter of the X-axis as a function of time and depth (Stein, 1985). The Sedigraph® sample suspension should have a particulate concentration of less than 2% volume. Too high concentrations will result in hindered settling of particles.

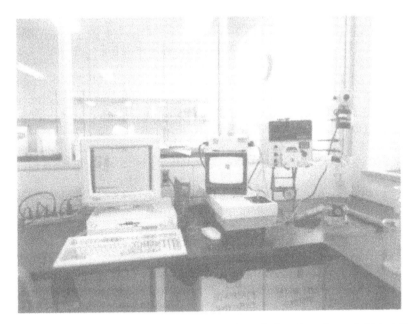

Figure 2-5 The Brinkman Particle Size Analyzer® showing, from right to left, the sample reservoir, peristaltic pump, video monitor, and computer. During analysis, images of particles can be viewed on the video monitor to ensure particle disaggregation.

2.4.2.3.3 Laser–time of transition theory (Brinkman PSA®). Direct particle measurement avoids sources of error inherent in many techniques that are dependant on secondary correlations, the index of refraction, or Brownian motion, etc. The Brinkman Particle Size Analyzer®, shown in Figure 2-5, measures the diameter of a particle directly, using the "time of transition theory" (Brinkman Instruments Manual). A laser scans the sample measurement field (e.g., sample cell) at a fixed velocity. The time of transit of the laser beam across a particle, multiplied by the laser velocity, gives a displacement that is equal to the diameter of the particle, such that

$$d = T \times V \tag{16}$$

where d = particle diameter (mm), T = time of transit across a particle (sec), and V = velocity of laser beam (mm/sec). The laser beam is rotated in a circle at a frequency fast enough that any motion of the particles relative to the beam is negligible. Particles are introduced into the measurement area, where they interact with the laser beam, producing a voltage pulse in a photodiode mounted behind the measurement area. The width of the pulse represents the time of interaction with a particle and is related to its size, whereas the rise time of the pulse is shortest for particles in the laser's focal plane and for those representing a perpendicular interaction with the laser (i.e., the diameter) (Walker and Reynolds, 1991). Particles out of focus or those with long pulses are disregarded.

Interchangeable sample cell modules allow particles to be measured in essentially any state — as liquids, powders, gases, or slurries. As a slurry, samples are suspended as a dilute solution and pumped by peristaltic action from a reservoir, through the flow-through cell, and back to the reservoir. Data accumulation continues until the confidence level (established before the analysis) is reached in each grain size range.

Extremely small samples may be suspended by a mechanical mixer in a cuvette cell. It is important to note that instruments that incorporate a magnetic stirrer to keep particles in suspension will draw magnetic particles from the solution to the stirrer paddles, effectively removing them from the measurement area.

A separate feature of this instrument is a video camera mounted in the measurement area at 90 degrees to the laser beam. It captures visual images of the particles and takes "still" pictures every 3 sec, while they are in dynamic flow. This visual representation gives direct information on particle shapes and aggregates and may be helpful in identifying particulate matter that has "signature" shapes (i.e., certain diatoms or phyllosilicate minerals).

2.4.2.3.4   Laser diffraction spectroscopy.   Laser diffraction analysis is based on the Fraunhofer theory, which states that the intensity of light scattered by a particle is directly proportional to the particle size (McCave et al., 1986). Large particles scatter light at high intensities through low angles, and small particles scatter light at low intensities through high angles (Walker and Reynolds, 1991). Again, particle sphericity is assumed. Below 0.007 mm, particles do not diffract light in the manner required for the application of the Fraunhofer theory because their diameter approaches that of the wavelength of light (de Boer et al., 1987; McCave et al., 1986). For particles in this size range, the mode of scattering is termed Mie scattering and depends upon particle size and differences in the refractive indices (Haley and Joyce, 1984). Combining these two scattering modes permits a particle size range from clay-sized through sand-sized. Instrument scattering inversion software compensates for "other" than Fraunhofer scattering (American Society for Testing and Materials, 1994b; standard method E1458-92).

Particles circulating through a cell are illuminated by a beam of monochromatic light, which is focused onto a multielement detector that senses the distribution of the scattered light intensity. A series of lenses and proprietary filters focuses the undiffracted light at the center of the detector, leaving only the surrounding diffraction pattern

(McCave et al., 1986). This angular distribution of light intensity is used to calculate size distributions.

  2.4.2.3.5   Image analysis.   Image analysis is used to characterize, classify, compare, and calculate geometric parameters of features in images (Petruk, 1989; Walker and LeCheminant, 1989). Light optically based image analysis involves input from a video camera, which may be linked to an optical microscope or macroviewer, to display and enhance a visual image of individual particles. Edge detection of individual particles is determined by developing sufficient contrast between the background and the projected particle image. This is achieved by the proper setting of the instrument's "grey-level" discriminators. Once the particle edges have been defined, any conceivable geometric parameter may be determined. Petruk (1989) stated that particle size distribution determined by image analysis relate closely to sieve analysis; both methods calculate the equivalent diameter based on the assumption that each particle is square. Overlapping grains, however, may not be seen as discrete particles, and this method may yield coarser particle size distributions.
  Electronic microprobe and scanning electron microscope technologies have enabled the development of electron beam-based image analysis, facilitating the characterization of extremely fine-grained sediments (Walker et al., 1989). When integrated with energy-dispersive X-ray analysis software, these systems provide detailed information on the physical and chemical properties of sediments (Walker and LeCheminant, 1989). An example of the type of results that can be determined using this system is shown in Table 2-3 and Figure 2-6. Ash particles were scanned using a Leica Cambridge Stereoscan S360 SEM. The mean diameter, number of particles, and percentage distribution were tabulated for 390 scanned features. The results from image analysis, unlike the other classical and instrumental methods, gave number distribution rather than volume distribution. Therefore, analysis of fine-grained sediments tended to be biased toward the fine particle size ranges.

2.4.3   Recommendations

Particle size analysis is an extremely complicated procedure in that most sediments are composed of a complex mixture of particles of varying sizes, shapes, densities, and matrices. All of the procedures described above make some assumption about particle sphericity, density, or light scattering ability. No two procedures will yield identical data for the same sample.
  This summary of methods has focused primarily on the analysis of fine-grained sediments, and most of the instrumental techniques offer good precision in the silt and clay size ranges (Table 2-4). Although

Table 2-3 Typical Results Obtained from
Image Analysis

Mean feret diameter (μm)	Frequency	Distribution (%)
<0.2	1	0.25
0.2–0.4	20	5.1
0.4–0.6	34	8.7
0.6–0.8	20	5.1
0.8–1.0	27	6.9
1.0–1.2	26	6.7
1.2–1.4	28	7.2
1.4–1.6	29	7.4
1.6–1.8	25	6.4
1.8–2.0	16	4.1
2.0–2.2	20	5.1
2.2–2.4	26	6.7
2.4–2.6	13	3.3
2.6–2.8	16	4.1
2.8–3.0	17	4.3
3.0–3.2	8	2.1
3.2–3.4	8	2.1
3.4–3.6	8	2.1
3.6–3.8	3	0.8
3.8–4.0	8	2.1
4.0–4.2	1	0.25
4.2–4.4	4	1.0
4.4–4.6	3	0.8
4.6–4.8	1	0.25
4.8–5.0	2	0.5
5.0–5.2	1	0.25
5.2–5.4	1	0.25
5.4–5.6	0	0
5.6–5.8	6	1.5
5.8–6.0	2	0.5
6.0–6.2	0	0
6.2–6.4	1	0.25
6.4–6.6	2	0.5
6.6–6.8	1	0.25
6.8–7.0	1	0.25
>70	11	2.8

Note: Frequency is the actual number of particles detected within each grain size range. Total number of grains detected was 390 in 28 scan-fields.

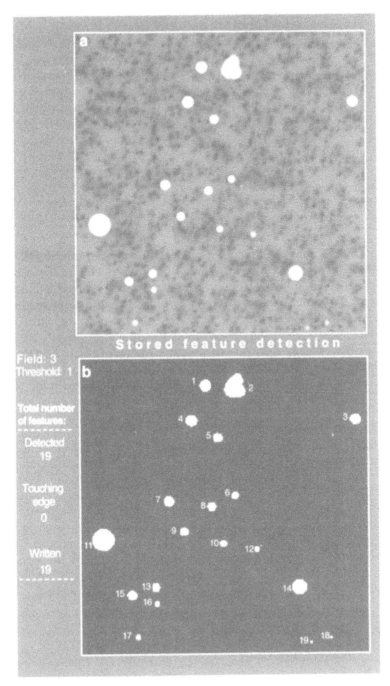

Figure 2-6 (a) Backscattered electron image of one scan-field in an image analysis automated run. (b) Example of the screen capture of the detected features in the scan-field (field width is 40 μm).

Table 2-4　Summary of Methods for Optimal Particle-Size Detection

Particle size analysis method	Particle size		
	Sand	Silt	Clay
Dry sieving	X	X	
Wet sieving		X	X
Pipette		X	X
Electrical sensing zone		X	X
X-ray sedimentation		X	X
Laser-time of transition		X	X
Laser diffraction		X	
Iconmetrics			X

Note: Sand: <2.0 mm, >0.063 mm particles; silt: <0.063 mm, >0.002 mm particles; clay: <0.002 mm based on Wentworth scale.

generally the manufacturers claim detection well into sand sizes, it is advisable to integrate another technique, such as sieving or the settling tube (Syvitski et al., 1991a) for reliable data on coarse-grained sediments. Normally, sample splits are much smaller for the instrumental methods, and equal representation of larger particles becomes a problem.

Sieving is a labor-intensive method, reproducible only into the fine silt ranges. Screens that are not properly cleaned can be a source of contamination. The pipette method has become outdated because of inherent sources of error in the methodology and its turnaround time, but it is still used in many laboratories where the cost of the instrumental methods is prohibitive. In addition to the capital expenditures required to purchase these instruments, some of them require extensive training for their operation, such as image analysis. However, this procedure is recommended only for detailed studies of fine-grained sediments.

Overall, the instrumental techniques are recommended. In most cases, integrated software applications perform automated, statistical handling of large data sets simultaneously with the analysis. The instrumental techniques have proven to be much more precise and far less susceptible to operator error than the classical pipette method. Standardization of sample preparation and evaluation of the specific requirements of the sediment monitoring objectives will aid in the selection of the most appropriate particle size analysis method.

2.5　Specific surface area

2.5.1　Introduction

Physical and chemical properties of sediments are controlled, to some extent, by surface area. Differences in surface area result from variations in texture, clay mineral type, and organic matter content (Sheldrick,

1984). The surface area is highly correlated with cation exchange capacity and is inversely proportional to particle size. Horowitz and Elrick (1987) suggested that the surface area plays an important role in sediment trace element concentrations in sediment, as the processes involved in the association of the trace elements with the sediments are related to surface reactions or surface chemistry.

The specific surface is defined as the surface area per unit mass of material (e.g., soil, clay, or sediment) and is expressed in square meters per gram (m^2/g). According to Jackson (1979), the specific surface area of spherical and plate-shaped particles can be calculated from simple geometry. This assumes that only the external surface is considered. Clay-sized particles contribute the most to the specific surface of an inorganic sediment, and some clay minerals have extensive internal surfaces. The specific surface for clay minerals varies from 5 to 20 m^2/g for kaolinite to 700 to 800 m^2/g for smectite-group minerals. These large differences reflect the relative contribution of the internal surfaces of the expandable clay minerals.

In a sediment, particles have variable size and shape, and determination of the specific surface area using size and shape distribution would be tedious. Indirect methods have been developed to measure a specific surface based on the principle that solid materials will adsorb a monolayer of molecules, such as nonpolar gas molecules or polar liquids, such as ethylene glycol or glycerol. Some of the more commonly used methods are described below.

2.5.2   Methods for determination of specific surface area

The classical multimolecular adsorption theory predicts that the quantity of gas adsorbed by a solid surface is a measure of the surface area (Langmuir, 1918). Brunauer et al. (1938) derived an equation (BET equation) that allowed for the calculation of the number of molecules of gas (usually N_2, but ethane, water, ammonia, Ar, K, and CO_2 have also been used; Aylmore, 1974) that form a monolayer of adsorbate. The BET equation makes two assumptions: (1) the heat of adsorption of all molecular layers subsequent to the first is equal to the heat of liquefaction, and (2) at equilibrium, the rate of condensation on the surface is equal to the rate of evaporation from the first or subsequent layer (Mortland and Kemper, 1965; Carter et al., 1986). Details of the BET method are given in Palik (1977) and ASTM (1992, method C1069-86).

The sample, placed in tube or holder, must be degassed thoroughly at low pressure and high temperature before analysis. The high temperature assures the collapse of expandable minerals. Temperatures used, in general, can vary from about 60°C to over 350°C for varying lengths of time and are instrument specific. In the standard method (American Society for Testing and Materials, 1992) 1 h at 150°C is recommended. Immediately after degassing, the sample and tube are immersed in liquid N_2 to enable adsorption of N_2 gas. The sample is purged with one (single point BET) or three (three point BET) different ratios of N_2/He gases. Following complete adsorption, the sample and the tube are immersed in water at room temperature to facilitate desorption of N_2 gas. A detector measures the changes in thermal conductivity as gases enter and leave the sample tube. The detector must be calibrated before or after the analysis. The specific surface is calculated, knowing the volume of gas needed to form the monolayer (derived from the BET plot of partial pressure vs. volume adsorbed) and the cross-sectional area of the adsorbate molecules (Palik, 1977). This method only measures the external surface.

2.5.2.1   *Ethylene glycol monoethyl ether method*

Simpler methods to determine a specific surface area have been developed using polar molecules such as ethylene glycol, glycerol, and methylene blue, to name a few. Dyal and Hendricks (1950) initially used ethylene glycol (EG) to form a monomolecular layer over the entire surface of clays. This method was modified by Carter et al. (1965) when they introduced ethylene glycol monoethyl ether (EGME) as the adsorbate. This proved to be more rapid and more precise than the EG method (Carter et al., 1986) and has become commonly used in soil studies.

In this method, preweighed (approximately 1 g each), air-dried samples in weighing tins are placed into a vacuum desiccator containing P_2O_5. The desiccator is evacuated over a period of specific time (e.g., 45 to 60 min). The sample is left in the dessicator overnight and then reweighed. This cycle (evacuation, stabilization, and weighing) is repeated until the weights are relatively constant (i.e., within 1 mg). A small amount (e.g., 1 to 3 mL) of the EGME solution is added to make a slurry. The slurry is placed in a vacuum desiccator containing $CaCl_2$ and allowed to equilibrate for about 30 min. Then the cycle described above is repeated until the weights become relatively constant. The surface area is calculated according to Carter et al. (1986):

$$SSA = M_g/(M_s \cdot 0.000286) \qquad (17)$$

where, SSA = specific surface area (m^2/g), M_g = mass of sample and EGME retained (g), M_s = mass of P_2O_5-dried sample (g), and 0.000286 = mass of EGME required to form a monomolecular layer on a square meter of surface (g/m^2).

Details for this method are given in Sheldrick (1984) and Carter et al. (1986). Although this method measures total surface area, it tends to underestimate vermiculite and hydrated halloysite. These minerals tend to adsorb only one layer of EGME molecules in the interlayer (Carter et al., 1986).

2.5.2.2   *Methylene blue method*

Methylene blue (MB; $C_{16}H_{18}ClN_3S$; MW = 319.85) has been used as an adsorbate to determine the surface area of clays (Phelps and Harris, 1967; Pham and Brindley, 1970) and cation exchange capacity (Nevins and Weintritt, 1967). In this method, fully hydrated material is exposed to progressively increasing concentrations of a dilute MB solution (e.g., 0.2% by weight, Faruqi et al., 1967). The amount adsorbed is determined indirectly, by spectrophotometrically measuring the amount of MB that remains in solution. The change in concentration of the MB solutions is plotted against the number of meq per 100 grams of material of MB added. The point where the line begins to curve corresponds to the formation of a monolayer or point of optimum flocculation. The specific surface area is calculated by

$$SSA = M_f \times A_m \times AV \qquad (18)$$

where, M_f = number of meq of MB adsorbed per 100 grams material, A_m = surface area of MB molecule (130×10^{-20} m^2/molecule), and AV = Avogadro's Number (6.023×10^{20} molecules per meq). This method is described in detail in Pham and Brindley (1970). The MB method measures the total surface area.

A variation of this method by Cuillé (1976) and Tran (1977) was used by Locat et al. (1984) in their study of sensitive clays from eastern Canada. As in the original method, varying amounts of a dilute MB solution were added to a soil-water mixture. After each addition, a drop of the MB-soil-water mixture was placed on chromatographic paper. The point of saturation was indicated when the MB diffused out of the drop of suspension into the paper.

2.5.3   *Recommendations*

If the sediments do not contain expandable clay minerals, each of these methods should yield similar results. As the BET method only measures external surface area, it would be a poor choice for analyzing sediments containing expandable clay minerals. The BET method is time-consuming and requires highly specialized equipment.

In a study that compared three methods (BET, EGME, and MB), Barran (1987) determined that the specific surface areas determined by the EGME and MB methods gave similar results, with the exception of one vermiculite-bearing soil sample. Both these methods were less susceptible to procedural error than the BET method. Barran (1987) suggested that the EGME was the simplest and least time-consuming of these three methods.

Other methods to measure specific surface given in the literature should not be ruled out. For example, Ristori et al. (1989) described a method that uses p-Nitrophenol (pNP) instead of EGME. Their results indicated good agreement with the N_2-BET sorption method. However, lower values were obtained relative to the EGME method. Ristori et al. (1989) believed that these lower values were more realistic for the soils examined. The method was simple, rapid, and reproducible. In contrast, Saluja et al. (1987) developed a rapid (<2 hours per sample) method to measure the surface area of powders. They derived the surface area from the enthalpy of saturation adsorption of n-butanol from a heptane solution using adsorption-desorption calorimetry. Their results were in good agreement with the BET method.

In selecting a method to determine the specific surface of sediments, it is important to characterize the sample well in terms of its mineralogy. The presence of abundant expandable minerals, or vermiculite, may preclude the selection of certain methods. Selection may also depend upon the commercial availability of the equipment, cost, and end use of the data.

Acknowledgments

The authors would like to acknowledge the general assistance given by M.M. Attard, R. Kelly, R. Lacroix, G. Lemieux, K. Mooney, P.M. O'Regan, and the staff of the Canadian Geoscience Information Centre of the Geological Survey of Canada (GSC). Discussions with T.J. Katsube, A. Tsai, and D.A. Walker of the GSC, and H. Kodama and G.C. Topp of the Land Resources Division of Agriculture Canada (AgCan) were most valuable and helped to keep us focused. We are grateful to T.W. Anderson, R.N.W. DiLabio, R.A. Klassen, G.M. LeCheminant, and J.A. Percival of the GSC and G.C. Topp (AgCan) for their helpful reviews of parts or of the entire manuscript. Thanks also to A. Mudroch, J.M. Azcue, and P. Mudroch, the editors of this book, for their assistance and encouragement.

References

Ackermann, F., A procedure for correcting the grain size effect in heavy metal analysis of estuarine and coastal sediments, *Environ. Technol. Lett.*, 1, 518, 1980.

American Petroleum Institute (API), Recommended Practices for Core-Analysis Procedure, API Recommended Practice 40, Washington, D.C., 1960.

American Society for Testing and Materials (ASTM), 1992 Annual Book of ASTM Standards (Sect. 15) General Products, Chemical Specialties, and End Use Products: Glass, Ceramic, Whitewares, Vol. 2, ASTM, Philadelphia, PA, 1992.

American Society for Testing and Materials (ASTM), 1994 Annual Book of ASTM Standards (Sect. 4) Construction: Soil and Rock (I), Vol. 8, ASTM, Philadelphia, PA, 1994a.

American Society for Testing and Materials (ASTM), 1994 Annual Book of ASTM Standards (Sect. 14) General Methods and Instrumentation, Vol. 2, ASTM, Philadelphia, PA, 1994b.

Aylmore, L.A.G., Gas sorption in clay mineral systems, *Clays Clay Miner.*, 22, 175, 1974.

Barbanti, A. and Bothner, M.H., A procedure for partitioning bulk sediments into distinct grain-size fractions for geochemical analysis, *Environ. Geol.*, 21, 3, 1993.

Barran, C., Specific Surface Area Determination of Soils: A Comparison of Three Methods, Honors Project, Department of Geography, Carleton University, Ottawa, Ontario, 1987.

Bengtsson, L., Chemical analysis, in Paleohydrological Changes in the Temperate Zone in the Last 15,000 Years, Subproject B, Lake and Mire Environments, Vol. II, Specific Methods, Berglund, B.E., Ed., Int. Geol. Corr. Prog., Proj. 158, 1979, 113.

Blake, G.R., Bulk density, in Methods of Soil Analysis — Part 1 — Physical and Mineralogical Properties, Including Statistics of Measurement and Sampling, Black, C.A., Ed., Agronomy Series No. 9, American Society of Agronomy, Inc., Madison, WI, 1965, 374.

Blake, G.R. and Hartge, K.H., Bulk density, in Methods of Soil Analysis — Part I — Physical and Mineralogical Methods, Klute, A., Ed., Agronomy Series No. 9, 2nd ed., American Society of Agronomy, Inc. and Soil Science of America, Inc., Madison, WI, 1986a, 363.

Blake, G.R. and Hartge, K.H., Particle density, in Methods of Soil Analysis — Part I — Physical and Mineralogical Methods, Klute, A., Ed., Agronomy Series No. 9, 2nd ed., American Society of Agronomy, Inc. and Soil Science of America, Inc., Madison, WI, 1986b, 377.

Boyce, R.E., Appendix I, Physical properties — methods, in *Initial Reports of the Deep Sea Drilling Project*, Edgar, N.T., Kaneps, A.G., and Herring, J.R., Eds., U.S. Government Printing Office, Washington, D.C., 15, 1973, 1115.

Boyce, R.E., Definitions and laboratory techniques of compressional sound velocity parameters and wet-water content, wet-bulk density, and porosity parameters by gravimetric and gamma ray attenuation techniques, in *Initial Reports of the Deep Sea Drilling Project*, Kaneps, A.G., Ed., U.S. Government Printing Office, Washington, D.C., 33, 1976, 931.

Brunauer, S., Emmett, P.H., and Teller, E., Absorption of gases in multi-molecular layers, *J. Am. Chem. Soc.*, 60, 309, 1938.

Carter, D.L., Heilman, M.D., and Gonzalez, C.L., Ethylene glycol monoethyl ether for determining surface area of silicate minerals, *Soil Sci.*, 100, 356, 1965.

Carter, D.L., Mortland, M.M., and Kemper, W.D., Specific surface, in Methods of Soil Analysis — Part I — Physical and Mineralogical Methods, Klute, A., Ed., Agronomy Series No. 9, 2nd ed., American Society of Agronomy, Inc. and Soil Science of America, Inc., Madison, WI, 1986, 413.

Carter, M.R. and Ball, B.C., Soil porosity, in *Soil Sampling and Methods of Analysis*, Carter, M.R., Ed., Can. Soc. Soil Sci., CRC/Lewis, Boca Raton, FL, 1993, 581.

Christensen, E.R. and Goetz, R.H., Historical fluxes of particle-bound pollutants from deconvolved sedimentary records, *Environ. Sci. Technol.*, 21(11), 1088, 1987.

Cuillé, C., Qualité des Sables Fins Auversiens et Stampiens de la Région Parisienne, Thesis, Université de Pierre et Marie Curie, France.

Culley, J.L.B., Density and compressibility, in *Soil Sampling and Methods of Analysis*, Carter, M.R., Ed., Can. Soc. Soil Sci., CRC/Lewis, Boca Raton, FL, 1993, 529.

Dadey, K.A. and Klaus, A., Physical properties of volcaniclastic sediments in the Izu-Bonin area, in *Proceedings of the Ocean Drilling Program, Scientific Results*, Maddox, E.M., Ed., College Station, TX, 126, 543, 1992.

Dadey, K.A., Janecek, T., and Klaus, A., Dry bulk density: its use and determination, in *Proceedings of the Ocean Drilling Program, Scientific Results*, Maddox, E.M., Ed., College Station, TX, 126, 551, 1992.

Dalton, F.N., Development of time-domain reflectometry for measuring soil water content and bulk soil electrical conductivity, in Advances in Measurement of Soil Physical Properties: Bringing Theory into Practice, Topp, G.C., Reynolds, W.D., and Green, R.E., Eds., Soil Science Society of America, Madison, WI, Special Publication 30, 1992, 143.

Damiani, V. and Thomas, R.L., The surficial sediments of the big bay section of the Bay of Quinte, Lake Ontario, *Can. J. Earth Sci.*, 11, 1562, 1974.

Day, P.A., Particle fractionation and particle size analysis, in Methods of Soil Analysis — Part 1 — Physical and Mineralogical Properties, Including Statistics of Measurement and Sampling, Black, C.A., Ed., Agronomy Series No. 9, American Society of Agronomy, Inc., Madison, WI, 1965, 545.

de Boer, G.B.J., de Weerd, C., de Thoenes, D., and Goossens, H.W.J., Laser diffraction spectrometry: Fraunhofer diffraction versus Mie scattering, *Particle Characterization*, 4, 14, 1987.

de Groot, A.J., Zschuppe, K.H., and Salomons, W., Standardization of methods of analysis for heavy metals in sediments, in *Sediment/Freshwater Interaction*, Sly, P.G., Ed., Developments in Hydrobiology 9, Dr. W. Junk Publishers, The Hague, 1982, 689.

DiLabio, R.N.W., Residence sites of trace elements in oxidized till, in Drift Exploration in the Canadian Cordillera, B.C. Min. of Energy, Mines and Petroleum Resources, Paper 1995-2, 1995, 139.

Dowding, C.H. and Pierce, C.E., Use of the time domain reflectometry to detect bridge scour and monitor pier movement, in *Time Domain Reflectometry in Environmental, Infrastructure, and Mining Applications*, U.S. Bureau of Mines, Minneapolis, MN, Special Publication SP 19-94, 1994, 579.

Durham, R.W. and Oliver, B.G. History of Lake Ontario contamination from the Niagara River by radiodating and chlorinated hydrocarbon analysis, *J. Great Lakes Res.*, 9(2), 160, 1983.

Dyal, R.S. and Hendricks, S.B., Total surface of clays in polar liquids as a characteristic index, *Soil Sci.*, 69, 421, 1950.

Elrick, D.E., Kachanoski, R.G., Pringle, E.A., and Ward, A.L., Parameter estimates of field solute transport models based on time domain reflectometry measurements, *Soil Sci. Soc. Am. J.*, 56, 1663, 1992.

Engstrom, D.R. and Wright, H.E., Jr., Chemical stratigraphy of lake sediments as a record of environmental change, in *Lake Sediments and Environmental History*, Haworth, E.Y. and Lund, J.W.G., Eds., University of Minnesota Press, Minneapolis, MN, 1984, 11.

Environment Canada, St. Lawrence Centre, Methods manual for sediment characterization, Ministère de l'Environment du Quebec, 1993, 145.

Faruqi, F.A., Okuda, S., and Williamson, W.O., Chemisorption of methylene blue by kaolinite, *Clay Miner.*, 7, 19, 1967.

Förstner, U., Accumulative phases of heavy metals in limnic sediments, *Hydrobiologia*, 91, 269, 1982.

Förstner, U., Inorganic sediment chemistry and elemental speciation, in *Sediments: Chemistry and Toxicity of In-Place Pollutants*, Baudo, R., Giesy, J.P., and Muntau, H., Eds., CRC/Lewis, Boca Raton, FL, 1990, 61.

Förstner, U. and Salomons, W., Trace metal analysis on polluted sediments, Assessment of sources and intensities, *Environ. Technol. Lett.*, 1, 494, 1980.

Förstner, U. and Wittmann, G., *Metal Pollution in the Aquatic Environment*, 2nd ed., Springer-Verlag, New York, 1981, 197.

Freeze, R.A. and Cherry, J.A., *Groundwater*, Prentice-Hall, Englewood Cliffs, NJ, 1979, 604.

Gardner, W.H., Water content, in Methods of Soil Analysis — Part I — Physical and Mineralogical Methods, Klute, A., Ed., Agronomy Series No. 9, 2nd ed., American Society of Agronomy, Inc. and Soil Science of America, Inc., Madison, WI, 1986, 493.

Gee, G.W. and Bauder, J.W., Particle-size analysis, in Methods of Soil Analysis — Part I — Physical and Mineralogical Methods, Klute, A., Ed., Agronomy Series No. 9, 2nd ed., American Society of Agronomy, Inc. and Soil Science of America, Inc., Madison, WI, 1986, 383.

Hakanson, L., Bottom dynamics in lakes, in *Sediment/Freshwater Interaction*, Sly, P.G., Ed., Developments in Hydrobiology 9, Dr. W. Junk Publishers, The Hague, 1982, 9.

Haley, M.P. and Joyce, I.H., Modern instrumental techniques for particle size evaluation, *J. Can. Cer. Soc.*, 53, 15, 1984.

Hamilton, E.L., Ocean basin ages and amounts of original sediments, *J. Sed. Petrol.*, 30, 370, 1960.

Harrison, W., Lynch, M.P., and Altschaeffl, A.G., Sediments of lower Chesapeake Bay, with emphasis on mass properties, *J. Sed. Petrol.*, 34, 727, 1964.

Hayhoe, H.N., Topp, G.C., and Bailey, W.G., Measurement of soil water contents and frozen soil depth during a thaw using time-domain reflectometry, *Atmosphere-Ocean*, 21, 299, 1983.

Horowitz, A.J. and Elrick, K.A., The relation of stream sediment surface area, grain size and composition to trace element chemistry, *Appl. Geochem.*, 2, 437, 1987.

Horowitz, A.J. and Elrick, K.A., Interpretation of bed sediment trace metal data: methods for dealing with the grain size effects, in *Chemical and Biological Characterization of Sludges, Sediments, Dredge Spoils, and Drilling Muds*, ASTM STP 976, Lichtenberg, J.J., Winter, J.A., Weber, C.I., and Fradkin, L., Eds., American Society for Testing and Materials, Philadelphia, PA, 1988, 114.

Horowitz, A.J., Elrick, K.A., and Hooper, R.P., The prediction of aquatic sediment-associated trace element concentrations using selected geochemical factors, *Hydrolog. Process.*, 3, 347, 1989.

Jackson, M.L., Soil Chemical Analysis — Advanced Course, 2nd ed., 11th printing, published by the author, Madison, WI, 1979.

Jenne, E., Kennedy, V., Burchard, J., and Ball, J., Sediment collection and processing for selective extraction and for total metal analysis, in *Contaminants and Sediments*, Baker, R., Ed., Ann Arbor Science, Ann Arbor, MI, 1980, 169.

Jones, B.F. and Bowser, C.J., The mineralogy and related chemistry of lake sediments, in *Lakes Chemistry Geology Physics*, Springer-Verlag, New York, 1978, 179.

Kachanoski, R.G., Pringle, E., and Ward, A., Field measurement of solute travel times using time domain reflectometry, *Soil Sci. Am. J.*, 56, 47, 1992.

Kemp, A.L. and Thomas, R.L., Cultural impact on the geochemistry of the sediments of Lakes Ontario, Erie and Huron, *Geosci. Can.*, 3, 191, 1976.

Krumbein, W.C. and Pettijohn, F.J., Principles and methods of size analysis, in *Manual of Sedimentary Petrography*, Mather, K.F., Ed., Appleton-Century-Crofts, New York, 1938, 91.

Langmuir, I., The adsorption of gases on plane surfaces of glass, mica, and platinum, *J. Am. Chem. Soc.*, 40, 1361, 1918.

Lavoie, D.L., Bruns, T.R., and Fisher, K.M., Geotechnical and logging evidence for unconsolidation of Lau basin sediments: rapid sedimentation vs. fluid flow, in *Proceedings of the Ocean Drilling Program, Scientific Results*, Maddox, E.M., Ed., College Station, TX, 135, 787, 1994.

Lewis, D.W., *Practical Sedimentology*, Hutchinson Ross Publishing Co., Stroudsburg, PA, 229, 1984.

Lindsay, P.J. and Shilts, W.W., A standard laboratory procedure for separating clay-sized detritus from unconsolidated glacial sediments and their derivatives, in Drift Exploration in the Canadian Cordillera, British Columbia Ministry of Energy, Mines and Petroleum Resources, Paper 1995-2, 1995, 165.

Locat, J., Lefebvre, G., and Ballivy, G., Mineralogy, chemistry, and physical properties interrelationships of some sensitive clays from eastern Canada, *Can. Geotech. J.*, 21, 530, 1984.

Lonardi, A., Measurement of chemical/physical properties, in *Initial Reports of the Deep Sea Drilling Project*, Wise, S.W., Jr., Ed., U.S. Government Printing Office, Washington, D.C., 36, 1033, 1977.

McCave, I.N., Bryant, R.J., Cook, H.F., and Coughanowr, C.A., Evaluation of a laser-diffraction-size analyzer for use with natural sediments, *J. Sed. Petrol.*, 56, 561, 1986.

McCave, I.N. and Syvitski, J.P.M., Principles and methods of geological particle size analysis, in *Principles, Methods, and Applications of Particle Size Analysis*, Syvitski, J.P.M., Ed., Cambridge University Press, New York, 1991, 3.

Mehra, O.P. and Jackson, M.L., Iron oxide removal from soils and clays by a dithionite-citrate system buffered with sodium bicarbonate, *Clays Clay Miner.*, 7, 317, 1960.

Mortland, M.M. and Kemper, W.D., Specific surface, in Methods of Soil Analysis — Part 1 — Physical and Mineralogical Properties, Including Statistics of Measurement and Sampling, Black, C.A., Ed., Agronomy Series No. 9, Amercian Society of Agronomy, Inc., Madison, WI, 1965, 532.

Mudroch, A., Particle size effects on concentration of metals in Lake Erie bottom sediments, *Water Poll. Res. J. Can.*, 19, 27, 1984.

Mudroch, A., Lake Ontario sediments in monitoring pollution, *Environ. Monitor. Assess.*, 28, 117, 1993.

Mudroch, A. and Azcue, J.M., *Manual of Aquatic Sediment Sampling*, CRC/Lewis, Boca Raton, FL, 1995, 219.

Mudroch, A. and Bourbonniere, R.A., Sediment preservation, processing, and storage, in *Handbook of Techniques for Aquatic Sediments Sampling*, 2nd ed., Mudroch, A. and MacKnight, S.D., Eds., CRC/Lewis, Boca Raton, FL, 1994, 131.

Mudroch, A. and MacKnight, S.D., Eds., *Handbook of Techniques for Aquatic Sediments Sampling*, 2nd ed., CRC/Lewis, Boca Raton, FL, 1994a, 236.

Mudroch, A. and MacKnight, S.D., Bottom sediment sampling, in *Handbook of Techniques for Aquatic Sediments Sampling*, 2nd ed., Mudroch, A. and Mac-Knight, S.D., Eds., CRC/Lewis, Boca Raton, FL, 29, 1994b.

Nevins, M.J. and Weintritt, D.J., Determination of cation exchange capacity by methylene blue adsorption, *Am. Ceram. Soc. Bull.*, 46, 587, 1967.

Palik, D.F., Specific surface area measurements on ceramic powders, *Powder Technol.*, 18, 12, 1977.

Petruk, W., Image analysis of minerals, in *Image Analysis Applied to Mineral and Earth Sciences*, Petruk, W., Ed., Mineral. Assoc. Can., Short Course, 1989, 6.

Pham, T.H. and Brindley, G.W., Methylene blue adsorption by clay minerals: Determination of surface areas and cation exchange capacities, *Clays Clay Miner.*, 18, 203, 1970.

Phelps, G.W. and Harris, D.L., Specific surface and dry strength by methylene blue adsorption, *Am. Ceram. Soc. Bull.*, 47, 1146, 1967.

Phogat, V.K., Aylmore, L.A.G., and Schuller, R.D., Simultaneous measurement of the spatial distribution of soil water content and bulk density, *Soil Sci. Soc. Am. J.*, 55, 908, 1991.

Ristori, G.G., Sparvoli, E., Landi, L., and Martelloni, C., Measurement of specific surface areas of soils by p-Nitrophenol adsorption, *Appl. Clay Sci.*, 4, 521, 1989.

Rootare, H.M., A review of mercury porosimetry, *Persp. Powder Metall.*, 5, 225, 1970.

Ross, D.A. and Riedel, W.R., Comparison of upper parts of some piston cores with simultaneously collected open-barrel cores, *Deep Sea Res.*, 14, 285, 1967.

Ryley, M.D., The Use of a Microwave Oven for the Rapid Determination of the Moisture Content of Soils, Road Research Laboratory, Report LR 280, 1969.

Saheurs, J.P.G., Wilson, W.P., and Sherwood, W.G., Sample preparation, in *Analysis of Geological Materials*, Riddle, C., Ed., Marcel Dekker Inc., New York, 1993, 65.

Salomons, W. and Förstner, U., *Metals in the Hydrocycle*, Springer-Verlag, Berlin, 1984, 349.

Saluja, P.P.S., Oscarson, D.W., Miller, H.G., and LeBlanc, J.C., Rapid determination of surface areas of mineral powders using adsorption calorimetry, in Proceedings of the International Clay Conference, Denver, 1985, Schultz, L.G., van Olphen, H., and Mumpton, F.A., Eds., The Clay Minerals Society, Bloomington, IN, 1987, 267.

Schultheiss, P.J., Physical and geotechnical properties of sediments from the northwest Pacific: deep sea drilling project leg 86, 1985, in *Initial Reports of the Deep Sea Drilling Project*, Turner, K.L., Ed., U.S. Government Printing Office, Washington, D.C., 86, 701, 1985.

Sheldrick, B.H., Analytical Methods Manual 1984, Research Branch, Agriculture Canada, LRRI Cont. No. 84-30, 1984.

Shilts, W.W., Geochemical partitioning in till, in *Drift Exploration in the Canadian Cordillera*, British Columbia Ministry of Energy, Mines and Petroleum Resources, Paper 1995-2, 1995, 149.

Soane, B.D., Dual energy gamma-ray transmission for coincident measurement of water content and dry bulk density of soil, *Nature* (London), 214, 1273, 1967.

Stein, R., Rapid grain-size analyses of clay and silt fraction by Sedigraph 5000D: comparison with Coulter Counter and Atterberg methods, *J. Sed. Petrol.*, 55, 590, 1985.

Syvitski, J.P.M., Asprey, K.W., and Clattenburg, D.A., Principles, design and calibration of settling tubes, in *Principles, Methods, and Applications of Particle Size Analysis*, Syvitski, J.P.M., Ed., Cambridge University Press, New York, 1991a, 45.

Syvitski, J.P.M., LeBlanc, K.W.G., and Asprey, K.W., Interlaboratory, interinstrument calibration experiment, in *Principles, Methods, and Applications of Particle Size Analysis*, Syvitski, J.P.M., Ed., Cambridge University Press, New York, 1991b, 174.

The Marine Geotechnical Consortium, Geotechnical properties of northwest Pacific pelagic clays: deep sea drilling project leg 86, hole 576A, in *Initial Reports of the Deep Sea Drilling Project*, Turner, K.L., Ed., U.S. Government Printing Office, Washington, D.C., 86, 723, 1985.

Topp, G.C., The measurement and monitoring of soil water content by TDR, in Soil Moisture Modelling and Monitoring for Regional Planning, Eley, F.J., Granger, R., and Martin, L., Eds., National Hydrological Research Institute, Saskatoon, Symposium No. 9, 1992, 155.

Topp, G.C., Soil water content, in *Soil Sampling and Methods of Analysis*, Carter, M.R., Ed., Can. Soc. Soil Sci., CRC/Lewis, Boca Raton, FL, 541, 1993.

Topp, G.C., Zegelin, S.J., and White, I., Monitoring water content using TDR: an overview of progress, in Time Domain Reflectometry in Environmental, Infrastructure, and Mining Applications, U.S. Bureau of Mines, Minneapolis, Minnesota, Special Publication SP 19-94, 1994, 67.

Tran, N.L., Un nouvel essai d'identification des sols l'essai au bleu de méthylène, *Bull. Liaison Laborat. Ponts et Chaussées*, No. 88, 1977, 136.

Walker, D.A. and LeCheminant, G.M., An integrated image and x-ray analysis system: description and techniques in a multiple use laboratory, in Image Analysis Applied to Mineral and Earth Sciences, Petruk, W., Ed., Mineral. Assoc. Can., Short Course, 1989, 46.

Walker, D.A., Paktunc, A.D., and Villeneuve, M.E., Automated image analysis applications: characterization of (1) platinum-group minerals and (2) heavy mineral separates, in Image Analysis Applied to Mineral and Earth Sciences, Petruk, W., Ed., Mineral. Assoc. Can., Short Course, 1989, 94.

Walker, L.K. and Reynolds, V.G., Particle Size Analyzer Comparison, Mineral Sciences Laboratories Division Report MSL 91-61 (TR), CANMET, Energy, Mines, and Resources Canada, 1991.

Wentworth, C.K., A scale of grade and class terms for clastic sediments, *J. Geol.*, 30, 377, 1922.

Wetzel, A., Interrelationships between porosity and other geotechnical properties of slowly deposited, fine-grained marine surface sediments, *Mar. Geol.*, 92, 105, 1990.

Whalley, B., Physical properties, in *Geomorphological Techniques*, Goudie A., Ed., George Allen & Unwin, London, 1981, 80.

Wroth, C.P. and Wood, D.M., The correlation of index properties with some basic engineering properties of soils, *Can. Geotech. J.*, 15, 137, 1978.

Zavarol, D.Z., Campanella, R.G., and Luternauer, J.L., Geotechnical properties of sediments on the central continental shelf of western Canada, in Current Research, Part H, Geol. Surv. Canada, Paper 89-1H, 141, 1989.

chapter three

Measurement of Eh and pH in aquatic sediments

Pierre Brassard

3.1   Introduction

The recognition of the importance of pH and Eh as "master variables" has been a great contribution to environmental science, particularly to aquatic chemistry. An accurate determination of these two parameters would enable the prediction of concentrations and chemical forms of the major and trace elements in water and aquatic sediments. Each combination of Eh and pH values uniquely adjusts the proportions of all dissolved species in solution. The appreciation of this potential has led investigators of biogeochemical processes to study and refine methods for measuring Eh and pH in the sediments. However, the natural aquatic environment is not a thermodynamic system at perfect equilibrium. Chemical reactions proceed at variable rates, and instruments suffer biases. To be useful, the measured values require careful interpretation. The following questions usually arise and need to be considered during the interpretation of the results obtained by measuring pH and Eh. Is the system at equilibrium, and if not, can we assume a predictable offset from the theory? What are the important reactions influencing the instrument(s) used in the measurements? Are these reactions relevant to the study? Is the system really homogeneous, or are we just assuming that it is? These and similar questions exemplify the problems associated with instruments based on rigorous theoretical requirements that are often not maintained during the measurement. In a typical field operation, Eh and pH electrodes are first calibrated with standard solutions for the purpose of adjusting the instrument(s) under some well-defined equilibrium conditions. The instrument is then used in the real environmental conditions (i.e., measurements of Eh and pH of water and sediments), and the values of the measured

1-56670-155-4/97/$0.00+$.50
© 1997 by CRC Press, Inc.

parameters are recorded. The belief that the measurements actually indicate the real environmental conditions is based on two implicit points. First, we assume that somehow the real conditions can be extrapolated from those existing in the standard solutions. Second, we also assume that the instrument response is similar and predictable under both conditions. In the end, the reality of the measurements largely depends on the extent to which these two assumptions are respected.

The purpose of this chapter is not to discourage the measurement and use of the above master variables in sediment studies, but to stress the importance of knowing the limits beyond which their use becomes questionable.

3.2   Measurement of Eh

3.2.1   General considerations

A quantitative interpretation of Eh measurements in natural aquatic systems is difficult because of problems associated with the measurement technique, the performance of the inert metal electrode, and the thermodynamic behavior of the environment (Whitfield, 1969). Critics regarding the use of Eh for determining the redox potential in marine sediments have pointed out that the absence of uniform measurement procedures, such as the choice and preparation of electrodes, their retention time in the sediment, etc., leads to different results (Rozanov, 1982). Whitfield (1974) pointed out several main physico-chemical objections to the use of Eh as an operational parameter. Some of the problems identified by him, as well as other problems observed during the attempts to measure Eh in bottom sediments discussed by Mudroch and Azcue (1995), were as follows:

- Difficulties associated with the disturbance of the sediment sample, resulting in release or adsorption of gases, particularly O_2 and H_2S, and reactions at the liquid junction of the reference electrode, such as precipitation of metal sulfides. The disturbance can occur during sampling but mainly originates during insertion of the electrode into the sediment.
- Instability and poor reproducibility of the measurements originate from low exchange current densities at the platinum surface and predominance of mixed potentials. Measurements obtained by the platinum electrode depend on the nature and rates of the reactions at the electrode surface. Therefore, a quantitative interpretation of the measured values is very difficult.
- Response of the platinum electrode to changes in the environment will depend greatly on the properties of the platinum

surface and the presence of adherent surface coatings. For example, upon inserting the electrode into fine-grained sediments containing a small amount of oxygen and reducing agents, a slow formation of platinum sulfide on the surface of the electrode may generate more negative potentials. On the other hand, in coarser sediments from oxygenated waters, such as mixtures of sand and silt in a near-shore area of a lake, the platinum may act as an oxide electrode and respond to pH rather than to the partial pressure of oxygen.

It is important to recognize that the platinum electrode is at best a transducer of a presumed equilibrium condition existing at the surface of the electrode. According to the Nernst relation, equilibrium is established when the chemical potential arising from differences in concentrations of the reduced and oxidized species exactly matches the electrical potential difference between the reference and platinum electrodes. At this point, the net electron flow is zero, and the reading of the platinum electrode indicates true redox potential.

However, actual values obtained by using platinum electrodes are biased by the measuring equipment. Since a small current must flow in the circuit to activate the meter, the potential at the electrode will be shifted toward oxidation or reduction in order to offset this bias current by an equal exchange current. If the concentrations in the redox system are small, there is little redox buffer capacity, or "poise" (Zobell, 1946), and the shift becomes excessive. Reliable measurements can only be made for an exchange current greater than 10^{-7} A/cm^2, which would correspond to the concentration of Fe ion in excess of 10^{-5} M (Stumm and Morgan, 1981), as measured by a high impedance, modern design meter. Kempton et al. (1990) found Nernstian responses for Fe systems at exchange currents well above the 10^{-7} A/cm^2 predicted by Stumm and Morgan (1981).

Another problem arises from mixed potentials. The calculations of equilibrium potential are based on the assumption that all participating redox reactions are coupled; i.e., they can share their electron. However, this is rarely the case in sediments since many reactions do not occur at the same rate. When two redox systems are not coupled with each other, the resulting potential is a composition of tendencies of the two reactions to transfer their electrons. The potential rests at a pseudoequilibrium, where the sum of all exchange currents is zero. Thus, equilibrium potentials based on the Nernst equation are, at best, operational. This problem is apparent in the study of Norrström (1994), who attempted to measure oxidation levels in wet soils. Ten platinum electrodes made from stainless rods and platinum wire were inserted in the soil in a circle about 10 cm in diameter, with a reference electrode in the center. The ten electrodes showed variations around 100 mV even after 24 h of equilibration. In addition, the drift rates of the values were different for each electrode. Similar results occurred in several soils of different humidity, granulometry, and organic matter.

Whitfield (1972) studied the problem from a different angle by measuring the charge/discharge profiles of natural cells made in the laboratory. The cell was left to achieve equilibrium Eh during the "charge" period. The "discharge" occurred when the anode and cathode of the cell were shunted with a resistor and the current measured. The cells were made by maintaining an oxidized layer of fresh water over a reduced layer of saltwater-saturated mud, amended with minced meat to promote a reduced condition. Gold and platinum electrodes were inserted in the layers. Generally, all cells behaved similarly for several cycles of charge/discharge, recharging always taking longer than discharge. The difference was attributed to adsorption on the surface of the platinum electrode during discharge. In all cases the calculated equilibrium current exceeded the lower value of 10^{-7} A/cm^2 (Morris and Stumm, 1967), except for the gold electrode. The author concluded that values obtained by the electrodes were reproducible in natural conditions provided that a small number of well-poised redox reactions controlled the equilibrium, such as the system S^0(rhombic)/HS$^-$(aq). It was suggested that a measure of equilibrium current in the field be used as a reliability indicator for Eh measurement.

Whitfield (1969) addressed the problem of drift and inhomogeneities by using two platinum electrodes immersed side by side in the sediment sample to ensure that reproducible potentials were being measured. The values of Eh measured by these two electrodes differed by 10 to 30 mV, due to the surface properties of the platinum and poisoning of the two electrode surfaces by an irreversible attack at different rates. To measure simultaneously Eh, pH, and sulfide activity (pS^{-2}), a total of five electrodes combined into a compound electrode were used, which enabled all working electrodes to be introduced in a single probe into the sediment. When this probe was used in conjunction with a remote junction reference electrode, stable and reproducible results were obtained, even on highly reduced samples. The Eh of each sample was monitored until a steady value (drift <1 mV/min) was observed. This value was recorded and the Eh measured by the second of the five electrodes was registered. The sulfide activity and pH were measured, and the Eh registered by the first electrode was checked again. When the results differed by more than 30 mV, a further 5 min were allowed for equilibration. The temperature was recorded before the sample was discarded. The whole cycle of readings was completed within approximately 10 min.

Some operational procedure is needed to bring uniformity to the bias of drift of the values caused by reactions occurring at different rates or by a slow fouling. A lower limit of stability of 0.1 mV/min or less may be arbitrarily set. Readings of values measured in sediments show that redox potentials stabilize within a few minutes in the oxidized layer but take up to several hours in the interface zone where bacterial population is high. The drift of the values is attributed to disturbances of sediment porewater, generated by movements of the probe and long equilibration times (Plante et al., 1989). To compensate for electrode biases, Dechev et al. (1974) have proposed an index of poisoning based on the ratio between the oxidized and the reduced zones in the sediment. The approach is operational and assumes poisoning to be similar at both sediment zones.

Generally, the measurement of Eh, although based on sound principles, fails as a general indicator because the sediment system does not conform to the ideal conditions required by equilibrium. An alternate approach is to infer redox potential by measuring the concentration of major constituents in the sediment and calculating the equivalent potential. Berner (1963) has demonstrated that a platinum electrode will record the potential of $S(s) + 2e^- = S^{-2}$ in H_2S-bearing sediments. However, there is no indication that the potential represents the ensemble of all other reactions and the extent to which they are coupled. Similarly, successful uses of Eh to estimate Fe^{3+} were reported in studies of aquifers because they were based on one dominant hydrous ferric oxide system (Macalady et al., 1990).

The measurements of Eh will thus be consistent if a sediment is overwhelmed by one dominant process. As a consequence, they should be used only in cases where the dominant redox couple is known and where the process under study is known to be related to that couple. This caveat severely limits the scope of applicability of Eh measurements. The major requirement for reliable Eh measurement is therefore a simple, well-defined, and stable Nernstian system. More often than not, such a condition can only be found in the laboratory. Up to the present, a reliable use of Eh for field applications is restricted to systems dominated by either sulfides, Fe^{3+}/Fe^{2+}, or Mn^{2+}/MnO_2 couples.

3.2.2 Methods used for field measurements of Eh

The problems listed above indicate that the potential measured by an Eh electrode arises both from fouling and nonideal redox systems. Studies published by different scientists on the meaning and problems with measuring and interpretation of sediment pH and Eh should be consulted prior to measuring these parameters in sediment samples (Mortimer, 1941, 1942; Berner, 1963; Morris and Stumm, 1967; Whitfield, 1969; Tinsley, 1979; Bates, 1981). Further, consideration of the excellent reviews of Berner (1981) and Thorstenson (1984) on the implications of electron activity and interpretations of redox processes in the sedimentary environment in natural waters is recommended. Some approaches are given here for those who, after considering the above limitations, decide to use Eh as an indicator of sediment processes.

It was found that cleaning the platinum electrodes by rubbing with a fine abrasive cloth was more efficient than rinsing with different chemicals and distilled water. The potential of the Eh electrode/calomel cell was measured occasionally in a Zobell solution (Zobell, 1946) (0.003 *M* potassium ferricyanide and 0.003 *M* potassium ferrocyanide in 0.1 *M* KCl) to check the performance of the liquid junction. The potential of the cell was adjusted to the standard hydrogen scale by adding 250 mV to the measured value to enable comparison with other data.

Bagander and Niemistö (1978) measured the Eh in marine sediment cores with a specially designed electrode inserting an attachment adaptable to a subsample slide used in the sectioning of the cores (Figure 3-1). This attachment allowed electrode measurements with minimal air contamination and disturbance of the sediment subsamples. Two different sets of electrodes were used. The first set consisted of a platinum wire fused into the end of a glass tube and an Ag/AgCl reference electrode. The second set was an Orion combination electrode.

Fenchel (1969) used redox potential discontinuities as an indicator of oxygenation and, indirectly, as an indicator of organic matter content in sediments. Measurements of Eh in a sediment profile must be carried out on a stable column of sediment, either *in situ* or in the laboratory. The former requires a very stable support mechanism anchored near the measured sediment column. The latter requires extracting a core from the sediments with precautions regarding maintenance of anoxic conditions. The core must be handled under an oxygen-free atmosphere at all times, which is a labor-intensive procedure.

Measurement of Eh is carried out with a platinum wire electrode mounted at the end of a thin tube to minimize disturbances. The reference electrode is held above the sediment in the overlying water. The platinum electrode is mounted on a stand equipped with a micrometer and is lowered in precise increments into the sediments. Stability is important because each reading requires from a few to 20 min equilibration time, as was found in the analysis of anoxic cave sediments (Fichez, 1990).

The need to leave electrodes in contact with an undisturbed sample has prompted some simple but ingenious constructions. In measurements of redox potential in wetland soils, Hargis and Twilley (1994) used a long, hollow, plastic tube. Each electrode consisted of an insulated copper wire welded to a small length of platinum wire. The wire was exposed along the exterior of the tube through a pair of holes drilled at the appropriate depth and sealed with epoxy. The bundle of copper wires inside the tube extended to the surface and was protected inside flexible latex tubing attached to the top of the plastic tube (Figure 3-2). The potential was measured against a calomel reference electrode

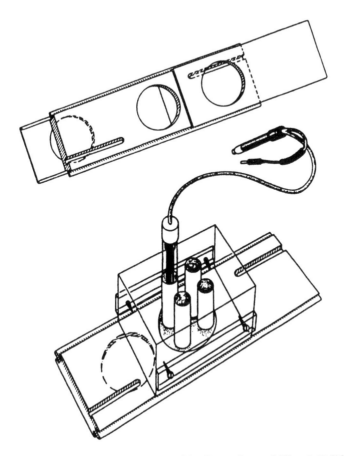

Figure 3-1   The sliding subsampler used by Bagander and Niemistö (1978). A slice of sediment is cut from the core by the sliding motion of two plastic sheets. The subsample is protected from air contamination. For measurement, a probe assembly is mounted on the top sheet as shown. (Modified from Bagander and Niemistö, 1978.)

inserted into the top soil layer. The method should be equally applicable to submerged sediments.

A portable Metrohm pH and Eh meter, with combination glass and platinum electrodes, was used for measurements in surficial sediments and sediment cores collected from the Laurentian Great Lakes (Kemp and Lewis, 1968; Kemp et al., 1971). Surface sediment samples were collected by a Shipek grab sampler. Upon retrieval, the sampling bucket with the collected sediments was placed on a special stand to keep the sediment surface horizontal. The Eh electrodes were calibrated using a Zobell solution (Zobell, 1946) and marked vertically at 0.5-cm intervals. They were then pushed into the sediment to the desired depth.

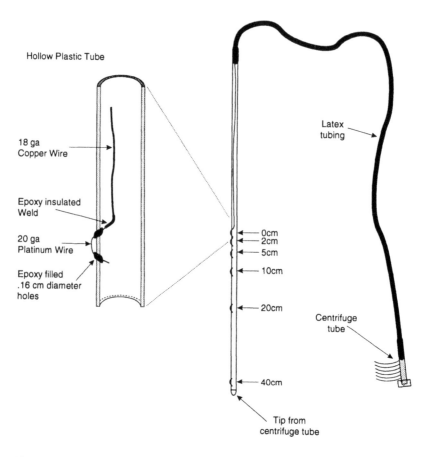

Figure 3-2 The simple probe used by Hargis and Twilley (1994) to obtain multiple Eh readings with minimal disturbance. (Modified from Hargis and Twilley, 1994.)

The pH electrodes were similarly marked and were calibrated using two buffer solutions of a known pH value (4.0 and 7.0). For measuring pH and Eh in sediment cores, the electrodes were inserted into the sediment after the core was placed on an extruder and uncapped, and water from the top of the core was siphoned off. For measuring the pH and Eh of the top 1 cm sediment layer, the electrodes were pushed 0.5 cm into the sediment. The pH reading was taken about 1 min after inserting the electrode into the sediment. However, it took approximately 10 min to stabilize the Eh value on the meter. After measuring the first subsample, the electrodes were removed from the sediment, cleaned with distilled water, and dried with soft paper tissue. The sediment layer, for which the measurements were made, was subsampled into a prepared container, and the electrodes were inserted into the next layer of the sediment core. After every five measurements, each electrode was recalibrated.

3.3   Measurement of pH

3.3.1   General considerations

As a master variable, pH is a more reliable parameter than Eh because the value of pH actually represents a real chemical species in solution, i.e., the concentration of H^+. If allowed sufficient time for diffusion, mixing, and reaction equilibrium, the concentration of H^+ in the system can be assumed to be the same for all participating reactions. In the case of Eh, this assumption is challenged because the electrode does not measure electron concentration but rather the potential to transfer electrons in the vicinity of the sensor. Treating the electron as a species is only a matter of convenience because the method of calculation becomes similar to a pH-equilibrium problem. In fact, the electron does not exist freely in solution due to its high reactivity (Thorstenson, 1984), and electrodes only render an indirect account of its transfer.

Problems, nonetheless, occur in pH measurements and are generally due to the complexity of sediment interactions. Four main equilibrium reactions are to be considered involving the following solid species in the sediment: sulfides, iron and manganese oxides, carbonates, and silicates. Except for carbonates, most proton and cation exchange capacity occurs on the surface of these solids. In addition, organic matter derived from biological degradation coats mineral surfaces and supplies most of the available exchange sites. Tessier (1995) recognized that sorption on oxic surfaces is mainly due to organic matter and hydrous oxides of Fe and Mn. Clays would only act as support of oxides and organic matter deposition. Cai and Reimers (1993) considered pH in sediments to be buffered mainly by carbonates, sulfides, and Fe/Mn oxides.

Since pH represents the equilibrium for reactions in the presence of many solids and surfaces, its value tends to be poised. During hundreds of pH measurements of estuarine and normal marine sediments, Berner (1981) never encountered pH values outside the range of 6 to 8. In more than 90% of the measurements he or his colleagues carried out, the pH was within the range of 6.5 to 7.5. The pH measured in lacustrine sediments, such as the Laurentian Great Lakes and other lakes in Canada, ranged mainly from 6.5 to 7.3 (Mudroch and Azcue, 1995).

Because changes in pH are expected to be small, measurements in the field must be done with instruments capable of 0.01 pH precision and directed at a specific chemical reaction. The readings are stable and reproducible. However, their small range cannot practically be used as a yardstick to classify sediment types. Other measurements, such as metal exchange, mineral composition, granulometry, and the concentration of organic matter, are more useful in the characterization of sediments.

In order to minimize disturbances to both sediment and electrode, pH measurement should be made *in situ* with a thin-bodied electrode. Reimers (1987) has developed a microprofiler for detail studies near the sediment surface. The device is essentially a motor-driven micrometer capable of moving an array of sensing electrodes vertically over a span of about 10 cm. The device is mounted on a tripod and lowered on the sediment surface from an anchored ship (Figure 3-3). Readings and profile movements are computer controlled. The sensing tip of the glass microelectrode is about 50 μm in diameter and 200 μm long (Cai and Reimers, 1993). The device is capable of responding to 99% of equilibrium value in less than 2 min. The stability of signal is less than 1 mv/5 h or 0.003 pH units.

In another study, Fisher and Matisoff (1981) showed that reaction profiles are much smaller in freshwater sediments than in the usually sampled 1-cm sediment increment. Vertical profiles of pH on a millimeter increments scale measured by microelectrodes followed delicate changes in sediment color. However, bioturbation was a major factor in levelling pH, and escaping gas bubbles caused a significant bias in reading of measured pH values in the sediment.

3.3.2   The suspension effect

Measurement of pH in sediments suffers from a problem that has yet to be resolved. The problem centers around the bias on the pH electrode when inserted into the sediment while the reference electrode is held in the overlying solution.

Jenny et al. (1950) used strong ion-exchange beads to simulate sediments. They noticed that an offset in pH potential occurred when the electrode passed from the supernatant into the sediment. If two reference electrodes are used between sediment and supernatant, a difference exists. However, there is no difference when two glass electrodes are used (Figure 3-4). They concluded that the bias was entirely due to changes of junction potential at the reference electrode when the electrode was exposed to the sediment. Cai and Reimers (1993) applied a similar experiment in their study of marine sediments. However, no significant differences were observed. It is possible that the high ionic strength of the sea water reduced the surface charge of the sediment to a small value. On the other hand, Brezinski (1983) examined cation exchange resins and concluded, contrary to Jenny et al. (1950), that the potential difference was not an artifact of junction

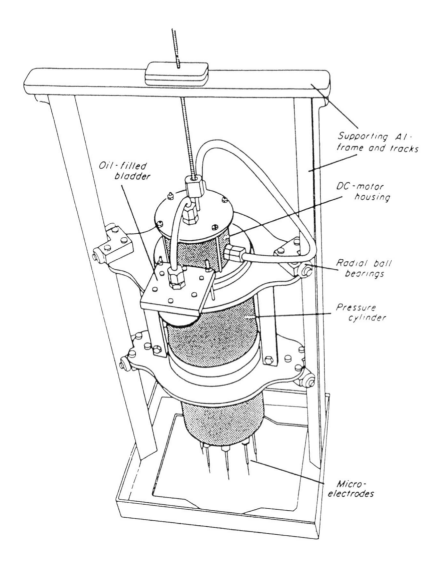

Figure 3-3 A microprofiler design for fine pH and pCO$_2$ work at the surface of oceanic sediments. (Modified from Reimers, C.E., *Deep Sea Res.*, 34, 2019, 1987.)

potential but of the equivalent potential due to a Donnan effect, where charge balance redistributes itself unequally between supernatant and sediment to compensate for the surface charge (Buck and Crabbe, 1986). Tschapek et al. (1989) have shown that the potential measured by two Ag/AgCl electrodes immersed respectively into an ion exchanger and the supernatant did not introduce a junction bias since the measured potential could be predicted solely by Boltzmann potential law.

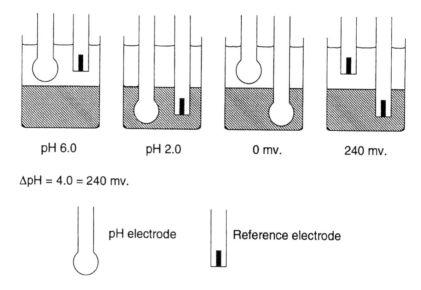

ΔpH = 4.0 = 240 mv.

Figure 3-4   Experimental set-up used by Jenny et al. (1950) to study the suspension effect. The four inserts represent all possible combinations for inserting a reference and a pH electrode in and above the sediment. (Modified from Jenny et al., 1950.)

The problem was again examined by Oman and Godec (1991). From the results of their thorough study, they concluded that the bias arose from the following two sources:

1. The potential shift of the indicator electrode depends on interactions between double layers of both indicator and charged sediment particles. This is the Pallman-Wiegner effect. This concept of double layer interaction makes sense, since the effect of the surface charge on minerals can influence the values obtained by glass electrodes. Stumm and Morgan (1981) estimated an apparent decrease from bulk solution of 2 pH units at the surface in fresh water. In sea water similar calculations would yield a 0.3 pH difference, which is in accordance with shrinking of the double layer at a higher ionic strength.
2. The effect of the Donnan potential occurs as an interaction with the reference junction when the reference solution contacts the sediment particles. However, a problem remains since careful measurement between two Ag/AgCl electrodes showed a potential difference near the interface of a charged sediment (Ji and Yu, 1985). The potential drops to zero if a conductor is introduced between the electrode and the interface, indicating the presence of an electrical field.

Although all the above options are probably making their contribution to the total potential difference, there is no consistent model to predict the behavior of reference electrodes in colloidal or polyelectrolyte media. For this reason, it is advisable to leave the reference electrode as far away from the sediment as possible without compromising conductivity and stability. As far as the indicating electrode is concerned, the consensus of double layer interaction appears to hold.

One way to alleviate the suspension effect is to calibrate the system *in situ* without moving the reference electrode. Yang et al. (1994) made microelectrodes mounted on a computer-driven profiler in a design similar to that of Reimers (1987). A drop of pH buffer was allowed to flow over the thin shaft of the pH microelectrode until it contacted the sensitive part and the soil surface.

3.4   General procedure for measurements of Eh and pH in bottom sediments

The pH and Eh in sediment samples should always be measured immediately upon retrieval of the samples. Therefore, in most cases, the measurements have to be carried out in the field. Mudroch and Azcue (1995) made the following suggestions for field measurements of Eh and pH in sediments.

3.4.1   Equipment and solutions used in the measurements

- A portable battery-operated pH/Eh meter, batteries, and a power cord for recharging the meter are needed.
- Glass and platinum electrodes or other electrodes suitable for measurements can be used. A set of separate reference and indicator electrodes is preferable since the reference should be kept away from the sediment in order to avoid the suspension effect. This point is less important in marine situations due to high ionic strength and high carbonate content.
- Plastic test tube-shaped containers with rubber sleeves or glass containers with a ground joint are needed for storing the electrodes in solutions during the transport and in the field. These containers are sometimes supplied by the manufacturer of the electrodes. They can be made from a test tube of a suitable size.
- Commercially available or laboratory-prepared pH buffer solutions (pH 4 and 7) in plastic bottles with lids should be used.
- A solution for calibration of the Eh electrode, such as a Zobell solution (0.003 M potassium ferricyanide, 0.003 M potassium ferrocyanide in 0.1 M KCl), should be freshly prepared in a

plastic bottle with a tight lid. The bottle with the solution **must** be labeled and handled according to the safety regulations for cyanides.

- A freshly prepared solution of saturated potassium chloride for storage of the electrodes and maintaining the solution inside the combination electrodes is needed. The solution should be stored in a plastic bottle with a lid.
- Other solutions necessary for the proper functioning of the electrodes during the field measurements, as outlined by the manufacturer of the electrodes, should be obtained.
- Distilled water and a wash bottle for storage and rinsing the electrodes between and after the measurements are important. The quantity of the distilled water will depend on the number of measurements.
- Small beakers for holding the buffer and calibration solutions and larger beakers or other containers for rinsing the electrodes after the measurements should be used. Plastic beakers or containers are more convenient to handle in the field than glass beakers or containers.
- Support stands, rods, connectors, and clamps to secure the electrodes in buffer and calibration solutions, in the sediment samples during the measurements, and during the cleaning of the electrodes should be available.
- Large plastic containers should be used for storage and transport of used buffers and Eh-calibration solutions. Used solutions need to be properly discharged according to the regulations for discharge of cyanide-containing liquids.
- A notebook and pens, as well as soft paper tissue, are necessary.
- Combination electrodes should be stored according to the manual, with information on handling of individual electrodes provided by the manufacturer. In most cases, combination glass electrodes should be stored in saturated or 4 M KCl solution when not in use to prevent the glass membrane drying out.

3.4.2   *Preparation of the equipment before the field trip*

- Check the batteries of the portable pH/Eh meter and recharge them, if necessary.
- Prepare the solutions for the calibration of the Eh and pH electrodes in the field.
- Check and test the pH and Eh electrodes; e.g., check their connections to the meter, remove the electrodes from the storage solution, wash them with distilled water, and calibrate them with the buffer and other solutions.

- Mark the electrodes vertically at desired intervals for insertion into the sediment samples.
- Store the electrodes, according to the manufacturer's instructions or in saturated KCl solution, for transport and use in the field.
- Pack all equipment and solutions for transport to the field.
- It is desirable to take spare electrodes to the field.

3.4.3   Measurements in the field

- It is recommended to allocate a space in the field where all pH and Eh measurements will be carried out. Within this space, all equipment should be assembled, checked for proper functioning, and prepared for measurements of the first sample. Electrodes must be connected, checked, calibrated, washed, and prepared for measurements of pH and Eh prior to the retrieval of the sediment sample. Electrodes must be calibrated under nitrogen if a nitrogen atmosphere is used. All solutions should also be equilibrated under nitrogen and kept there.
- Grab sampler and sediment corers with recovered sediment need to be placed in such way that they will remain steady without disturbing the sediment samples during the measurements.
- Electrodes must be carefully inserted into the undisturbed sediment samples to avoid any air contamination, particularly around the Eh electrode. Care must be taken not to generate open space between the electrode and the sediment. Proper insertion of the electrode without disturbing the sediment is the most important step in measuring the Eh.
- The electrodes are inserted into the sediment to the depth marker. The pH/Eh meter is switched to the pH scale and the value recorded within 1 min after inserting the electrode into the sample. The meter is then switched to the mV-scale for recording the Eh value. The potential usually drifts considerably over the first 10 to 15 min, and after this period becomes stabilized. After stabilization, the value on the mV-scale should be recorded. When measuring the Eh of sediments from waters with low ionic strength, such as most freshwater bodies, it is recommended to "acclimatize" the electrodes in the water prior to measurement, particularly the electrodes that were stored in saturated KCl solution. This will reduce drifting of the potential after inserting the electrode into the sediment. This researcher's measurements were made using a Zobell's solution for the calibration of the platinum electrode. The solution has an Eh value of +430 mV at

25°C. The Eh value of the Zobell solution recorded by the platinum electrode was usually around +250 mV. Therefore, the difference (i.e., +180 mV) was always added to the values measured in sediment samples by the electrode to standardize the measurements. However, the Eh was measured mainly to obtain general information on the oxic or anoxic conditions in the sediments, with no further attempts to interpret the measured values.

- Both electrodes are then removed from the sample, washed with distilled water to remove all adhering sediment particles, and dried gently with a soft paper tissue.
- The electrodes should be calibrated after each fifth measurement. However, in measuring pH and Eh in a sediment core, the electrodes may need a less frequent calibration.

3.5 Reference electrode

Reference electrodes are made separately or as part of a combined Eh or pH electrode. In either case, they accomplish the same function. The reference electrode is an important part of the equipment for measuring pH and Eh. It provides a reference half-cell voltage against which measurements are made. Reference electrodes are more likely to be contaminated than pH or Eh electrodes because they must have direct solute contact with the bulk solution. The major cause of improper readings in voltametric measurement can be traced back to improper care and use of the reference electrode.

The reference electrode should normally be a standard hydrogen electrode. However, this is not practical, and calomel or AgCl half-cells are used instead. In its simplest form, the electrode is made of a metal, such as Hg or Ag, and coated with a metal chloride salt of low solubility. The electrode is immersed in a saturated KCl solution. At equilibrium, the saturated solution fixes the concentration of the chloride ion around the metal chloride salt. The metal chloride salt dissolves slightly until saturation, at which point the concentration of the free metal ion (i.e., Ag^+ of Hg^{++}) also becomes fixed and establishes the cell potential.

The construction of the reference electrode is a compromise between exposing the half-cell to bulk solution potential and maintaining a saturated solute concentration in contact with the half-cell. Typically, a small fritted plug acts as a salt bridge and allows a slow leak of the reference solution outside, just enough to establish electrical contact but not enough to disturb the sample. The frit restricts movement of solutes to a few microliters per hour, and generates its own potential (the junction potential) due to differential mobilities of crossing ions. Contamination, precipitation, and deposition of bulk sub-

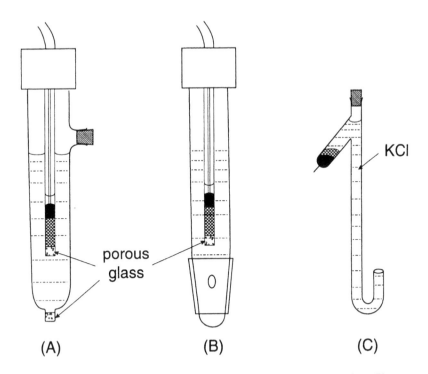

porous
glass

KCl

(A)   (B)   (C)

Figure 3-5 Designs of reference electrode junction. A) porous plug; B) cone and sleeve junction; C) "J" junction, which relies on the heavier density of saturated KCl. (Modified from Hills, G.J. and Ives, D.J.G., in *Reference Electrodes Theory and Practice*, Ives, D.J.G. and Janz, G.J., Eds., Academic Press, New York, 1961, 127.)

stances on the frit alter the junction potential and cause drift of the reference potential. Different types of junctions are available to accentuate the problem of low junction potential vs. acceptable leakage (Figure 3-5). Small surface junctions have low leakage rates but produce high junction potentials that are prone to vary with fouling and sudden changes in ionic strength. Large surface junctions can be cleaned easily. They have low and stable junction potentials, but the leakage rate is high. A "J"-shaped reference electrode maintains an interface between the heavier saturated KCl solution and the bulk solution. The interface acts as a junction of low potential and can be easily cleaned by purging with fresh solution. Because settling of particles on the junction alters junction potential, these electrodes should be kept above the sediment. Hills and Ives (1961) and Janz (1961) have reviewed the theory and construction of reference electrodes.

Proper care of the reference electrode can be summarized by the following:

- The reference solution must be kept at the proper concentration. Some cells require saturated KCl, others a 4 M KCl solution. Long-term use of the electrode dilutes the reference solution. For saturated solutions, precipitated salts should always be present within the body of the electrode.
- The reference solution should be replenished and rinsed with the proper solution, not water, and never be allowed to dry.
- Storage of the reference electrode can introduce a bias. In theory, the junction should be saturated with the reference solution with a leakage rate sufficiently high to prevent solutes from entering the junction (Brezinski, 1983). In practice, low flow reference electrodes are often used in long-term measurements. In such cases, it is better to condition the electrode to the bulk solution before measurement starts, with the assumption that the introduced bias is constant. Thus, for a long-term measurement, when left in solution, the reference electrode should be "acclimatized" to a solution of ionic strength similar to that in the measured medium. A time period of between a few hours to a day is required to stabilize the flow through the junction and equilibrate temperature changes. For storage between uses, the reference electrode should be kept in 4 M KCl solution.
- The calibration of pH electrodes is usually carried out with standard buffers at an ionic strength of 0.05. Transfer of the reference junction to a low ionic strength medium will introduce a bias. Preparing secondary pH standards at the working ionic strength or bringing the sample up to 0.05 ionic strength prior to the measurement is the only way to correct for this difference. The first method requires the use of a large surface junction to ensure stability in the very low ionic strength of fresh water; the second alters the sediment-porewater interactions by causing a partial collapse of charged double layers on the surface of particles, resulting in altered surface exchanges. Adequate pH calibration at a low ionic strength can be achieved *in vitro* with water saturated with solid $CaCO_3$ at the adequate ionic strength and temperature. Equilibrium pH can be calculated from first principles (Stumm and Morgan, 1981) and used as the standard.
- The junction remains the most fragile part of the electrode and should be kept from fouling. In sediment studies it is better to use the reference electrode in the water overlying the sediment.

3.6 Conclusions

Measurements of pH and Eh in sediment studies are meaningful only if proper conditions exist in the sediment. Although fouling of reference and indicator electrodes can be compensated for by repeated and consistent maintenance and use, the major problem in measurement arises from the excessive demands placed on the natural sample to conform to a theoretical ideal state. Therefore, the electrodes only measure the extent to which the sample conforms to this ideal state.

In practice, natural sediments approach the thermodynamic requirements when the redox system is well poised by one major reaction, usually involving sulfides or ferrous/ferric compounds in aquatic sediments. Measurements are more stable for pH but require an instrument with 0.01 pH unit precision and careful handling to be meaningful. In all cases measurements carried out with Nernstian or Nernstian-assumed systems should always be accompanied by equilibrium calculations to determine how far the system is off the mark. The main problem in this type of study remains not so much the instrument error, which can always be minimized by use of better electronics and electrode design, but the difficulty in associating obtained measurements with meaningful interpretation of the physico-chemical conditions in the sediment.

References

Bagander, L.E. and Niemistö, L., An evaluation of the use of redox measurements for characterizing recent sediments, *Estuarine Coastal Mar. Sci.*, 6, 127, 1976.

Bates, R.G., The modern meaning of pH, *CRC Crit. Rev. Anal. Chem.*, 10, 247, 1981.

Berner, R.A., Electrode studies of hydrogen sulphide in marine sediments, *Geochim. Cosmochim. Acta.*, 27, 563, 1963.

Berner, R.A., A new geochemical classification of sedimentary environments, *J. Sed. Petrol.*, 51, 359, 1981.

Brezinski, D.P., Kinetic, static and stirring errors of liquid junction reference electrodes, *Analyst*, 108, 425, 1983.

Buck, R.P. and Crabbe, E.S., Electrostatic and thermodynamic analysis of suspension effect potentiometry, *Anal. Chem.*, 58(9), 1938, 1986.

Cai, W.J. and Reimers, C.E., The development of pH and pCO_2 microelectrodes for studying the carbonate chemistry of porewaters near the sediment water interface, *Limnol. Oceanogr.*, 38(8), 1762, 1993.

Dechev, G.D., Matveeda, E.G., and Yordanov, S.D., Determining the characteristics of sea water and bottom sediments as well as their interrelation by means of the redox potential, *Comptes Rendus de l'Academie Bulgare des Sciences*, 27(4), 3, 1974.

Fenchel, T., The ecology of marine microbenthos IV, *Ophelia*, 6, 1, 1969.

Fichez, R., Absence of redox potential discontinuity in dark submarine caves sediment as evidence of oligotrophic conditions, *Estuarine Coastal Shelf Sci.*, 31, 875, 1990.

Fisher, J.B. and Matisoff, G., High resolution vertical profiles of pH in recent sediments, *Hydrobiologia*, 79(3), 277, 1981.

Hargis, T.G. and Twilley, R.R., A multi depth probe for measuring oxidation reduction (redox) potentials in wetland soils, *J. Sed. Res., Sect. A, Sed. Petrol. Proc.*, 64(3), 648, 1994.

Hills, G.J. and Ives, D.J.G., The calomel electrode and other mercury-mercurous salt electrode, in *Reference Electrodes Theory and Practice*, Ives, D.J.G. and Janz, G.J., Eds., Academic Press, New York, 1961, 127.

Janz, G.J., Silver halide electrodes, in *Reference Electrodes Theory and Practice*, Ives, D.J.G. and Janz, G., Eds., Academic Press, New York, 1961, 179.

Jenny, H., Nielsen, T.R., Coleman, N.T., and Williams, D.E., Concerning the measurement of pH, ion activities and membrane potentials in colloidal systems, *Science*, 112, 164, 1950.

Ji, G.L. and Yu, T.R., Long range effect on electrical charge of soils on liquid junction potential of salt bridge, *Soil Sci.*, 139(2), 166, 1985.

Kemp, A.L.W. and Lewis, C.F.M., A preliminary investigation of chlorophyll degradation products in the sediments of Lakes Erie and Ontario, in Proc. 11th Conf. Great Lakes Res., International Association of Great Lakes Research, Ann Arbor, MI, 1968, 206.

Kemp, A.L.W., Savile, H.A., Gray, C.B., and Mudrochova, A., A simple corer and a method for sampling the mud-water interface, *Limnol. Oceanogr.*, 16, 689, 1971.

Kempton, J.H., Lindberg, R.D., and Runnells, D., Numerical modeling of platinum Eh measurements by using heterogeneous electron-transfer kinetics, in *Advances in Chemistry Series, Chemical Modeling of Aqueous Systems*, 1990, 339.

Macalady, D.L., Langmuir, D., Grundl, T., and Elzerman, A., Use of model generated Fe^{3+} ion activities to compute Eh and ferric oxyhydroxide solubilities in aerobic systems, in *Advances in Chemistry Series, Chemical Modeling of Aqueous Systems*, 1990, 350.

Morris, J.C. and Stumm, W., Redox equilibria and measurements in the aquatic environment, *Adv. Chem. Ser.*, 67, 270, 1967.

Mortimer, C.H., The exchange of dissolved substances between mud and water in lakes — I, *J. Ecol.*, 29, 280, 1941.

Mortimer, C.H., The exchange of dissolved substances between mud and water in lakes — II, *J. Ecol.*, 30, 147, 1942.

Mudroch, A. and Azcue, J.M., *Manual of Aquatic Sediment Sampling*, CRC/Lewis, Boca Raton, FL, 1995, 219.

Norrström, A.C., Field measured redox potentials in soils at the groundwater surface-water interface, *Eur. J. of Soil Sci.*, 45(1), 31, 1994.

Oman, S. and Godec, A., Effect of dispersed particles on the response of indicator electrode in suspension, *Electrochim. Acta*, 36(1), 59, 1991.

Plante, R., Alcolado, P.M., Martinez-Iglesias, J.C., and Ibarzabal, D., Redox potential in water and sediments of the Gulf of Batabanó, Cuba, *Estuarine Coast Shelf Sci.*, 28, 173, 1989.

Reimers, C.E., An in-situ profiling instrument for measuring interfacial porewater gradients: methods and O_2 profiles for the North Pacific, *Deep Sea Res.*, 34, 2019, 1987.

Rozanov, A.G., Pacific sediments from Japan to Mexico: some redox characteristics, in *The Dynamic Environment of the Ocean Floor*, Fanning, K.A. and Manheim, F.T., Eds., Lexington Books, Lexington, MA, 1982, 239.

Stumm W. and Morgan J.J., The solid-solution interface, in *Aquatic Chemistry*, Wiley-Interscience, New York, 1981, 621.

Tessier, A., Sorption of trace elements on natural particles in oxic environments, in *Environmental Particles*, Buffle, J. and van Leewen, H.P., Eds., CRC/Lewis, Boca Raton, FL, 1995, 425.

Thorstenson, D.C., The Concept of Electron Activity and Its Relation to Redox Potentials in Aqueous Geochemical Systems, U.S. Geological Survey Open-File 84-072, 1984, 45.

Tinsley, I.J., *Chemical Concepts in Pollutant Behavior*, Wiley-Interscience, New York, 1979, 92.

Tschapek, M., Wasowski, C., and Sanchez, R.M.T., Liquid junction potential at the salt bridge in heterogeneous systems, *Ber. Bunsen Ges. Phys. Chemie*, 93(10), 1139, 1989.

Whitfield, M., Eh as an operational parameter in estuarine studies, *Limnol. Oceanogr.*, 14, 547, 1969.

Whitfield, M., The electrochemical characterization of natural redox cells, *Limnol. Oceanogr.*, 17(3), 383, 1972.

Whitfield, M., Thermodynamic limitations on the use of the platinum electrode in Eh measurements, *Limnol. Oceanogr.*, 19, 857, 1974.

Yang, J., Blanchar, R.W., and Hammer, R.D., Calibration of glass microelectrodes for in-situ measurement, *Soil Sci. Soc. Am. J.*, 58(2), 354, 1994.

Zobell, C.E., Studies on the redox potential of marine sediments, *Bull. Am. Assoc. Petrol. Geol.*, 30, 1946.

chapter four

Rapid and cost-effective analysis for aquatic sediment samples by X-ray fluorescence spectrometry

Ken F. Maley

4.1   Introduction

This chapter will address the application of X-ray fluorescence spectrometry (XRF) in the context of an aquatic sediment sampling program where total as opposed to acid-extractable analyses are required. The technique will be described in general terms, emphasizing those aspects common to academic and commercial settings. These include the client's need for accuracy, speed, and economy. The decision to choose XRF for a specific analytical task is usually based on some balance between these often conflicting obligations.

The appeal of XRF, as a broadly-defined array of methods, may stem from the axiom that if one can irradiate some material or object and observe the behavior of the energy absorbed, then the structure or composition can be predicted. The utility of such data will be proportional to the amount of information available regarding the measurement conditions. The latter is the basis of the need for some theoretical exposure, not only for analysts but for managers and clients as well. Applications for XRF abound, including analyses of various solids (e.g., rocks, metals, soils), and also of slurries (Ereiser et al., 1981), solutions (Davidson et al., 1994), and even fumes (Carsey, 1982). This review will focus on practical considerations for sediment samples received as solids.

1-56670-155-4/97/$0.00+$.50
© 1997 by CRC Press, Inc.

4.2 Theory of X-ray fluorescence spectrometry

Electromagnetic radiation of the X-region (keV) has been used analytically for many years. Moseley (1913) demonstrated a relationship between X-ray spectral line wavelength and intensity for brass. By 1948, Friedman and Birks (1948) had designed and built the first prototype commercial spectrometer. In a sense, XRF may be viewed as a mature technology. Comprehensive references by Jenkins and De Vries (1967) and Bertin (1975) are still consulted today. More significant has been the rapid increase in computing power at diminishing cost that has enabled the XRF analyst to utilize theoretical data to produce more accurate results. Newer texts by Tertian and Claisse (1982) and Lachance and Claisse (1995) reflect this shift.

4.2.1 Excitation

Simply stated, X-rays are produced by an atom when it has absorbed a quantity of energy greater than the binding energy of "inner" shell electrons. Electrons ejected from inner shells are replaced by transitions from shells further out. These de-excitation steps release the difference in energy levels as X-ray photons. Since the atomic structure is different for each element, it follows that:

1. The binding (critical) energies are characteristic for each shell and energy sublevel by element.
2. The energies of the emitted photons are characteristic for the transitions possible between energy levels of each element.
3. The probability of each transition occurring, called the fluorescent yield (ω), determines the intensity ratios of the various emissions (lines), which will vary with atomic number (Z).

Bertin (1975) illustrates five excitation mechanisms that give rise to electron vacancies. Primary (direct) excitation refers to bombardment of matter by high speed electrons (e.g., in an X-ray tube or electron microprobe), or by other particles (α, protons) from radioisotopes or accelerators. Secondary (fluorescent) excitation results from photoelectric absorption of X-rays or γ-emissions by a sample. Internal conversions from γ or β-decay as well as electron capture also produce characteristic X-rays.

Typical instrumentation uses an X-ray tube for excitation. Electrons are focused onto an anode target at sufficient potential for primary radiation to be produced. The excitation potential corresponds to the critical energy in kV. Tube sources are polychromatic, consisting of a broad continuum of Brehmstrahlung (braking) emissions, in addition to the characteristic lines of the target element.

Table 4-1   Partial K and L X-ray Spectra for Ba

Transition: IUPAC notation	Binding energies (keV)	Line energy[a] (keV)	Wavelength (Å)	Seigbahn notation	Relative intensity[b]
K-L$_3$	37.441–5.247	32.194	0.385	Kα$_1$	100
K-L$_2$	37.441–5.634	31.807	0.390	Kα$_2$	50
K-M$_3$	37.441–1.062	36.379	0.341	Kβ$_1$	15
K-M$_2$	37.441–1.137	36.304	0.342	Kβ$_2$	5
L$_3$-M$_5$	5.247–0.781	4.466	2.776	Lα$_1$	100
L$_3$-M$_4$	5.247–0.796	4.451	2.786	Lα$_2$	11
L$_2$-M$_4$	5.634–0.796	4.838	2.568	Lβ$_1$	53

[a] Bearden and Burr (1967) energy data adapted from Lachance, G.R. and Claisse, F., *Quantitative X-ray Fluorescence Analysis*, John Wiley & Sons, Chichester, 1995, 402.

[b] Philips X40 4.1 on-line tables.

4.2.2   Lines and notation

An intuitive grasp of the above principles and a set of tables are all that is needed to interpret X-ray spectra for qualitative purposes. Using Table 4-1, for example, if the analyst observes Ba K-lines from a specimen, then the L lines must also be present, due to the lower critical energy of the L-shells. If the source energy is known (e.g., tube potential), then the presence of either can be verified. In conjunction with the relative intensities, possible matches from other elements can be eliminated. By convention, XRF spectroscopists prefer to deal in angstrom units (10⁻⁸ cm), probably to avoid using exponential notation. The older Seigbahn line notation is still favored, but the IUPAC system is used in publications.

For purposes of quantification, the intensity of the analyte lines is a function of (but rarely directly proportional to) the composition. The interaction of radiation with matter in a specimen must be considered before a strategy for converting intensities to concentrations can be developed.

4.2.3   Absorption

Lachance and Claisse (1995) provide a succinct review of X-ray absorption relationships. In brief, the extent to which a material absorbs radiation can be expressed by a linear attenuation coefficient (μl), which represents the fraction of intensity absorbed by an irradiated cross-section in cm⁻¹. By incorporating density (ρ) in g cm⁻³, a more useful mass attenuation coefficient (MAC) is obtained:

$$MAC = \frac{\mu l}{\rho} \qquad (1)$$

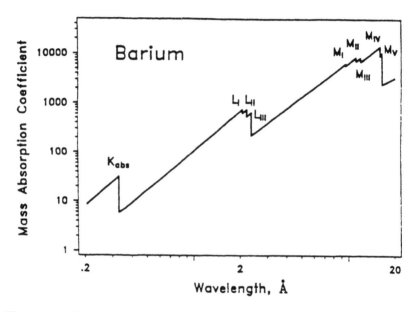

Figure 4-1   X-ray absorption curve for barium showing K, L, and M absorption edges. (From Willis, J.P. and Duncan, A.R., Course on Theory and Practice of WRF Spectrometry, Course Notes Week 1, University of Western Ontario, Department of Earth Sciences, 1995.)

The MAC, in cm^2g^{-1}, is an index of a substance's "stopping power", which can be used for direct comparison between materials and the wavelength of X-rays. From a table of MAC values (Bertin, 1975), the following trends are observed:

- an increase in MAC at a given wavelength with atomic number (Z);
- an increase in MAC of an element with wavelength;
- a series of abrupt discontinuities in the above trends (Figure 4-1).

These anomalies in the expected trend, referred to as "absorption edges", correspond to the critical energies in the various shells for fluorescent emission in terms of wavelength. At an incident wavelength that is shorter than an absorption edge of an element,

- all lines of that series (e.g., K) will be emitted;
- all lines of lower critical energy (e.g., L1, L2, L3) will be emitted;
- these emissions will contribute to the fluorescence of other elements in the specimen with absorption edges at longer wavelengths (enhancement).

Spectrometer design, as well as sample composition, are also important. The total or effective MAC (μ^*), which accounts for absorption of both the incident and emergent radiation, is expressed by:

$$\mu^* = \mu_p \, csc\psi' + \mu_s \, csc\psi'' \tag{2}$$

where, ψ' is the incident angle between the source and specimen, ψ'' is the takeoff angle between specimen and optics, μ_p is the sample MAC at the primary (incident) wavelength, and μ_s is the sample MAC at the secondary (analyte) wavelength, and at their respective wavelengths,

$$\mu \, (_p \, or \, _s) = \Sigma \, \mu_i \, C_i \tag{3}$$

where, C_i are the weight fractions of the component i elements.

4.2.4   Scatter

A proportion of the incident radiation will be scattered from collisions with atoms in the sample. Although the terms "mass attenuation" and "mass absorption" are often interchanged, a MAC can be considered to be a sum of photoelectric absorption and scatter coefficients (Lachance and Claisse, 1995). Coherent or Rayleigh scatter is observed at the same wavelength as the incident photons. Incoherent or Compton scatter results from a loss of energy in the collision (recoil) and is observed at a longer wavelength. Scattered intensity is an important consideration, since

- Rayleigh and Compton peaks from the excitation source (Figure 4-2) will appear in the measured spectrum;
- the intensity of Compton peaks is inversely proportional to the average Z of the scattering medium, and therefore the total MAC. This relationship can be used in calibration to empirically estimate MAC (Feather and Willis, 1976; Willis, 1991);
- scattered continuum from the source contributes to the background intensity which must be subtracted from the analyte intensity. Compton peaks can also be employed in background corrections (Feather and Willis, 1976).

4.3   Practical application for analysis of environmental samples

4.3.1   Detection and counting

The task of dispersion and detection of X-ray emissions is accomplished by two main categories of instruments, a) wavelength dispersive (WDS)

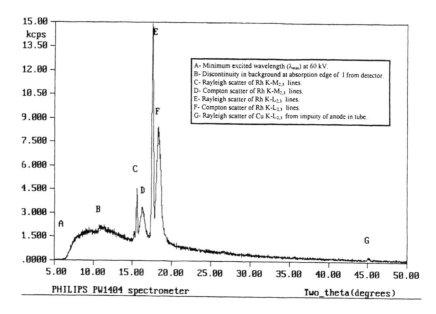

Figure 4-2 Wavelength scan of a blank sample using a rhodium tube operated at 60 kV with NaI scintillation and Ar/CH₄ flow detection.

and b) energy dispersive (EDS) spectrometers. Wavelength dispersive systems employ flat or curved crystals to separate the spectrum by diffraction according to Bragg's law (Bertin, 1975). Such systems typically combine scintillation (e.g., NaI) and gas-flow (e.g., Ar-CH₄) detectors for short and long wavelength regions (>2 Å), respectively. Sequential or simultaneous configurations are available depending on the need for flexibility or speed. A sequential wavelength dispersive spectrometer is shown in Figure 4-3.

Energy dispersive units typically use a high-resolution semiconductor detector (e.g., SiLi) with a multichannel analyzer sorting the lines by energy pulse amplitude. In comparison to WDS, EDS offers the advantage of simultaneous detection with lower capital and installation costs, but sacrifices light element performance and high intensity counting capacity (Willis and Duncan, 1995). Hybrid (multidispersive) instruments are now available for specialized simultaneous applications.

For either class of system, the same sources of systematic error must be addressed:

- Elemental matrix effects.
- Physical effects (particle size, mineral segregation).
- Spectral effects (background, line overlap).
- Detector nonlinearity (dead time, pulse pile-up).

Figure 4-3 Philips PW1480. A sequential wavelength dispersive spectrometer fitted with a 72 position sample changer. (Photo by S.T. Caverly.)

4.3.2 Sample preparation

The treatment of sediment samples prior to analysis was described in detail by Mudroch and Bourbonniere (1994). Generally, samples for analysis by XRF need relatively simple preparation, such as drying, grinding, etc., in comparison to sample preparation for analysis by dissolution-based instrument techniques. The theoretical considerations outlined above dictate three main requirements:

1. Infinite thickness: If the specimen is thicker than the depth through which the shortest wavelength of X-ray photon under consideration can emerge, then the actual thickness or weight of the preparation (usually a disk) is not a variable. This will be inversely proportional to MAC and limited by the sample width and spectrometer geometry.
2. Homogeneity: As the MACs for the elements that compose the sample differ, the effective penetration depth will vary accordingly. Since the measurement, which will be particularly shallow for the long wavelength lines of low Z elements, must represent the entire specimen, homogeneity is ideal. This is usually accomplished by fusion. For such analyses in powders, fine grinding (e.g., –325 mesh) is essential. Small particle size also reduces sampling error, assists fusion, and improves the binding of pellets.

Table 4-2   Examples of Elemental Analyses and Typical Lower Limits
of Detection Reported by XRF

Element or oxide	Preparation method	Detection limit (g/t)
SiO_2, Al_2O_3, Fe_2O_3, CaO, K_2O, P_2O_5, TiO_2, MnO	Borate fusion	10–100
Na_2O, MgO	Borate fusion	100–500
La, Ce, Nd	Low dilution borate fusion	3–9[16]
Mo, Nb, Zr, Y, Sr, Rb	Pressed powder	1–5
Cu, Pb, Zn, Ni, Co, Sb, Cd, As, U, Th	Pressed powder	5–10
Mn, Fe, Cr, Ba	Pressed powder	10–20

Reported by Eastell, J. and Willis, J.P., *X-ray Spectrom.*, 19, 3, 1990.

3. Flat surface: A consequence of expression (2) is that the incident
   and takeoff angles must be consistent across the entire measurement area. Certainly, any theoretical approach to matrix correction will be based on this assumption. Perhaps more importantly,
   a sample intruding toward the source and detectors will exhibit
   a high bias vs. flat standards. Likewise, a rough surface will have
   an effect analogous to coarse particles, particularly with respect
   to long wavelength lines.

Geological samples are typically submitted to our XRF laboratory
for major element characterization and trace analyses (Table 4-2). For
major "whole rock" elements, high-dilution (e.g., 10 to 1) fusion with
lithium tetraborate ($Li_2B_4O_7$) may be applied to a wide variety of oxide
minerals over broad concentration ranges. Borate fusions, usually performed in platinum crucibles, can be easily automated for high
throughput (Figure 4-4). Flux mixtures may include up to 50% lithium
metaborate ($LiBO_2$) for high levels of acidic oxides such as silica, and
nonwetting agents such as an iodide or bromide salt added to facilitate
casting into molds. Highly-contaminated sediments containing metal-
lics, sulfides, or elements such as arsenic or copper may also require
pretreatment with an oxidant (e.g., $LiNO_3$) to prevent damage to the
costly platinum vessels. The advantages of fusion are numerous
(Claisse, 1995). The borate glass bead is homogeneous, stable, and may
be stored for future investigations. Particle size, mineral, and chemical
effects are eliminated. Hence, calibration standards can be made from
a diversity of reference materials that need not match the samples.
Elemental matrix effects are suppressed, allowing relatively simple
calibration models to be used.

Figure 4-4 Claisse *Fluxy* three burner fusion system. Each unit produces about 60 fused disks per 8-h shift. (Photo by S.T. Caverly.)

The dilution does, however, elevate detection levels. For routine trace analyses, briquettes of finely ground sample mixed with a binding agent can be produced at low cost. In a few cases, loose powder in a cell suspended on a thin film window will produce satisfactory results. The strong matrix effects present in powdered samples can lead to large analytical errors, especially for low-Z elements. As a result, major element analysis is usually not practical on briquettes. Alternatively, a low-dilution fusion technique (Eastell and Willis, 1990) may be employed to obtain high analyte line intensities from a homogeneous specimen to combine major and trace packages. Unfortunately, some minerals may be difficult or impossible to fuse at low dilution (Eastell and Willis, 1990) ratios, and more comprehensive calibration protocols are required to correct for elemental matrix effects (Eastell and Willis, 1993).

4.3.3   Calibration

The process of calibrating an XRF system involves two steps:

1. Optimum measurement conditions are selected to minimize random errors. These include line selection, excitation parameters

(e.g., tube voltage, current, and target material), as well as detection settings (e.g., crystal options, pulse height selection). Whenever possible, these decisions should be quantitatively evaluated by a Figure of Merit (FOM) calculation (Tao et al., 1985) and not left to instinct or the instrument manufacturer. The benefits of a properly configured XRF system include minimized counting time requirements for a given precision level, as well as lower detection levels and reduced occurrence of unanticipated interferences.

2. The proper correction for systematic errors is made to produce accurate, unbiased results. Since XRF matrix effects are well understood theoretically, they should not be a barrier to good analyses. Rather, the accuracy requirements of a test program will determine the level of sophistication for calibration needed and reflect upon the cost and speed accordingly.

Most XRF calibration models follow the general form:

$$C_i = D + E\, R_i\, M \tag{4}$$

where, C_i is the concentration of element i, D is a constant if there is a calibration intercept, E is a method-specific calibration constant (slope), R_i is the net intensity of the selected X-ray line for element i, and M is a term for matrix correction.

The net intensity R_i can be calculated by subtracting background intensity (R_b) from the peak (P_i) either by mathematical interpolation from peak offsets (Lachance and Claisse, 1995) or by a ratio method (Feather and Willis, 1976). Spectral overlaps (R_o) can be subtracted as a function of a reference intensity, derived from blanks containing the interferent. These deconvolution steps, which may be iterative in a multielement program, are of critical importance for trace analysis.

$$R_i = P_i - (\Sigma\, R_b + R_o) \tag{5}$$

In some major element applications, background offset measurement can be omitted for expediency, and the calculation of an intercept (D) by regression or from a suitable blank may suffice. Nonzero intercepts may also result from scattered instrument impurities (e.g., Cu, Cr, Fe), from contaminants in flux (e.g., Si, Al), or from empirical calibrations that have failed to correct for MAC.

The term M may take several forms. In the simplest case of calibrating with matched in-type standards or at a very high-dilution fusion, M = 1 will provide satisfactory results, especially for semiquantitative purposes. Other possible matrix correction methods used for sediments are:

- M = a ratio of the analyte to an internal standard. Usually an element line of an adjacent element with a similar MAC. This approach is useful in powder preparations for volatile elements, which may be lost in fusion (e.g., As), or for situations where infinite thickness cannot be achieved.
- A correction for MAC using Compton scatter or a ratio of Compton to Rayleigh scatter (Willis, 1991). This works very well for high-Z elements between the scatter lines (e.g., Rh Z = 45) and the first major constituent absorption edge encountered, which is usually iron (Z = 26). This method, which corrects for absorption only, is used predominantly on briquettes.
- A binary (α) influence coefficient model. Coefficients are used to express changes in MAC and account for enhancement effects. Corrections are applied as constants in the form M = 1 + (Σ $\alpha_{ij}W_j$), where α_{ij} is the effect of matrix element j on analyte i and W_j the weight fraction. Although alphas may be calculated empirically by multiple regression analysis, a large number and compositional range of standards are required to avoid gross errors due to cross-correlations (Tao et al., 1985). It is preferable to derive alphas from theoretical principles. In practice, this involves using specialized software (Jenkins and Gilfrich, 1992; Rousseau, 1990a,b) to generate the coefficients from fundamental parameters such as tube voltage and anode material, spectrometer geometry, fluorescent yields, flux composition, and absorption edge data. The Lachance-Traill (Lachance and Traill, 1966) and the Claisse-Quintin (Claisse and Quintin, 1967) (which has a third element term) alpha models work very well for moderate- to high-dilution fusions. Correction must also be made for volatile elements, usually by the determination of loss on ignition (LOI) or loss on fusion (LOF).
- A variable coefficient fundamental algorithm. Although equation (2) above assumes monochromatic excitation, this will rarely occur in analytical contexts. Claisse and Quintin (1967) demonstrated that influence coefficients vary as a function of specimen composition using polychromatic incident sources. Subsequently, it was observed that the semiempirical treatment of constant, single alpha factors over the entire analyte range was inadequate for undiluted or low-dilution specimens (Eastell and Willis, 1993). Models such as the COLA (Lachance and Claisse, 1995) expression employ multiple influence coefficients, which can be varied with analyte concentration. Computer programs (Jenkins and Gilfrich, 1992; Rousseau, 1990a,b), which iterate each sample composition starting with a modified Claisse-Quintin algorithm, make this task possible for routine work. One drawback, particularly for sequential systems, is that

the total specimen composition must be analyzed or the unmeasured components known. The operating software sold with new instruments will now often include a fundamental parameter option.

The use of theoretical matrix corrections significantly reduces the number of standards required for calibration by limiting the calculating parameters to essentially a sensitivity constant (E). Isolating matrix effects permits the analyst to concentrate on spectral corrections and the quality of the standard sample preparation. Although it is not necessary to calibrate with actual sediment standards in the case of fused samples, a selection of such reference materials is desirable for method verification. For trace analyses, the use of certified reference materials is also more practical than the preparation of synthetic mixtures, which may involve very high dilution. Numerous lake and stream sediment reference materials are available. Resources such as the *Geostandards Newsletter* (Lynch, 1990) and compilations by Potts et al. (1992) are essential tools for compiling and assessing sediment standard concentration data.

4.4   Discussion

4.4.1   Advantages and limitations

The X-ray spectrum, of which K, L, and occasionally M lines only are used routinely, is relatively simple. The inner-shell transitions are independent of chemical state, except for low-Z elements. That problem is overcome by fusion. The irradiation is nondestructive for geological matrices. Sample preparation is simple and adapts well to automation with minimal reagents. Consequently, the routine operation of an XRF system does not require extensive, specialized training. The instrumentation is generally robust and very stable. As a result, frequent recalibration is not necessary, and routine maintenance can be conducted by the user. The potential for speed, particularly when simultaneous detection is employed, compares favorably with other techniques. It is possible to complete a multielement analysis within minutes of sample receipt.

XRF techniques are very accurate when applied correctly. There is no theoretical foundation for XRF measurements to be less accurate than other techniques. Optimum setup and calibration does, however, require skilled analysts and is time-consuming. The simplicity of the spectrum often dictates the use of spectral lines, which must be corrected for one or more interferences due to lack of alternatives. It is becoming common practice for laboratories to outsource these tasks and then operate the system internally. The capital costs, especially for

WDS, can be prohibitive, making it attractive to send samples to a custom laboratory for nonproduction facilities.

Sensitivity for light elements (i.e., $Z < 15$) is relatively poor due to low fluorescent yield and absorption by detector windows. Although specialized instruments may have the ability to measure boron ($Z = 5$) (Willis and Duncan, 1995), the lightest element for routine determination by a typical spectrometer is usually sodium ($Z = 11$). Lower limits of detection in the order of 0.01% for these elements may be unacceptably high for some sampling programs. The use of fusion does make measurement of volatile elements (e.g., S, As, Cl) problematic. For example, even when oxidants were added, retention of sulfur was found to be semiquantitative (Eastell and Willis, 1990). Flux mixtures and fusion techniques are a major focus for XRF method development because the full benefits of cost and speed can only be realized if a wide range of elements and concentrations can be determined from a single preparation.

In summary, XRF is an ideal technique for rapid, nondestructive total-metal analysis of solid materials for elements with $Z = 11$ (sodium) to 92 (uranium). The simplicity of sample preparation and virtual lack of recovery problems for most elements of XRF remain very attractive. However, the perception of simplicity itself can be a pitfall. It is very easy to carry out bad X-ray work. Independent training in the principles of X-ray fluorescence is essential for XRF analysts (Willis and Duncan, 1995). This is particularly important when one is tempted to apply an empirical model that may not be valid, or a theoretical formalism that is not understood.

Acknowledgments

The author would like to thank Mr. R.W. Calow, Dr. J.R. Johnston, and S.T. Caverly for their critical review of the manuscript. Figure 4-1 appears by courtesy of Dr. J.P. Willis from the University of Cape Town, South Africa.

References

Bearden, J.A. and Burr, A.F., Reevaluation of X-ray atomic energy levels, *Rev. Mod. Phys.*, 125, 1967.

Bertin, E.P., *Principles and Practice of X-ray Spectrometric Analysis*, 2nd ed., Plenum Press, New York, 1975, 467.

Carsey, T.P., X-ray fluorescence analysis of welding fume particles, *Adv. X-ray Anal.*, 25, 209, 1982.

Claisse, F., Glass Disks and Solutions by Borate Fusion for Users of Claisse Fluxers, Corporation Scientifique Claisse Inc., Saint-Foy, Quebec, 1995.

Claisse, F. and Quintin, M., Generalization of the Lachance-Traill method for the correction of the matrix effect in X-ray fluorescence analysis, *Can. Spectros.*, 12, 129, 1967.

Davidson, R.A., Walker, E.B., Barrow, C.R., and Davidson, C.F., On-line radio-isotope XRF analysis of copper, arsenic, and sulfur in copper electrolyte purification solutions, *Appl. Spec.*, 48, 796, 1994.

Eastell, J. and Willis, J.P., A low dilution fusion technique for the analysis of geological samples. 1 — Method and trace element analysis, *X-ray Spectrom.*, 19, 3, 1990.

Eastell, J. and Willis, J.P., A low dilution fusion technique for the analysis of geological samples. 2 — Major and minor element analysis and the use of influence/alpha coefficients, *X-ray Spectrom.*, 22, 1993.

Ereiser, J., Frank, G., Guy, J.W., and Litchinsky, D., On-stream analysis of float process slurries by SRF, *Adv. X-ray Anal.*, 24, 297, 1981.

Feather, C.E. and Willis, J.P., A simple method for background and matrix correction of spectral peaks in trace element determination by X-ray fluorescence spectrometry, *X-ray Spectrom.*, 5, 41, 1976.

Friedman, H. and Birks, L.S., Geiger-counter spectrometer for X-ray fluorescence analysis, *Rev. Sci. Instrum.*, 19, 323, 1948.

Jenkins, R. and DeVries, J.L., *Practical X-ray Spectrometry,* Springer-Verlag Inc., New York, 1967, 183.

Jenkins, R. and Gilfrich, J.V., Figures of merit, their philosophy, design, and use, *X-ray Spectrom.*, 263, 1992.

Lachance, G.R. and Traill, R.J., A practical solution to matrix problem in X-ray analysis, *Can. Spectrosc.*, 11, 43, 1966.

Lachance, G.R. and Claisse, F., *Quantitative X-ray Fluorescence Analysis,* John Wiley & Sons, Chichester, 1995, 402.

Lynch, J., Provisional elemental values for eight new geochemical lake sediment and stream sediment reference materials, LKSD-1, LKSD-2, LKSD-3, LKSD-4, STSD-1, STSD-2, STSD-3, STSD-4, *Geostand. Newsl.*, 14, 153, 1990.

Moseley, H.G.J., High-frequency spectra of the elements, *Phil. Mag.*, 27, 703, 1913.

Mudroch, A. and Bourbonniere, A., Sediment preservation, processing and storage, in *Handbook of Techniques for Aquatic Sediments Sampling,* 2nd ed., Mudroch, A. and MacKnight, S.D., Eds., CRC/Lewis, Boca Raton, FL, 1994, 131.

Potts, P.J., Tindale, A.G., and Webb, P.C., *Geochemical Reference Material Compositions: Rocks, Minerals, Sediments, Soils, Carbonates, Refractories, and Ores Used in Research and Industry,* Whittles Publishing, Latheronwheel, U.K., 1992, 313.

Rousseau, R.M., CiLT, Software for Chemical Composition Calculations Using the Lachance-Traill Algorithm in X-ray Fluorescence Analysis, Ver. 4.1, Les Logiciels R. Rousseau Inc., Cantley, Quebec, 1990a.

Rousseau, R.M., CiROU, Software for Chemical Composition Calculations Using the Fundamental Algorithm in X-ray Fluorescence Analysis, Ver. 4.1, Les Logiciels R. Rousseau Inc., Cantley, Quebec, 1990b.

Tao, G.Y., Pella, P.A., and Rousseau, R.M., NBSGSC-FORTRAN Program for Quantitative X-ray Fluorescence Analysis, NBS Tech. Note 1213, National Institute for Standards and Technology, Gaithersburg, MD, 1985.

Tertian, R. and Claisse, F., *Principles of Quantitative X-ray Fluorescence Analysis,* Heyden, London, 1982, 485.

Willis, J.P., Mass absorption coefficient determination using Compton scattered tube radiation: applications, limitations, and pitfalls, *Adv. X-ray Anal.,* 34, 243, 1991.

Willis, J.P. and Duncan, A.R., Course on Theory and Practice of XRF Spectrometry, Course Notes Week 1, University of Western Ontario, Department of Earth Sciences, 1995.

chapter five

Determination of trace elements in sediments

Gwendy E.M. Hall

5.1 Introduction

Today the geoanalyst is extremely fortunate to have a plethora of analytical techniques and methods available; in fact, the choice is perhaps becoming daunting. The gravimetric and colorimetric methods dating mainly from the first half of this century have been superseded by the introduction of instrumental techniques, such as atomic absorption spectrometry (AAS), X-ray fluorescence (XRF), instrumental neutron activation analysis (INAA), inductively coupled plasma atomic emission spectrometry (ICP-AES), and ICP mass spectrometry (ICP-MS). An historical perspective of this instrumental revolution can be gleaned from Potts et al. (1993), who discuss major events, discoveries, and publications related to the development of modern inorganic geochemistry. In this chapter, the so-called "wet-chemical" techniques of AAS, ICP-AES, and ICP-MS are reviewed as they relate to the determination of trace elements in aquatic sediments. The literature cited comes from the traditional journals read by the geoanalyst working in the fields of environmental studies and exploration geochemistry, disciplines that merge in their common goal of gaining further understanding of the geochemical processes occurring in the surficial environment.

It should be noted that the term *analytical technique* pertains to the instrumentation employed and associated principles, while the term *analytical method* describes the complete scheme used to obtain the result, including not only the instrumentation but the sample decomposition and any other procedures involved prior to measurement. Some of the pertinent questions that the analysts must ask themselves are the following: Do I need to decompose the sample, or is there a direct analytical technique with sufficient sensitivity? Does the diges-

Table 5-1 Glossary of Acronyms Used in This Chapter

Acronym	Definition
AVS	Acid-volatile sulfide
APDC	Ammonium pyrrolidine dithiocarbamate
CCD	Charge coupled device
CCRMP	Canadian Certified Reference Materials Project
CID	Charge injection device
CVAAS	Cold vapor atomic absorption spectrometry
DCP	Direct current plasma
DIN	Direct injection nebulizer
DSI	Direct sample insertion
ETV	Electrothermal vaporization
FAAS	Flame atomic absorption spectrometry
FI	Flow injection
GFAAS	Graphite furnace atomic absorption spectrometry
GSC	Geological Survey of Canada
HEN	High efficiency nebulizer
HGAAS	Hydride generation atomic absorption spectrometry
HG-ICP	Hydride generation inductively coupled plasma
ICP-AES	Inductively coupled plasma atomic emission spectrometry
ICP-MS	Inductively coupled plasma mass spectrometry
IP	Ionization potential
MIBK	Methyl isobutyl ketone
NIST	National Institute of Standards and Technology
PMT	Photomultiplier tubes
PN	Pnuematic nebulizer
REE	Rare earth element
RSD	Relative standard deviation
SRM	Standard reference material
TDS	Total dissolved salt
ppm; ppb; ppt; ppq	$\mu g/g$; ng/g; pg/g; fg/g

tion suit the project's objectives for all elements? Which elements are susceptible to loss via volatilization or precipitation during digestion? Is the analytical technique to be employed sensitive enough and the level of contamination by reagents low enough for adequate accurate and precise measurement of the analytes of interest? What are the potential interferences to be aware of? Which reference material(s) (international or in-house) should be inserted in the batch of samples in order to evaluate accuracy? How much replication is needed to satisfactorily estimate precision? Although the analytical community has made great strides in instrumentation, the tendency has been to overlook sample preparation and decomposition until recently. No matter how accurate and precise the measurement step is, the result is only as good as all the preparation that has gone before.

The heyday of flame AAS, first introduced by (Sir) Alan Walsh in 1955, is certainly over, though it is used for specialized purposes; the companion technique of graphite furnace AAS (GFAAS) is enjoying a revival. The robust and inexpensive technique of flame AAS remains in use in developing countries and in small-scale laboratories where large, multielement data sets are not required. The high sensitivity of GFAAS and the discovery of "universal" matrix modifiers, such as palladium, to minimize interferences have ensured this technique a continuing place in the modern geoanalytical laboratory. Methods published in the 1960s demonstrated the strength of FAAS in the determination of the major elements, first row transition series, Rb, Sr and Ba in geological samples.

The first ICP emission spectrometer was brought to the marketplace in 1975 and subsequently became the workhorse of laboratories with a tradition of wet chemical analysis. Its advantages over the established technique of FAAS included simultaneous multielement measurement (i.e., speed) and long linear dynamic range (i.e., 10^5); a major disadvantage, however, was its much more complex spectrum, with greater chance of line overlap. Research is continuing today into ways of improving the performance of ICP-AES, though the instrumentation is far more automated and refined than it was in the first generation. This research into such areas as understanding the many processes taking place in the plasma and designing superior modes of sample introduction has benefited the younger technique of ICP-MS, introduced commercially in 1983. Hailed as the most sensitive and versatile technique in our repertoire today, ICP-MS can be thought of as providing detection limits (3σ) in solution of ca. 0.01 to 0.1 µg/L for most elements, while those by ICP-AES generally fall into the range 1 to 10 ppb. These values, however, are moving targets, due to the rapid progress being made in instrument design in both techniques. Certainly, manufacturers are now quoting ppt (i.e., 0.001 ppb) detection limits by ICP-MS for elements such as Au and Ag, but these are "ideal" and apply to pure solutions rather than to acid leachates, for example. Although ICP-MS is superior in sensitivity to ICP-AES, it is far less tolerant of total dissolved salt (TDS) content and thus requires greater dilution of leach solution to a maximum TDS of ca. 0.1%. The two ICP techniques should be employed in geoanalysis in a complementary fashion, with ICP-AES covering major, minor, and higher abundance trace elements, and ICP-MS being used for the heavier and less abundant elements. Figure 5-1 illustrates the range in pure solution (i.e., ideal) detection limits of these techniques. At a glance, it is clear that ICP-MS has a leading role to play in the analysis of sediments for ultratrace elements.

Further details on the techniques themselves, not necessarily pertaining to sediment applications, can be found in books authored by

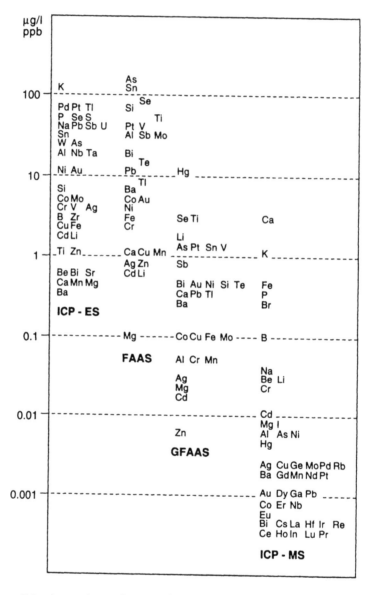

Figure 5-1   Approximate detection limits in pure solution associated with the techniques under discussion.

Slavin (1984, GFAAS), Thompson and Walsh (1989, primarily ICP-AES), Date and Gray (1989, ICP-MS), and Potts (1987, multitechnique). The *Journal of Analytical Atomic Spectrometry* (*JAAS*) annually publishes an extremely useful and extensive review of the applications of atomic spectrometry to the chemical analysis of environmental samples, sediments being listed under "geological materials" (e.g., Cresser et al., 1995). Also, a regular review in *JAAS* is an update of activities in the techniques themselves, leaning more toward research in instrumentation and theory than to application. Volume 44 of the *Journal of Geochemical Exploration*, a special issue published in 1992 entitled "Geoanalysis" and edited by Hall, was written for the geochemist (environmental and exploration) rather than the analyst. It was intended to provide an appreciation for the state-of-the-art in this field, covering each prominent analytical technique and methodology in widespread use today together with discussion on reference materials, sample decomposition, and statistical considerations of data quality. There, as here, the less common techniques, such as electrochemical (e.g., anodic stripping voltametry, ion-selective electrodes) and atomic fluorescence, have been omitted from discussion. Atomic fluorescence is not offered by manufacturers as a multielement instrument, though the technique is used for specific elements, such as Hg, where sensitivity is superior to that by AAS (Moreales-Rubio et al., 1995). Detection limits quoted throughout this chapter refer to that concentration of element in the original sediment sample, not in a digestate (i.e., the method's detection limit); exceptions are noted.

5.1.1   Standard reference materials

The number of international sediment standard reference materials (SRMs) is becoming larger rapidly, as the need for quality control and assurance (QC/QA) is realized in more and more programs. Thirty SRMs, together with the trace elements for which they are certified and their associated organizations, are shown in Table 5-2, which is by no means an exhaustive list. For example, the National Institute for Environmental Studies (NIES) in Japan, the Commission of the European Communities Measurements and Testing Programme (BCR) in Brussels, and the National Water Research Institute (NWRI) of Environment Canada all provide a variety of sediment SRMs. Organizations such as NIST and NRCC focus on materials suitable for environmental studies in industrial river, estuarine, and marine sedimentary environments, whereas sediments issued under the IGGE and GSJ were developed for geochemical exploration purposes. The certified values refer to total element concentration, obtained either as a result of an extensive round-robin among reputable laboratories (e.g., CCRMP) or by in-house analyses using independent techniques (e.g., NIST, NRCC). Recommended (i.e., certified), proposed, and informational values, in order of degree of certainty, for the SRMs listed in Table 5-2 can be found in a very useful compilation by Govindaraju (1994), which is also available digitally. The lake and stream sediment series, LKSD 1-4 and STSD 1-4 prepared by the Geological Survey of Canada (GSC) and marketed

Table 5-2 International Sediment Standard Reference Materials (SRM). Parentheses
Indicate Not All the SRMs in That Series Are Certified for That Element

SRM	Source	Trace elements certified
MESS-2, marine	NRCC	Ag As Be Cd Co Cr Cu Li Mo Ni Pb Sb Se Sn Sr V Zn
BCSS-1, marine	NRCC	As Be Cd Co Cu Ni Pb Sb Se V Zn
PACS-1, marine	NRCC	As Cd Co Cr Cu Hg Mo Ni Pb Sb Se Sn Sr V Zn
BEST-1, marine	NRCC	As Cd Co Cr Cu Hg Mo Ni Pb Sb Se Sn Sr V Zn
LKSD 1-4 series, lake	CCRMP	Ag As Au B Ba Be Br Ce Co Cr Cs Cu Dy Eu F Hf La Li Lu Nb Nd Ni Pb Rb Sb Sc Sm Sn Tb Th U V Y Yb Zn Zr
STSD 1-4 series, stream	CCRMP	Ag As Au B Ba Be Br Ce Co Cr Cs Cu Dy Eu F Hf La Li Lu Nb Nd Ni Pb Rb Sb Sc Sm Sn Tb Th U V Y Yb Zn Zr
MAG-1, marine mud	USGS	B Ba Be Cd Ce Co Cr Cs Cu Dy Eu F Ga Gd Hf Ho La Li Lu Nd Ni Pb Rb Sb Sc Se Sm Sr Tb Th Tm U V Y Yb Zn Zr
SL-1, lake	IAEA	As Ba Br Cd Ce Co Cr Cs Cu Dy Hf La Nd Ni Pb Rb Sb Sc Sm Th U V Yb Zn
SL-3, lake	IAEA	As Br Ce Cs Dy Eu Hf La Lu Nd Rb Sb Sc STa Tb Th U Yb
GSD 1-8, stream	IGGE	Ag As B Ba Be Bi Cd Ce Co Cr Cs Cu Dy Er Eu F Ga Gd Ge Hg (In) La Li (Lu) Mo Nb Nd Ni Pb Pr Rb S Sb Sc (Se) Sm Sn Sr (Ta) Tb Th Tl Tm U V W Y Yb Zn Zr
GSD 9-12	IGGE	Ag As B Ba Be Bi Cd Ce Co Cr Cs Cu Dy Er Eu F Ga Gd Ge Hf Hg I In La Li Lu Mo Nb Nd Ni Pb Pr Rb Sb Sc Se Sm Sn Sr (Ta) Tb Th Tl Tm U V W Y Yb Zn Zr
SARM 51, 52, stream	MINTEK	Co Cr Cu Nb Ni Pb Rb Sr V Y Zn Zr
1645, river	NIST	As Cd Co Cr Cu Eu F Hg La N Ni Pb Sb Sc Se Th Tl U V Zn
1646, estuarine	NIST	As Be Cd Ce Co Cr Cs Cu Eu Ge Li Mo Ni Pb Rb Sb Sc Se Te Th Tl V Zn
2704, river	NIST	As Ba Br Cd Ce Co Cr Cu Dy Eu Ga Hf Hg La Li Lu Ni Pb Rb Sb Se Sm Sn Sr Th Tl U V Yb Zn Zr
Jsd 1-3, stream	GSJ	As Be Ce Co Cr Cs Cu (Er) (Eu) La Li Nd Ni Pb Rb Sc (Sm) Sr V Y Yb Zn Zr
GSMS-1, marine	IRMA	As B Ba Ce Cl Co Cs Cu Dy Er Eu Ga Gd Ho La Li Lu Mo Nd Ni Pb Pr Rb Sb Sc Sm Sr Tb Th Tm U V W Y Yb Zn Zr
OOK 201-204	RIAP	Ag (As) Au B Ba (Be) (Bi) (Cd) (Ce) Co Cr Cs Cu Ga Ge La Li Mo Nb Ni Pb Rb Sc Sn Sr (Th) V (W) Y Yb Zn Zr

Note: NRCC: National Research Council of Canada; CCRMP: Canadian Certified Reference
Materials Project; USGS: United States Geological Survey; IAEA: International Atomic
Energy Agency, Vienna; IGGE: Institute of Geophysical and Geochemical Exploration,
China; MINTEK: Council for Mineral Technology, South Africa; NIST: National Institute
of Standards and Technology, U.S.A.; GSJ: Geological Survey of Japan; IRMA: Institute
of Rock and Mineral Analysis, China; RIAP: Research Institute of Applied Physics,
Irkutsk, Russia.

under the CCRMP, now has a wealth of data derived from partial leaches, such as *aqua regia*, and these values will soon be added to those originally provided (Lynch, 1990).

In large programs, insertion of SRMs at an adequate frequency to monitor accuracy (and sometimes precision) would be prohibitively expensive. Ideally, an international SRM should only be used to verify new methodology, or a new laboratory and secondary in-house control samples should be employed for routine evaluation of accuracy and precision. It must be remembered that SRMs do not necessarily reflect the accuracy associated with the samples under study; they are exceptionally well homogenized and ground to a fine fraction (usually <74 μm), and almost certainly have a different mineralogical structure. Today there are numerous laboratories, governmental and private, that are offering the service of preparing reference materials to suit a particular project. Government agencies, such as the NWRI in Canada, carry out interlaboratory quality control studies in which an individual laboratory can participate (for a small fee) in analyzing the prepared material and can then evaluate itself against others (e.g., Cheam et al., 1988).

5.2   *Decomposition techniques*

The reader is referred to three excellent books on decomposition methods; those of a general nature are discussed by Dolezal et al. (1968) and Bock (1979), and the book by Gorsuch (1970) focuses on destruction of organic matter. Kingston and Jassie (1988) have summarized methods based on microwave-assisted digestion in a book that will need updating soon in view of the high level of activity in this field today. Another valuable reference concerning decomposition methods applied to geological materials is a review paper by Chao and San-zolone (1992).

After preparation of the sample to a fineness of at least 177 μm (i.e., <80 mesh), preferably 63 μm, the elements of interest must be brought into solution for analysis by techniques such as ICP-AES, ICP-MS, ICP-AFS, AAS, and electrochemical methods. In general, decompositions are based on two modes of breaking up the crystal lattice: the use of acids and of fusions. In choosing the most appropriate decomposition, the following criteria must be addressed: the chemical and mineralogical properties of the sample; the elements of interest; the constraints imposed by the analytical technique(s) to be employed (e.g., interferences, especially from major constituents); the precision and accuracy limits acceptable to meet the objectives of the program; the selection of a partial or total attack to suit the information required; productivity; and cost.

5.2.1   Acid digestion

The advantages enjoyed by acid as opposed to fusion decomposition of sediments include the following: organic material is easily volatilized by oxidation as CO_2; extraneous salts are not added to the analyte solution, thus avoiding potential interferences and high blank levels; acids can be obtained at a high degree of purity; Si can be volatilized with HF, thus reducing salt content; and procedures are adaptable to large scale production. Acids high in oxidizing strength include HNO_3, $HClO_4$ and H_2SO_4, while those nonoxidizing in their action are HCl, HF, HBr, H_3PO_4 and dilute $HClO_4$ and H_2SO_4. Two acid digestions in common use today are *aqua regia* and $HF–HClO_4–HNO_3–HCl$, though many combinations can be employed for different objectives. In North American laboratories, there has been a trend away from employing the $HClO_4–HNO_3$ attack, requiring stainless steel rather than wooden fume hoods, for the decomposition of organic- and sulfide-bearing material in favor of the simpler and less hazardous *aqua regia* procedure. However, destruction of organic matter by the $HClO_4–HNO_3$ attack is superior to that by *aqua regia*; thus, the objectives of the program must be considered in method selection. These oxidizing decompositions are employed in preference to the more rigorous and time-consuming HF-based methods when elements of interest (e.g., Cu, Pb, Zn, Cd, Co, Ni) are sorbed onto clay minerals or are present in other readily solubilized phases (e.g., as humic or fulvic complexes or in Fe, Mn oxides), rather than within resistant silicate lattices.

5.2.1.1   Acid properties

Nitric acid is a widely used acid for the destruction of organic matter, reacting readily with both aromatic and aliphatic groups. The acid boils at about 120°C, a factor that assists in its removal after oxidation but that correspondingly limits its effectiveness. Hot, concentrated HNO_3 is used to decompose organics, sulfides, selenides, tellurides, arsenides, sulfoarsenides and phosphates through oxidative degradation (S is oxidized to SO_4, C to CO_2). Nitric acid dissolves the majority of metals occurring in nature, with the exception of Au and Pt. Iron sulfides and molybdenite dissolve easily, but HNO_3 is not as effective as HCl in dissolving Fe and Mn oxides. Practically all O-containing primary uranium minerals are decomposed with concentrated HNO_3. Nitric acid can be used in the presence of $HClO_4$ which continues the oxidation after HNO_3 has been removed. Care must be exercised in using $HClO_4$ as it is such a powerful oxidizing agent, and explosions have occurred. Nitric acid is added far in excess of $HClO_4$ (e.g., at a ratio of 4:1), so that much of the oxidation is carried out before the action of perchloric is initiated. The powerful oxidizing and dehydrating properties of hot, concentrated $HClO_4$ are extremely effective in decomposing organic matter and sulfides. Oxidizing power is lost upon dilution of the acid with water. All the perchlorates formed after decomposition with $HClO_4$ are soluble in water, with the exceptions of those of K, Rb, and Cs. Its high boiling point (203°C for the 71.6% acid) makes it useful in driving off HF and more volatile acids such as HNO_3 and HCl, which have boiling points in the range 110 to 120°C. Although H_2SO_4 has similar properties to $HClO_4$, it has not found such widespread application, probably due to interference effects created by SO_4 in analytical techniques such as AAS and ICP-AES, and to the low solubility of alkaline earth and Pb sulfates. The improved purity of hydrogen peroxide (H_2O_2) and the lack of residual products (H_2O only) have promoted its popularity in recent years, particularly in an effort to avoid $HClO_4$.

Aqua regia has become a popular replacement for the HNO_3–$HClO_4$ attack. The mixture of 3 parts HCl to 1 part HNO_3 also has a strong oxidizing power due to the formation of nascent chlorine and nitrosyl chloride, and thus the organic component of a sediment is efficiently "wet-ashed":

$$HNO_3 + 3HCl \rightarrow NOCl + 2H_2O + 2(Cl)$$

$$NOCl \rightarrow NO + (Cl)$$

Hot *aqua regia* is an efficient solvent for numerous sulfides (e.g., those of As, Se, Te, Bi, Fe, Mo), arsenides, selenides, tellurides, sulfosalts, and native Au, Pt, and Pd. The minerals belonging to the group of simple oxides and their hydrates (e.g., Fe–Mn) are completely decomposed with *aqua regia*. Natural U-oxides, Ca-phosphates, and most sulfates (except barite) are solubilized, as are some silicates, such as the zeolites. The oxidizing strength of *aqua regia* can be enhanced by adding bromine. Evaporation to dryness with concentrated HCl converts salts to chlorides, ready for final solubilization in dilute (0.5 to 1 M) mineral acid, which is compatible with the analytical technique. *Aqua regia* digestion for As, Sb, Bi, Se, and Te should not be taken to dryness to avoid loss of analytes via volatilization. Strong oxidizing acid mixtures are required to convert all forms of Hg to Hg^{2+}. Sometimes HF is added to *aqua regia* to decompose silicates.

The complexing ability of F^- is important in the mechanism of decomposition by HF, though the degree of dissociation is quite low (about 10% at a strength of 1 M). Hydrofluoric acid is most effective in breaking up the strong Si–O bond to form SiF_4, which volatilizes upon heating. Fluorides of As, B, Ti, Nb, Ta, Ge, and Sb may be lost to varying extents upon heating. Hydrochloric acid, a strong acid, is effective in the dissolution of carbonates, phosphates, borates, and sulfates (except barite), and has become an almost universal solvent suitable for most techniques, with the possible exception of ICP-MS due to the formation of Cl-molecular species. Its capacity to attack Fe and Mn oxides is superior to that of HNO_3, due to its reducing and complexing properties. It solubilizes acid-volatile sulfides (AVS), consisting of poorly crystallized secondary sulfides and monosulfides (e.g., ZnS, FeS, PbS), but its action on pyrite (FeS_2) is limited. Hydrofluoric acid is customarily used with oxidizing mineral acids to effect decomposition of organics, oxides, and sulfides, as well as silicates. Fluoride is usually removed by evaporation with $HClO_4$, thereby preventing the precipitation of insoluble fluorides (e.g., Ca and the rare earth elements [REE]) later in the digestion. As HF attacks glassware, polytetrafluorethylene (PTFE or Teflon) or platinum dishes are employed, and the absence of HF in the analyte solution makes it suitable for passage through glass nebulizers, spray chambers, and torches. Nitric acid is added to moderate the action of $HClO_4$ on organic material, which could be explosive. There are many variations of the $HF–HClO_4–HNO_3–HCl$ procedure, but normally the mixed acids are evaporated to dryness and the residue dissolved in HCl (0.5 to 1 M) for analysis. Although this attack is often referred to as total, it does not dissolve the refractory minerals cassiterite, wolframite, chromite, spinel, beryl, zircon, tourmaline, magnetite, and high concentrations of barite, among others. Ignoring this fact can lead to misinterpretation of data and subsequently to incorrect hypotheses (Hall and Plant, 1992).

5.2.1.2   Open systems

The effectiveness of extraction naturally depends upon parameters, such as sample to reagent volume ratio, grain size distribution of the sample, strength of acids, temperature, pressure, and time of contact between sample and acid mixture. Open vessel decomposition has the advantage of being amenable to large-scale batch type operation. Decomposition with *aqua regia*, with or without water added, is usually carried out in borosilicate test tubes or beakers placed in or on a heating block, hot plate, or water bath. Decomposition involving HF is carried out in Teflon or Pt beakers on a hot plate so that reaction products can be fumed off and the sample taken to insipient dryness prior to solubilization in the final acid medium. Fuming to white fumes of $HClO_4$ not only removes insolubles fluorides (e.g., those of the REEs and Ca) but also eliminates Si (as SiF_4), which would otherwise hydrolyze and precipitate on standing, thereby removing some trace elements from solution. To achieve this end, repetitive fuming with $HClO_4$ is preferable to single action rapid evaporation (Croudace, 1980). Some procedures specify the addition of HF alone initially, while others call for the three acids to be added at once. This order of addition and attack can have an impact on loss of elements by volatilization. For example, the presence of $HClO_4$ can inhibit volatilization by oxidizing elements, such as As and Sb, to their less volatile fine-valency state.

The U.S. Environmental Protection Agency (U.S. EPA) Method 3050A for acid digestion of sediments, sludges and soils, is based on the addition of HNO_3 and H_2O_2 to 1 to 2 g of sample. Initially, the sample is refluxed with repetitive additions of HNO_3, and after approximately 1.5 h the solution is evaporated to 5 mL. Refluxing is continued with the addition of up to 10 mL of 30% H_2O_2 until effervescence has subsided. The sample may be dissolved in HCl for analysis by FAAS or ICP-AES after filtration or centrifugation, or alternatively the acid-peroxide digestate can be further reduced to ca. 5 mL and taken up in H_2O to 100 mL for analysis by GFAAS or ICP-MS. The digestion specified for stream sediments under the National Geochemical Reconnaissance Program (NGRP) at the GSC is based upon HCl and HNO_3. To 1 g of sediment in test tubes, 3 mL of HNO_3 are added and allowed to stand overnight. The test tubes are then heated to 90°C for 30 min after which 1 mL of HCl is added and the digestion continued for 90 min with intermittent mixing of sample and acid. After cooling, the solution is diluted to 20 mL for analysis. The NGRP protocol for lake sediments involves the addition of 6 mL of 4 M HNO_3–1 M HCl to 1 g of sediment and, after standing overnight, heating at 90°C for 2 h, followed by dilution to 20 mL after cooling.

The extensive round-robin exercise, prior to certification of the CCRMP series of lake and stream standard reference sediments (LKSD 1-4 and STSD 1-4), allowed for comparison of data obtained by (1) concentrated *aqua regia*; (2) *aqua regia* diluted with water; (3) HNO_3–$HClO_4$; (4) and "total" digestions (Lynch, 1990; Lynch, personal communication, 1995). Results for all eight reference sediments were identical by all these digestions (including HF) for Ag, Cd, Cu, Mo, and Pb. There was a tendency for As, Co, and Ni values to be slightly lower by attacks (1) to (3), whereas these acid mixtures resulted in distinctly low values for Ba, Cr, Sr, and V. Results for Sb in the lake sediments agreed well between concentrated *aqua regia* and total digestion, but values obtained by *aqua regia* were distinctly low compared to total values for the stream sediments (e.g., 3.6 vs. 7.3 ppm for STSD-4). Interestingly, the Mn value was significantly low, outside the limits set by the standard deviations, by attacks (1) and (2) for LKSD-1 and -4, and for STSD-2 and -4 (e.g., 670 ± 80 cf. 1060 ± 60 ppm in STSD-2). One participating laboratory provided precise replicate data obtained by refluxing the sediments in HNO_3 alone. Their data were generally lower than the accepted *aqua regia* values, even for elements such as Ni, Co, and Pb, and in turn were much lower than total values, especially for the stream sediments.

5.2.1.3   Closed systems

A closed system, as the term implies, involves carrying out the decomposition in a vessel closed to the atmosphere, such as a pressure bomb. The appeal of this methodology is multifaceted: volatilization loss is minimized or negated; atmospheric contamination is minimized (important for ubiquitous elements, such as Zn, Pb, Fe); the volume of reagents required is reduced, hence creating lower blank levels and dilution factors; and the efficiency of decomposition is increased by the action of higher temperatures and pressure. The negative aspect is cost, in terms of both capital outlay for the vessels and lower productivity. Vessels range from platinum-lined nichrome crucibles housed in a metal casing, made to withstand temperatures up to 425°C and pressures up to 6000 psi, to simple screw-capped polypropylene and polycarbonate bottles and test tubes. Hale et al. (1985) demonstrated that extraction of Cu, Pb, Zn, Mn, Fe, Co, and Ni in a suite of sediments and soils using HCl and HNO_3 (at a ratio of 4:1) in sealed borosilicate glass tubes for 1 h at 175°C was equivalent to the performance of the more corrosive and noxious mixture of HNO_3 and $HClO_4$ (4:1) with fuming. This approach became popular in geochemistry laboratories, but in the interests of safety and to accommodate the use of HF, borosilicate glass gave way to various organic polymers. Different materials have their own properties, Teflon being able to withstand temperatures up to 250°C, whereas a maximum temperature of 130°C is appropriate for polypropylene and polycarbonate. While Teflon-lined pressure vessels are expensive, their use can ensure efficient dissolution of minerals such as beryl, kyanite, staurolite, tourmaline, garnet, ilmenite, chromite, rutile, tantalite, pyrite, chalcopyrite, and pyrrhotite (Langmyhr and Sveen, 1965). For routine high-production analyses, inexpensive screw-capped Teflon vials of 15 to 50 mL capacity have become popular. In order to complex fluoride in solution, boric acid is normally added following the digestion.

Closed system digestion is increasingly being used, with the energy source being microwave radiation rather than heat. The advantages include more complete dissolution in much less time (minutes rather than hours), lower volume of acids required, and less exposure to toxic fumes. The magnetron produces an electromagnetic wave with a frequency of 2450 MHz, which rapidly generates heat in samples by energizing polar molecules; a chemical reaction, however, is still necessary to complete the digestion. Kingston and Jassie's landmark paper (1986) described the interaction of mineral acids with the microwave field and proposed quantitative expressions relating dynamic system pressure and temperature to microwave power and sample load. The first commercial microwave digestion system, released in the mid 1980s, was the Model MDS-81D from CEM (Matthews, NC, U.S.A.). Twelve Teflon vessels of 120 mL capacity were loaded in a carousel situated in a Teflon-lined oven of variable power up to 600 W. A pressure relief valve on each vessel facilitated the venting of vapors to a collection unit if pressures exceeded about 85 psi. A number of explosions have been reported by those using sealed containers without pressure relief mechanisms (Gilman and Grooms, 1988). Using the CEM equipment, Nakashima et al. (1988) found results for As, Cd, Co, Cu, Fe, Mn, Ni, Pb, and Zn in MESS-1 to be in good agreement with certified values; results for Cr, however, were low (this requires multiple additions of $HClO_4$–HNO_3). Their scheme involved the addition of 3 mL of HNO_3, 3 mL of HF, and 1 mL of $HClO_4$ to 0.2 to 0.6 g of sediment and application of full power until the pressure reached 60 to 65 psi, where it was maintained for 20 min. The ensuing solution was then evaporated on a hot plate to remove HF and the residue dissolved in HCl. Using similar equipment but employing a mixture of HF, HCl, and HNO_3, Lamothe et al. (1986) reported no loss of volatile elements (e.g., B, Cr, Pb, Si) during microwave digestion of 51 geological samples. However, recoveries were low for Al, Cr, Li, Pb, Si, and Zr in many samples, including marine sediments MAG-1 and MESS-1 and the Chinese stream sediment GSD-series, owing to the presence of refractory mineral phases. In contrast to these two papers, Bettinelli et al. (1989) reported full recovery of elements, such as Cr in NIST River Sediment 1645 and Marine Sediment 1646. The key differences here are the higher amounts of acid added (e.g., about 20 mL of HF–HNO_3–HCl to 0.25 g of sediment), and the much longer time of digestion (approximately 25 min) of their program. *Aqua regia* extraction using the Model 81D was compared with conventional refluxing in a Pyrex test tube and was found to produce equivalent results for Cd, Cr, Cu, Fe, Mn, Pb, and Zn in IAEA Lake Sediment SL-1, though the conditions selected (e.g., digestion time of 64 min) did not demonstrate an advantage to the microwave approach (Nieuwenhuize et al., 1991). The status of microwave dissolution up to 1989 was reviewed by Matusiewicz and Sturgeon (1989).

The first flurry of applications of microwave technology for decomposition was followed by the realization that the advantage of shorter time of digestion itself was counterbalanced by disadvantages, such as the tedious and repetitive manipulation of capping and uncapping vessels and the lengthy cool-down time. One mining company in Canada, Kidd Creek, designed a robotic system to handle samples for each microwave, but this was a costly venture and did not prove sufficiently rugged. The number of science-oriented microwave oven manufacturers increased, and research was carried out into developing temperature and pressure feedback mechanisms and vessels capable of withstanding higher pressures. The U.S. Environmental Protection Agency Method 3051, describing the microwave-assisted digestion of sediments, affirmed the importance of temperature control with time: "any sequence of power [may be used] that brings the sample to 175°C in 5.5 minutes and permits a slow rise to 180°C during the remaining 4.5 minutes". The latest generation of microwave systems is completely computer controlled and designed to work at up to 230°C and 625 psi for particularly refractory samples. Applications directories are built in, but the analyst is advised to customize and validate procedures to suit the samples at hand. An interesting concept from Prolabo (Paris, France) incorporates focused microwaves directly onto the sample (at ambient pressure) rather than dispersing the energy through an oven cavity. Two such systems are offered, one accommodating a single sample at a time and the other sixteen samples under fully automated and programmable treatment. Clearly this open design avoids the nuisance of capping manipulations but at the expense of low pressure conditions and, hence, decreased efficiency of attack. One model from Prolabo, the Maxidigest 401, is intended for decomposition of larger samples than normal (1 to 8 g cf. 0.1 to 1 g) which would be beneficial for inhomogeneous samples or those where analyte content is extremely low (Liu et al., 1995).

The new U.S. EPA Microwave Method 3052 is based on the digestion of up to 0.5 g of sample (sediment, soil, sludge, etc.) in 9 mL of HNO_3 and 3 mL of HF for 15 min at $180 \pm 5°C$; different acid combinations, including HCl and H_2O_2, are recommended for certain matrices. The method is said to be for total values of Al, Sb, As, B, Ba, Be, Cd, Ca, Cr, Co, Cu, Fe, Mg, Mn, Hg, Mo, Ni, K, Se, Ag, Na, Sr, Tl, V, and Zn, and allows for either boric acid addition to complex F^- or evaporation and dissolution in 2% HNO_3. Hardware performance compliance with U.S. EPA Method 3052 implies such criteria as precise temperature sensing (to within $\pm 2.5°C$) and heating rates; a minimum of 435 psi capability pressure; a rotating turntable for even interaction; pressure relief; and a ventilated and corrosion-resistant cavity. Milestone's latest model 1200 MEGA system more than matches these specifications, with vessel ratings of up to 1595 psi. This design is based on a versatile rotor module accommodating different configurations of 3 to 12 vessels. The attractive feature of this system is the ability to evaporate off acids such as HF using the accessory FAM-40 module without transferring sample solution. By simply changing rotors, one can change from the act of drying to digestion to evaporation.

An obvious evolution of the somewhat clumsy batch-type approach is automated flow-through microwave digestion, and several systems have recently been offered commercially. The CEM Corporation has developed the SpectroPrep continous-flow system. Sturgeon et al. (1995) have evaluated this equipment for the decomposition of the marine sediment BCSS-1, among other samples. A sample of 0.25 g was weighed into a conical centrifuge tube, and 25 mL of a mixture of 20% HF, 50% HNO_3, 10% HCl, and 20% H_2O_2 were added to form a 1% m/v slurry. With automated agitation, a 10-mL slurry aliquot was pumped at 5 mL/min through the microwave operating at a power of 360 W (65%). The solution was filtered on-line prior to delivery to an autosampler. Internal standards were used to compensate for an average trace element recovery of 90%, and results for As, Cd, Co, Cu, Mn, Ni, Pb, and Zn agreed well with certified values. However, recovery of Cr was low (88.7 ppm cf. 123 ppm) and efficiency of oxidation of biological matrices was only 64%. It was stressed that production of a uniform and fine slurry is mandatory, and whereas this is a relatively easy task for reference materials it would not be so for real sediment samples. Currently under evaluation and development in the author's laboratory is the AutoPrep-Q5000 from Questron, an automated, discrete flow, high-pressure microwave digestion system. Up to 150 samples can be loaded for preprogrammed digestion at up to 200 psi and 200°C; decomposition of highly organic samples is aided by integrated predigestion with HNO_3 or other reagent. To date, excellent results have been obtained for the CCRMP lake and stream sediments for Zn, Cu, Pb, Ni, Co, As, Mo, and V extracted in various mixtures of HNO_3 and HCl. Doubtless these systems will find wide application once the "bugs" (mostly mechanical) are overcome, and laboratory personnel will feel justified in accepting the price of the equipment.

Other less mature systems are currently under development and will certainly rival these two current commercial instruments for a share of the vast potential market. Though not yet developed past the prototype stage, the automated stop-flow capsule-based microwave digestion system of Legere and Salin (1995) offers the possibility of digesting at least five capsule-contained samples in an oven simultaneously. The ambitious work of De la Guardia et al. (1993) deserves mention and may well be the configuration of the future. Their design comprises direct injection, in a water carrier flow, of dispersions of solid samples in HNO_3; merging of these slurries with 30% H_2O_2; microwave digestion in a Teflon coil of 100 cm length; cooling and degassing of the solution in an appropriate interface; and finally analysis by FAAS. A sample injection frequency of 180/h was achieved, together with good results for Pb, Cu, and Mn, but stronger extraction conditions were required for Zn in sewage sludge. In a less sophisticated but still fully automated way, Torres et al. (1995) employed a robotic station to manipulate soil samples through digestion using the Prolabo Microdigest 301 and analysis by FAAS for Fe, Cu, and Zn. Clearly, the scientific community is now poised to conquer this major limitation to efficient and accurate analysis sample decomposition. While progress in hardware and software development is occurring at an astonishing rate, caution is warranted against applying designated methods, such as U.S. EPA 3052, without thorough optimization for the particular sam-

ples under study and the objectives of the program. An elegant dem-
onstration of an orthogonal array design for such customization was
recently described by Zhou et al. (1995) for trace metals in sediments.
Five variables were studied: the addition of HF in the digestion acids
(HCl and HNO_3); the ratio between HCl and HNO_3; the maximum
pressure setting; output power; and digestion time. Excellent results
were found for Zn, Se, Ni, Co, Cu, Pb, and Cd, but Cr in one of three
sediment standard reference samples (NIES CRM 2 from Japan) was
low, probably reflecting its mineralogy.

5.2.2 Fusion

Fusions are not often carried out on sediments, as most elements of
interest can be brought into solution by the means discussed above.
However, elements such as Zr, Nb, Hf, Ta, and B are probably not fully
recovered by an acid attack and require a more rigorous breakdown
with a flux, such as lithium metaborate ($LiBO_2$). The great efficiency of
fusion compared to acid attack is due to the effect of the high temperature (500 to 1100°C). Heterogeneous reactions taking place in the melt
are of two types: acid-base and oxidation-reduction. Alkaline flux
reagents include Na and K carbonate (and bicarbonate), Na and K
hydroxide, and sodium tetraborate; acid fluxes include Na and K
hydrosulfate, Na and K pyrosulfate, boron trioxide, and hydrofluoride.
Oxidative reagents comprise largely Na_2O_2, KNO_3, and $KClO_3$, while
carbonaceous substances such as flour and starch are added to flux
mixtures for a reducing action. The drawbacks of a fusion are (1) the
potential addition of contaminants due to the high flux to sample ratio
(3:1 to 10:1), (2) the high salt concentration introduced and subsequent
need for a higher dilution factor (typically 500 to 1000, cf. 20 to 100 by
acid attack), and (3) difficulty in streamlining the operation for high
throughput.

   Although classical methods of silicate analysis relied heavily upon
the Na_2CO_3 flux, borate fusion, particularly $LiBO_2$ rather than $Li_2B_4O_7$,
now dominates in geoanalysis due to its widespread applicability and
effectiveness at low flux material to sample ratios (3:1 or greater). Flux
material and sample are simply mixed together in a graphite or Pt
crucible, fused at 1000°C for 30 min with gentle swirling, allowed to
cool, and the melt dissolved in HNO_3 or HCl. This fusion is employed
for major and minor element determination as well as for traces where
a dilution factor of ca. 500 suffices. Lithium metaborate can be replaced
by an equal molar mixture of Li_2CO_3 and B_2O_3 because in the fusion
process the following chemical reaction generates $LiBO_2$:

$$Li_2CO_3 + B_2O_3 \rightarrow 2LiBO_2 + CO_2$$

In order to ensure full decomposition and yet maintain a relatively low dilution factor (i.e., 100) beneficial for trace elements, the GSC routinely employs a method suggested by Thompson and Walsh (1989). The sample is first digested in $HF–HClO_4–HNO_3$, and the residue is then fused with $LiBO_2$. The small amount of residue necessitates only a small weight of flux material, and the two resulting solutions, reasonably low in dissolved salts, may be recombined for analysis.

Just as in acid decomposition, if not more so, automation of the fusion procedure is vital for adequate productivity. Govindaraju of the Centre de Recherches Petrographiques et Geochimiques (CRPG) in Nancy, France, has led the way in automating the $LiBO_2$ fusion (Govindaraju and Melvelle, 1987). The early designs of the 1970s evolved into the "LabRobStation", which is built around the Zymark robot and controlled by microcomputers. The system integrates peripherals for weighing, fusion, dissolution of melt, dilution, and finally ion exchange where needed (e.g., for REEs). About 8000 samples were processed between 1982 and 1986 in this way, providing, for example, ICP-AES results for the REEs and Y with an RSD of 2 to 6%. A conveyor belt approach to automation of the Na_2O_2 sinter process was adopted by the Bureau of Mineral Research in Orleans, France, again providing thousands of preparations annually without operator attendance. This procedure is particularly attractive because it can be carried out with a low flux material to sample ratio (e.g., 2:1 or greater) and at a temperature of only 500°C. Furthermore, the sole concomitant added to the analyte solution is Na, which can be accommodated in most analytical techniques. Schemes, based on automated flow injection technology, to preconcentrate trace elements in solutions derived from attack by $LiBO_2$ and Na_2O_2, will undoubtedly proliferate and make these decompositions more desirable where total values for many elements are required.

Full recovery of the precious metals from sediments demands reductive fusion with litharge (PbO), the classical lead fire assay procedure, or fusion with NiS. Details of these procedures may be found in the review article by Hall and Bonham-Carter (1988). Lead fire assay with collection in Ag is most suitable for Au, Pt, and Pd, while fusion with NiS is required for the rarer elements, such as Ru, Os, and Ir, which would otherwise be lost in the oxidative cupellation stage following Pb fusion. High sample weights of >10 g are normally taken for two reasons: first, to ensure sample representivity, as these elements are known to be inhomogeneously distributed in nature, and second, to be able to detect these elements, of low abundance in the low ppb or ppt range. Acid mixtures, such as *aqua regia*, have been used extensively to extract Au and, to a lesser degree, Pt and Pd, but results cannot be assumed to be the total values of the elements (Hall et al., 1989).

5.3   Atomic absorption spectrometry

Atomic absorption spectrometry (AAS) was rapidly adapted to geoanalytical applications in the early 1960s following the availability of commercial instrumentation, and it remained the leading technique in the analysis of solutions until the advent of ICP-AES in the late 1970s. The reason for its decline in favor of ICP-AES is due mainly to the fact that only one element at a time can be determined by AAS; hence, it cannot compete in speed with multielement techniques. Furthermore, the short linear dynamic range of AAS necessitates dilution for the more highly concentrated analytes, which leads to reduced productivity and greater error. The principal advantages of AAS include its specificity, simplicity, low capital outlay, ruggedness, and relative freedom from interferences. Today, probably the most important geological applications of AAS lie in the specialized areas where flame atomization is replaced by graphite furnace or vapor generation-quartz tube modifications. Flame AAS is advantageous to employ when only three to four elements are required in the analysis of sediments. Two useful books focusing on the application of AAS to the analysis of geological materials have been written by Angino and Billings (1967) and Van Loon (1980). Insight into the present role of the technique in geochemistry can be gleaned from Viets and O'Leary (1992).

5.3.1   Principles and instrumentation of atomic
absorption spectrometry

An AA spectrometer consists of a light source (usually a hollow cathode lamp or HCL), which emits a sharp line of the analyte; a chopper to eliminate emitted light from the cell; an atomization cell (flame, furnace, or quartz tube); a monochromator to select the line of interest; a photomultiplier detector (PMT); and a read-out system (Figure 5-2). Thus, conventionally, a solution containing the element of interest is nebulized into an air–acetylene (2200°C) or nitrous oxide–acetylene (2900°C) flame, which dissociates compounds into atoms, and absorption of light at the resonant wavelength(s) emitted by the specific HCL takes place. The concentration of analyte is proportional to the absorbance measured (Beer's Law), and calibration is made against known standard solutions. Double-beam instruments (Figure 5-2) compensate for drift in lamp intensity; one manufacturer offers two monochromators for simultaneous detection of two analytes. Options range from a basic single beam unit to a fully automated microprocessor-controlled instrument.

   Interferences fall broadly into the following categories: spectral; chemical; ionization; and viscosity effects. Spectral overlap, that is, the inability to resolve adjacent absorption lines of two elements, is rare;

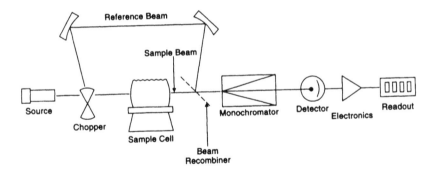

Figure 5-2   A double-beam AA spectrometer.

nonspecific broad-band absorption by molecules such as CaOH is subtracted by correction procedures employing deuterium arc, Zeeman,
or Smith-Heiftje measurement. Background correction should be
employed routinely for elements such as Pb, Ag, Ni, and Co. Chemical
interference occurs when the analyte is bound in a stable compound
in the flame rather than being present as a free atom. For example,
Ca–O–P and Al–O–Mg bonding is relatively stable in an air–acetylene
flame but is broken down in the hotter nitrous oxide–acetylene environment. Alternatively, a releasing agent, such as La, is added in solution, which preferentially binds with oxygen. Alkali and alkaline-earth
elements suffer from ionization effects (low ionization potential, IP)
where the population of ground state atoms is depleted by formation
of their ions and hence the measured absorbance of resonance light is
depressed. Addition of an excess of another alkali of low IP (e.g., Cs)
creates a large population of free electrons in the flame, thus suppress-
ing analyte ionization. Differences in viscosity or surface tension properties and hence nebulization efficiency can be overcome by matrix
matching or the method of standard additions. However, this interference is rare.

5.3.2   Applications of atomic absorption spectrometry

5.3.2.1   Flame atomic absorption spectrometry
Direct FAAS methods are the most rapid and least complex of AAS
procedures, but they render detection limits sufficiently low for only a
relatively small number of elements in sediments. Assuming a dilution
factor of 20 to 30 after sample decomposition by acid attack, the following elements can be measured to crustal abundance levels (* indicates nitrous oxide–acetylene flame): Al*, Ba*, Ca*, Co, Cr*, Cu, Fe, K,
Li, Mg, Mn, Mo*, Na, Ni, Pb, Si*, Sr*, V*, and Zn. The list of elements
where FAAS can be employed to adequately analyze sediments without

resorting to preconcentration is similar to that for ICP-AES, with the exceptions of B, S, and P included in the latter's repertoire. Clearly, this list is short on the number of trace elements adequately determined by FAAS. In addition to interferences caused by ionization and compound formation effects, some specific interelement nonspectral interferences have been reported in FAAS and must be avoided. For example, the extent of Fe interference in the determination of Cr depends upon the stoichiometry of the flame (Ajilec et al., 1988).

Solvent extraction has been the popular method by which to pre-concentrate a suite of trace elements to enhance detection limit capa-bility by FAAS-based methods. Various separation methods have been developed based on the extraction of metal iodides into solutions of Aliquat 336, a quaternary alkylammonium chloride, in methyl isobutyl ketone (MIBK). Viets (1978) developed a method, using a $KClO_4$–HCl digestion, which extracts Ag, Bi, Cd, Cu, Pb, and Zn in the presence of ascorbic acid and KI into a 10% Aliquat 336-MIBK organic phase to attain detection limits in the 0.1 to 1 ppm range. This method was later modified to include Mo, Sb, and As and to accommodate various acid digestions as well as potassium pyrosulfate fusion (O'Leary and Viets, 1986). Detection limits as low as 0.05 ppm were obtained for Ag and Cd in the determination of these nine elements using an HCl–H_2O_2 digestion to liberate nonsilicate-bound metals in sediments and soils (O'Leary and Viets, 1986). Hall and Bouvier (1988) reported results for a large suite of geological SRMs to limits of 0.2 ppm for Cd, Ag, and Pb by FAAS using this extraction after an HF–$HClO_4$–HNO_3 decompo-sition. An important advantage of this method, as compared to chelate extraction methods (e.g., with APDC), is that the metal halide com-plexes, with the exception of Zn, are extracted both as ion pairs with the MIBK solvent and by replacing the chloride of the Aliquat 336. The metals are therefore extracted over a broad range of acid concentration, and time-consuming dilution and pH adjustment are avoided.

5.3.2.2   *Quartz tube atomic absorption spectrometry*

Cold-vapor (CV) AAS remains the analytical technique of choice for the determination of Hg in sediments (Figure 5-3). Mercury ions, present in an acid leachate (H_2SO_4–HNO_3; HCl–HNO_3; or HNO_3–$HClO_4$) of the sample, are reduced to the metal (with sodium borohydride or stannous sulfate), which is then swept in vapor form with air and liberated H_2 into an absorption cell located in the optical axis of an AA spectrometer. Measurement of absorbance at 253.7 nm results in detection limits in the range of 5 to 15 ppb, using the GSC method based on the decomposition of a 1-g sample. Interference via the formation of stable Hg compounds with precious metals, Se, and Te is minimal due to the low natural abundance of these elements and to their incomplete digestion by common acid attacks. An alternative

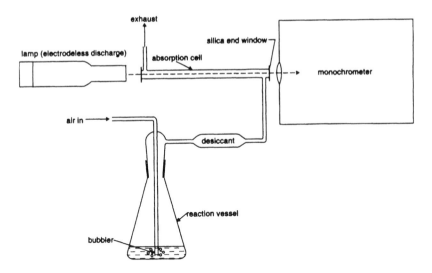

Figure 5-3   Apparatus for the CVAAS measurement of Hg.

procedure eliminates the digestion stage by combustion of the sample at 1000°C, capture of the evolved Hg vapor on Au foil, and subsequent release of Hg into an absorption cell by rapid heating. This thermal release method encounters problems with organic matter-rich samples, such as lake sediments. Volatile organics may condense on the collector and affect its efficiency by occluding Hg; furthermore, broadband absorption is significant when they revolatilize. Background correction and maintaining the collector at 150 to 170°C reportedly obviate these limitations. Some laboratories use a thermal release method for many routine samples, and the wet chemical method for organic matter- and sulfide-rich samples.

Most digestions for Hg are based on the use of acids HNO_3 and HCl (Puk and Weber, 1994). Work by Landi and Fagioli (1994) explored a two-step application of the oxidizing $K_2Cr_2O_7$–H_2SO_4 mixture at its boiling point and found excellent results for sediment SRMs. The U.S. EPA Hg Method 7471, based on detection by CVAAS, calls for *aqua regia* decomposition at 95°C followed by oxidation with $KMnO_4$ at 95°C to ensure organically bound Hg is converted to Hg^{2+} for reduction. A recent report by Scifres et al. (1995) of the U.S. EPA suggests that they are turning away from CVAAS to CVAFS for ultratrace detection in the 1 to 300 ppb range. The method involves microwave decomposition in *aqua regia*, oxidation of Hg with BrCl, prior reduction with $NH_2OH \cdot HCl$, and reduction with $SnCl_2$. Their results, for 30 marl sediments, are slightly higher and certainly more precise than those by U.S. EPA Method 7471.

As for Hg, instrument manufacturers offer front-end generation systems to evolve the hydrides of As, Sb, Se, Te, Bi, Sn, and Ge. The first three are commonly determined in geoanalytical laboratories by hydride generation (HG) AAS. Similar to Hg, the most sensitive absorption lines of these elements lie in the 200 nm region, where air and normal flame gases themselves absorb strongly. General practice involves formation of the hydride by the addition of a strong reducing agent, such as sodium borohydride ($NaBH_4$), to an acid leachate ($HCl-HNO_3$; $HF-HClO_4-HNO_3$) and transfer of the vapor in a flow of argon into a heated (electrically or by air–acetylene flame) quartz tube situated in the light path of an AA spectrometer. Detection of the analytes to levels of approximately 0.1 ppm is practical, but care must be taken to avoid interferences, which can be numerous, and whose magnitude is specific to the operating conditions and generation system employed. Interferences exist in both the liquid phase during hydride generation and in the vapor phase in the quartz cell during atomization. Transition elements, such as Cu, Ni, Co, and precious metals, suppress the signal by their reduction and subsequent adsorption and destruction of the analyte hydrides or by precipitation of metal selenides, tellurides, etc. This type of interference can be minimized by such procedures as separation using coprecipitation of analytes with $La(OH)_3$ (Tao and Hansen, 1994, using FI); and complexation with masking agents, such as EDTA, tartrate, cysteine, DTPA (e.g., Brindle et al., 1989; Wickstrom et al., 1995a,b). Mutual interferences in the quartz tube have been studied in detail by Barth et al. (1992), and they probably sound the death knell for HGAAS in favor of HG-ICP-AES or HG-ICP-MS. These interferences are particularly problematic for low abundance elements, such as Se and Te, in the presence of higher concentrations of As and Sb. A wealth of literature exists describing procedures for the minimization of interferences in HGAAS, as is attested to by the 287 references cited in a review of the subject by Yan and Ni (1994).

The analyst must ensure that the element is in its reactive valency state after sample decomposition for complete reduction with $NaBH_4$ to the gaseous hydride. For example, Se (VI) does not form the hydride, Se (IV) does; As (III) and (V) both form the hydride but at different rates. Prereduction is often carried out to reduce As and Sb (V) to (III) by addition of KI and, less frequently, with NaI. Selenium requires KBr or heating in 2 to 6 M HCl, as KI would reduce Se to the unreactive Se (0) state. Fortunately, Se and Te are in their (IV) valency state following the widely used digestion with *aqua regia*, and As and Sb-valency can be reduced from (V) to (III) on-line with KI. About 3000 sediment samples have been analyzed annually by this method at the GSC since the mid 1970s. Separation after *aqua regia* digestion by coprecipitation with $La(OH)_3$, followed by dissolution of the precipitate in 4 M HCl, has become routine and produces a detection limit of 0.1 ppm for As, Sb, Bi, Se, and Te (when As concentration is <100 ppm). Similar detection capability was reported for As and Se by Chan and Sadana (1992) in sediments using 1,10-phenanthroline to mask interferences from Cu and Ni. Elements such as Pb and Ge require different conditions of hydride generation, such as higher pH, and a less severe reducing environment. However, the expanding literature suggests that this method is gaining acceptance for these elements also. In fact, a recent review of HGAAS for Pb proposes that this technique is superior to GFAAS for low levels of Pb (Madrid and Camara, 1994).

Ultratrace detection of As, Sb, Se, Sn, and Pb by a combination of techniques — HGAAS and GFAAS — has been demonstrated by Sturgeon et al. (1987). Their method comprises generation of the hydride into a preheated graphite furnace at 600 to 800°C, where the gases are trapped and undergo thermal decomposition, leading to deposition of the analytes on the graphite surface. The furnace is then rapidly heated to atomize the elements. A detection limit of 30 ppb Se in sediment was achieved upon taking 500 μL of sample digestate having a dilution factor of 100 (i.e., 0.5 g sample to 50 mL). A measurement RSD of 10% was reported for BCSS-1 at 0.46 ppm Se (Willie et al., 1986).

5.3.2.3   *Graphite furnace atomic absorption spectrometry*

In the 1950s, L'vov, in Russia, pioneered the graphite furnace and in 1970, Perkin-Elmer introduced the first commercial GFAAS instrument. Graphite furnace atomic absorption spectrometry enjoys superior sensitivity over FAAS by 1 to 3 orders of magnitude but suffers from more complex interferences and lower productivity. The last 8 to 10 years have seen a rebirth of this technique due to the following: improved furnace design (e.g., STPF, stabilized temperature platform furnace, introduced in 1984) to achieve atomization into an isothermal atmosphere to minimize gas phase interferences; matrix modifiers (e.g., Pd, $Mg[NO_3]_2$), proven to be of widespread application in negating chemical interferences; accurate background correction facilities; and sophisticated electronics for signal integration and output. Two early books written by Fuller (1977) and Slavin (1984) describe different furnace configurations, interferences, and applications. Tantalum and tungsten filaments formed alternative atomization cells in the early days of electrothermal AAS, but these have given way almost exclusively to the graphite furnace, with its superior heating characteristics and developments in interference minimization.

The L'vov style of graphite furnace is constructed of a graphite tube 5 cm or less in length with an inside diameter of less than 1 cm. It is open at both ends and has a hole in the center for sample introduction. The tube attaches to a pair of water-cooled electrical contact rings located at the ends. The graphite components are sheathed inter-

Table 5-3   Comparison of GFAAS Results with Accepted Values
for BCSS-1; Values in ppm

| Element | Standard additions | | L'vov platform | |
	Direct	Chelation	Direct	Accepted
Cd	0.24 ± 0.04	0.26 ± 0.10	0.23 ± 0.03	0.25 ± 0.04
Pb	22.0 ± 3.0	22.0 ± 10.1	20.0 ± 5.4	22.7 ± 3.4
Cu	15.0 ± 3.4	16.2 ± 2.2	14.6 ± 2.5	18.5 ± 2.7
Ni	54.7 ± 5.1	56.6 ± 5.0	53.0 ± 2.2	55.3 ± 3.6
Co	12.7 ± 3.6	11.9 ± 6.9	10.7 ± 1.3	11.4 ± 2.1
Be	1.1 ± 0.2		1.1 ± 0.1	1.3 ± 0.3
Cr	122 ± 7			123 ± 14

Modified from Sturgeon, R.E., Desaulniers, J.A.H., Berman, S.S., and Russell, D.S., *Anal. Chim. Acta*, 134, 283, 1982.

nally and externally by a stream of Ar or N_2 gas to prevent oxidation of the sample and incineration of the tube. The gas flow also purges vapors generated by the sample during the heating cycle. A sample aliquot (5 to 50 µL) is automatically injected into a graphite furnace (± matrix modifier), whereupon a progressive heating program is applied to dry (100 to 150°C), ash (400 to 1000°C), and atomize (1000 to 3000°C) the sample. Typically, each analysis requires 2 to 3 min. Unlike in using FAAS, it is desirable to avoid HCl in the final sample matrix to negate Cl vapor phase interferences. Therefore HNO_3 and, to a lesser extent, H_2SO_4 are preferred. Besides the addition of modifiers, such as ascorbic acid, nickel, and palladium, gases such as methane, hydrogen, and nitrogen have been injected during the drying, ashing, and volatilization steps to negate interferences. The advantage of the L'vov platform in minimizing interferences was demonstrated by Sturgeon et al. (1982) for trace metal analysis of estuarine sediments. Table 5-3 shows their results for BCSS-1, compared to the much lengthier methods of calibration by standard additions (rather than with aqueous solutions) and chelation-extraction with APDC-MIBK.

The poor efficiency of GFAAS has been recognized as a major limitation among manufacturers, and consequently, new instrumentation has recently become available that is capable of determining two to six elements simultaneously, or can be programmed to run eight elements sequentially. Some models reduce drying time by injecting the sample solution onto a preheated graphite surface. As with FAAS, the degree of automation in GFAAS is high and the design versatile, allowing for pre- and postinjection of various matrix modifiers and facility for recalibration should excessive drift occur. Halls (1995) has reviewed recent developments in fast furnace technology and suggests that, with a study of the function and time of each stage in the program, many determinations can be reduced from the typical 2 to 3 min to 30 sec. Features associated with present-day models are discussed by Wach (1995). Typical instrumental detection limits are 0.03 to 0.1 ppb for Pb and 0.1 to 0.6 ppb for Cr. L'vov (1990), known as the father of GFAAS, is optimistic that we are on the road to *absolute* analysis (i.e., elimination of matrix effects) with proper use of the STPF, improved quality of pyrolytic coating of the graphite, and standardization of spectrometer design.

No other modifier has been demonstrated to perform with comparable success for such a large number of elements as the Pd-Mg system. It has been extensively studied by Welz and co-workers in a series of papers (e.g., Welz et al., 1992). In a thorough investigation of 21 elements, they found, for example, that NaCl at a concentration of 10 g/L had no influence on the majority of elements. Sulfate matrices, however, caused problems in the determination of some elements (e.g., Se, Te, Sb) because of gas-phase interferences or coexpulsion with the matrix. Furthermore, the Pd-Mg modifier did not perform equally well for all elements in raising the pyrolysis temperature to expel concomitants prior to analyte atomization without decreasing sensitivity. Thus, preliminary research into optimal conditions and potential interferences is required for the set of elements and the matrix under study. Most published methods concerning the application of GFAAS to the analysis of sediment digestates incorporate a separation step prior to analysis, owing to the complex nature of marine and terrestrial sediments. However, a recent paper by Sen Gupta and Bouvier (1995) reports the direct determination of Ag, Cd, Pb, Bi, Cr, Mn, Co, Ni, Li, Be, Cu, and Sb in solutions derived from a microwave HF–*aqua regia* decomposition of 23 environmental geological SRMs. They show good agreement with accepted values, down to levels as low as 0.1 ppm for Ag and Bi. Earlier work by Bettinelli (1983) also demonstrated good accuracy in the GFAAS determination of these trace elements in an HCl solution obtained after a $Li_2B_4O_7$ fusion of geological SRMs. The flux was thought to act as a buffer or modifier to minimize interferences.

In the author's opinion, analysis of sediments by GFAAS is best accomplished following some type of analyte separation from potential interferences in the matrix. Continuing earlier work on extraction with Aliquat 336, Clark (1986) used back-extraction from the organic phase and GFAAS to determine a large suite of elements, including Hg, Ga, Tl, Te, and Se. The greatest weakness of the Aliquat 336 method, that extraction efficiency is drastically diminished if HNO_3 is used in the digestion, was overcome by Rubeska (1987) in adding sulfamic acid or urea. This modification was used in an elegant integrated scheme by Macalalad et al. (1988), where 16 elements were determined from an *aqua regia* digest using FAAS, CVAAS, and GFAAS (Figure 5-4). Much of the analysis carried out for certification of geological SRMs under the NRCC program has been accomplished with GFAAS using

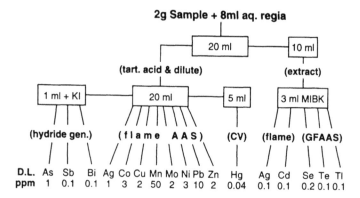

Figure 5-4   Determination of 16 elements by AAS. (Schematic adapted from Macalalad et al., *J. Geochem. Explor.*, 30, 167, 1988.)

separation techniques where deemed necessary (Sturgeon, 1989). Gold is widely determined in sediments by GFAAS following digestion with *aqua regia* (Hall et al., 1989) or HBr–Br$_2$ (Meier, 1980), and extraction of the halide anion into MIBK. High sample weights and preconcentration factors can enable detection limits of 0.2 ppb to be obtained. Detection limits in the range 5 to 10 ppb for a 1-g sample are achievable for Ag, Au, Pt, Pd, and Rh by GFAAS following separation by coprecipitation on Te (Hall et al., 1987). The advent of flow injection technology, where separation procedures can be carried out on-line in an automatic mode coupled with the excellent sensitivity of GFAAS ensure a continuing, important role for this technique in the environmental geological laboratory.

The discovery of Pd as a universal modifier, enabling the use of similar rather than specific atomization programs for a suite of elements, has fostered research into *simultaneous multielement* GFAAS using a high intensity continuum lamp with a high resolution polychromator capable of wavelength modulation background correction (Littlejohn et al., 1991). A somewhat similar goal is associated with another research direction: FANES or Furnace Atomic Nonthermal Excitation Spectrometry. Here, a sample is injected, dried, and charred at atmospheric pressure, as in conventional GFAAS. Then the atomizer is evacuated, a pressure of 5 to 40 Torr of argon is established, and a hollow cathode discharge is formed with the tube as cathode. The sample is vaporized into the discharge, and analyte emission is induced and measured. Thus, the technique derives the benefits of electrothermal atomization and nonthermal excitation. If a radio frequency plasma is formed in the graphite tube at atmospheric pressure, the technique is known as FAPES (Furnace Atomization Plasma ES). These techniques, now in their infancy, could well lead to a new generation of flexible GFAAS instruments.

5.3.2.4   Slurry atomic absorption spectrometry

The preparation of a slurry, or suspension, to facilitate the introduction of material for direct analysis by AAS has received considerable attention since the mid 1980s. Slurry introduction is clearly advantageous to and more representative than the direct weighing-in approach using graphite cups, where extremely small amounts of sample (20 to 500 μg) are analyzed by GFAAS directly, and the associated errors are consequently high. The advantages of this technique are obvious: (1) time-consuming digestion can be omitted; (2) the risks of contamination and analyte loss are much reduced; (3) the dilution effect of dissolution is minimized; and (4) selective amounts of small sample weights can be handled. Grinding of the sediment sample is performed first to prepare a powder, which is then weighed (1 to 50 mg/mL of diluent) into a measured volume of diluent, such as 0.1% HNO_3, and finally injected (typically 25 μL) into the furnace. For example, the Perkin-Elmer instrument manufacturer provides an ultrasonic slurry sampling accessory to ensure a uniform distribution of particles just prior to injection; this is a useful alternative to a hand-held mixing probe or vortex device. Optimal conditions for furnace cycling are similar to those in analysis of solutions and, indeed, calibration is often carried out with solution standards. Controversy seems to exist in the requirement for small particle size (commonly <10 μm) to ensure accuracy. This has probably limited widespread adoption of the methodology, the thought being that one sample preparation procedure (i.e., acid decomposition) is just being traded for another (i.e., grinding). Naturally, care must be taken to avoid contamination during grinding; hence, stainless steel parts in the grinding should be avoided. Zirconium oxide, silicon nitride, and boron carbide beads have all been used successfully. Particle size effect is probably more important in slurry analysis by ICP-AES than by GFAAS. Miller-Ihli (1992) has demonstrated accurate analyses using particles in excess of 100 μm, at least for the more volatile elements. A slurry dispersant, such as Triton X-100, is often added to prevent particle aggregation.

An inherent problem with direct solid analysis, particularly for geological samples, is the uncertainty of adequate sample representation when taking only mg-sized aliquots. Most of the variability in slurry analysis of sediments is undoubtedly due to sample inhomogeneity. Hence, proponents of the technique advocate taking about five readings of five different slurry preparations (i.e., weights), but this adds substantially to analysis time. Calibration with matrix-matched SRMs, rather than with aqueous standard solutions, is likely a prerequisite for sediment analysis, not a desirable feature of the technique

(Sturgeon, 1989). These potential drawbacks of slurry sampling may be turned to advantage if its role among the profusion of analytical methods available is viewed differently, as a means by which to study element distribution in sediments. As a bulk analysis tool, slurry GFAAS is clearly more suited to biological applications, as demonstrated by the papers cited in a review of the technique (Bendicho and de Loos-Vollebregt, 1991).

5.4   *Inductively coupled plasma atomic emission spectrometry for sediment analysis*

The pioneering work of Greenfield et al. (1964) and Wendt and Fassel (1965) in demonstrating the analytical potential of ICP spectrometry in the mid 1960s led to the introduction of the first commercial ICP emission spectrometer in 1975. Though FAAS, also a solution-based technique, had served the scientific community well in the 1960s and early 1970s, its pre-eminence was gradually eroded in view of the numerous advantages of ICP-AES. These comprise (1) the ability to measure 20 to 60 elements simultaneously in a cycle time of 2 to 3 minutes, (2) a long linear dynamic range of 4 to 6 orders of magnitude so that major, minor, and trace elements could be determined in the same solution (i.e., no dilution), (3) superior sensitivity for such elements as B, P, S, the refractories (e.g., Al, Mo, Nb, Ti, Zr), and the REEs, and (4) much reduced chemical interferences in the hotter environment of the argon plasma. The low-cost sequential ICP emission spectrometer, based on a computer-controlled scanning monochromator, probably does compete on a more even footing with FAAS. However, it is the *simultaneous* ICP spectrometer, of direct reading design, that is firmly entrenched as a major workhorse in today's geoanalytical laboratory. A comprehensive treatise on the application of ICP-AES to the analysis of geological and environmental materials has been written by Thompson and Walsh (1989); other books containing relevant information include those by Montaser and Golightly (1987), Boumans (1987), and Van Loon and Barefoot (1989). An excellent review of ICP-AES in the analysis of geological materials has been carried out by Jarvis and Jarvis (1992).

5.4.1   *Principles and instrumentation of ICP-AES*

A schematic diagram of a conventional ICP simultaneous spectrometer is shown in Figure 5-5. The common means of sample introduction is the pneumatic nebulizer of concentric or cross-flow design, but the efficiency of transfer to the ICP is only about 2%, most of the larger droplets (>10 μm) being removed in the spray chamber. The plasma is a gas in which atoms are present in an ionized state. When a high

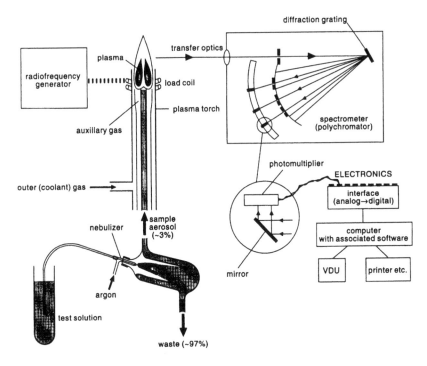

Figure 5-5   Conventional simultaneous ICP atomic emission spectrometer.

frequency current flows in an induction coil, it generates a rapidly varying magnetic field within the coil. The interaction or *inductive coupling* of the oscillating magnetic field with flowing ionized gas generates the ICP flame. The inductive heating of the flowing gas maintains plasma temperatures of 6000 to 10,000°K, much hotter than the nitrous oxide–acetylene or air–acetylene flames used in FAAS. Argon is chosen as the plasma gas because (1) it is inert and likely to suppress chemical interferences, (2) it is transparent in the UV-visible region where most of the ion and atom lines lie, (3) it has a high ionization potential (IP of 15.75 eV), permitting detection of all elements that can be excited to emit lines in the UV-visible spectra, and (4) it has a moderately low thermal conductivity. The function of the ICP is to vaporize, dissociate, atomize, and excite the sample, thereby promoting atomic and ionic line spectra as photons are emitted in energy transfer reactions. The torch comprises three accurately aligned concentric tubes of silica. The outer coolant argon flow-rate is usually 10 to 20 L/min, while the sample carrier flows at ca. 1 L/min. The auxiliary gas is employed to lift the plasma off the torch, to prevent salt build-up on the torch tip, and to nebulize organics. Two generator frequencies are offered: 27.12 or 40.68 MHz. The normal analytical zone, where emission is measured, is in a 4- to 5-mm window some 12 to 20 mm above the load coil. In some spectrometers, the ICP is replaced by the direct current plasma (DCP), where an argon plasma discharge is maintained between two tungsten anodes and a cathode. Direct current plasma-AES has found limited application, probably owing to its generally inferior sensitivity and susceptibility to chemical interferences.

Spectrometers are capable of either simultaneous (polychromator) or sequential (monochromator) measurement. The simultaneous spectrometer is a direct reader design, usually based on the Pashen-Runge mounting. Light from the ICP is focused on a fixed diffraction grating on the circumference of the Rowland's circle, where fixed exit slits are arranged to coincide with the diffracted lines of a suite of predetermined elements. Light passing through these slits is focused on individual photomultiplier tubes (20 to 60 in number). The sequential spectrometer offers the flexibility of line choice but at the expense of productivity. A scanning monochromator can be programmed to move rapidly from one line to another under computer control. It has been demonstrated that the analysis of 30 samples for 30 elements takes 2 h on a simultaneous system compared to 6 h on a sequential one. Manufacturers offer combined simultaneous-sequential spectrometers arranged at 90° to each other, viewing the same ICP; this provides maximum flexibility and speed. In order to measure spectra at wavelengths below 200 nm, the path of the spectrometer is either evacuated of air (absorbs in the low-UV) or purged with nitrogen. Measurements are made from this region up to the Cs line at 852.12 nm.

A new generation of ICP-AES instrumentation has evolved during the past few years, which is decreasing the sales of the conventional design described above. There have been significant improvements in detector design. Conventional photomultiplier tubes (PMT) provide excellent sensitivity, but their physical size limits the number of lines available on a simultaneous spectrometer. Photodiode array detection combined with an echelle grating, as on the instrument marketed by Leco, allows for the simultaneous detection of many lines (or pixels), but sensitivity of the array is inferior in the low UV, where many lines of interest are located. New solid-state detectors, known as charge coupled devices (CCD) and charge injection devices (CID) are purported to match or better the photometric performance (quantum efficiency, noise, dynamic range) of the PMT. The CCD design is incorporated in the Perkin-Elmer Optima 3000 ICP emission spectrometer. Each silicon-based detector consists of 224 discreet subarrays, which have from 20 to 80 photosensitive pixels per subarray. These are strategically arranged to take advantage of the most sensitive emission lines for all elements and, consequently, cover over 5000 lines. The impressive amount of spectral information obtainable by this revolutionary design has resulted in an unusually high number of initial sales for an instrument just graduated from the prototype stage; its performance will be interesting to monitor. Similarly, the rival firm of Thermo Jarrell Ash is offering the CID-based 'Iris' ICP emission spectrometer, in addition to the more conventional PMT-based Model 61E. The optics in these instruments are configured to view the plasma axially rather than side-on, an arrangement now optional from at least seven major manufacturers. Axial or end-on viewing is aimed at improving the efficiency of observation of the most useful zone in the plasma, the central channel, while avoiding the surrounding plasma, which produces an intense and unwanted background. Technological modifications are required to achieve this. Because of the formation of a highly unstable recombination zone at the tip of the plasma, an interface is necessary to permit the removal of the hot air–argon interaction zone and the efficient observation of the central channel. Improvements in detection limits of two- to tenfold have been realized; these depend on resolution, the aperture of the observation beam, and the length of the central channel viewed. As always, there is a down-side to this advance, which is that the ICP system appears to be less robust, that is, more susceptible to matrix effects (Mermet and Poussel, 1995). In reviewing figures of merit associated with today's instrumentation, Mermet and Poussel noted that there have been major advances recently in terms of resolution, long-term stability, and detection limits. High resolution has been achieved by using high grating line densities (>4200 mm^{-1}), orders up to four, or echelle gratings. Narrow entrance and exit slits (<20 μm) have been employed, along with better control of optical aberrations. Several ICP-AES systems exhibit long-term (i.e., 2- to 3-h) stability below 1% RSD. These various improvements have resulted in a detection limit for Pb, for example, of about 1 ppb in a clean solution compared to about 20 to 40 ppb a decade ago.

5.4.2   Optimization in ICP-AES

Critical parameters to be optimized in ICP-AES include carrier gas flow-rate, plasma power, and observation height in the flame. As many elements are usually determined simultaneously, a set of compromise conditions is chosen, normally weighted to facilitate greatest detection capability for the trace rather than major/minor elements. The group of elements adversely affected by such compromise is the alkali metals: they are very efficiently ionized in the argon plasma, but their most sensitive lines are atomic, which are prominent higher in the ICP flame. A procedure now employed routinely to optimize a set of operational interdependent variables is known as the Simplex method (Jarvis and Jarvis, 1992). The goal of optimization procedures is not only to obtain maximum sensitivity of measurement of the trace elements of interest but also to minimize interferences from matrix elements; hence, a compromise is sometimes necessary.

Chemical and ionization interferences are of less impact than in AAS, but spectral overlap is of greater concern in the absence of the lock-and-key effect created by the hollow cathode lamp. Hence, spectrometer resolution (generally <0.01 nm) becomes important, especially in the analysis of geological samples, which are complex matrices with Fe as a major element (emits >1000 lines between 200 and 300 nm). There are now several comprehensive line atlases available for the ICP spectroscopist, as for example the atlas published by Winge et al. (1985). Where spectral interference cannot be avoided by judicious selection of an atomic or ionic line, mathematical correction factors for a particular instrument are computed using known standard solutions. There is general consensus that numerical uncertainty in these measured values is the limiting factor in ICP-AES geoanalysis for trace elements. Background emission interference is subtracted by measurement of intensity on either or both sides of the analyte line; this is trivial with flexible sequential systems while manufacturers of simultaneous units differ in design. Commonly, the polychromator entrance slit can be moved ±0.5 nm on either side of its normal position, thereby shifting measurement at each detector off the analyte peak. The new CCDs and CIDs detectors allow for true simultaneous measurement of background, negating the need to move off-line when a change in ICP conditions may have occurred. This new generation of instruments should provide improvements in accuracy and precision, particularly where background correction is significant. Aluminium, for example, a major element in sediments, emits continuum radiation in the region 190 to 220 nm, which must be corrected for while measuring in this part of the spectrum.

Physical/matrix interferences are caused by changes in nebulization efficiency due, for example, to differences in salt content or acid concentration. Matrix-matching is only expedient when analyzing a group of similar samples; therefore, other methods of circumventing interferences have been established. Normalization to the intensity of an added internal standard (e.g., Y, Sc) compensates for nebulization efficiency changes as long as that element behaves in the identical fashion to the analyte(s). An alternate approach is to calibrate using solutions directly prepared from suitable SRMs. These are digested in an identical fashion to the unknowns, permitting automatic matching of the matrix. The disadvantage is the uncertain confidence that must be placed on the recommended values for SRMs, particularly for some ultratrace elements.

5.4.3 Alternatives to conventional nebulization

Sample introduction, via a pneumatic nebulizer (PN) and spray chamber, is considered the major weakness of ICP-AES, as only about 2 to 3% of the sample solution reaches the plasma. Meinhard concentric nebulizers are widely employed and provide highly reproducible and reliable performance for the analysis of dilute aqueous solutions. Cross-flow nebulizers use larger capillary sizes and are easier to manufacture from inert materials such as Teflon; hence, they enable the introduction of HF-containing solutions. The Babington V-groove nebulizer is capable of handling samples containing high TDS contents and particulate matter and are uniquely suited to the aspiration of slurried samples. Although much research was carried out in the 1980s into ultrasonic nebulization (USN) with desolvation, the early systems were plagued with such inadequacies as short- and long-term instability, lack of reliability, larger interelement effects, cross-contamination and memory effects, excessive clean-out time, and drift in aerosol generation efficiency (Fassel and Bear, 1986). However, these negative features were eventually overcome, and improvements in detection limits of about tenfold have been achieved with this superior transfer of analyte to the plasma. Using the CETAC U-5000, Brenner et al. (1992) demonstrated its excellent performance in the ICP-AES analysis of dilute sediment leachates, waters, and high salt solutions when combined with a sheath gas attachment and wide injector tube. They suggested that USN-ICP-AES could negate the need to turn to hydride generation or other means by which to improve the measurement sensitivity of PN-ICP-AES. Several manufacturers now offer an ultrasonic nebulizer, though sales of the CETAC model U-6000 certainly dominate this field. Other nebulizers have been developed for purposes other than improvement of detection power. The direct injection nebulizer (DIN), also marketed by CETAC, is well suited to chromatographic applications (in line with ICP-AES or -MS) for speciation studies and bypasses the spray chamber, thus reducing memory effects for elements, such as Hg and B. Pneumatic high efficiency nebulizers (HEN), capable of delivering only 50 µL/min, are useful when sample size is limited and are now equivalent in detection limit to PN-ICP-AES operating at 1 mL/min (Olesik et al., 1994). The spray chamber is also progressing in design: a new model has been demonstrated to clear out memory from a 500 ppm solution of Fe and Zn to <10 ppb in 20 sec (Legere and Salin, 1994).

Electrothermal vaporization (ETV), using modified commercial graphite furnaces to deliver µL amounts of solution efficiently to the plasma, has been demonstrated to enhance sensitivity significantly. The technique, however, has not transferred out of the research laboratory to high production or commercial facilities, perhaps because of the longer analysis time and the fact that equipment manufacturers have not provided this tool routinely until relatively recently with ICP-MS instrumentation. In a review of the technique, Matusiewicz (1986) cited 100 papers, none of which reported an application to sediment analysis. Similarly, the coupling of the Direct Sample Insertion (DSI) device to the spectrometer is used in only a handful of laboratories, despite its advantages over nebulization. The DSI takes the form of a computer-controlled graphite cup mounted on a rod, which is driven to various positions along the injector tube of the torch and into the ICP. Instead of aqueous solutions, solids, for example, in the form of sample mixed with graphite (Karanassios et al., 1991) or analyte-loaded polydithio-carbamate resin (van Berkel et al., 1990), have been analyzed by ETV- and DSI-ICP-AES. Some reasons for the lack of success of these techniques include inadequate speed of measurement electronics for true capture of the signal, necessity to measure background simultaneously and with the same speed, and interferences.

Gaseous rather than liquid introduction is used in HG-ICP-AES for determination of the elements As, Sb, Se, Te, Bi, Ge, and Sn. The advantages of HG over nebulization are essentially three-fold: transport efficiency to the plasma is close to 100%, the analytes are separated from their matrix in solution, and background emission is low. Introduction as the hydride leads to improvements in detection capability over nebulization of up to 500-fold for some elements, reaching lower limits of 0.1 to 1 ppb in solution. Many manufacturers now supply their own design of hydride generator, such as the GSC model shown schematically in Figure 5-6. Sodium borohydride at about 1% (stabilized in NaOH) is used almost exclusively as the reductant, forming hydrogen when coming into contact with the acidified sample solution or leachate. Membrane separation is being examined currently, but the gas-liquid phase separator shown still predominates. The complete analysis is under computer control, negating the need for constant operator attention. Indeed, whatever the mode of sample introduction, analysis by ICP-AES is fully automated today, with control by a 486 microcomputer and a built-in library of wavelength information and protocols for calibration, interference corrections, and quality control.

5.4.4   Applications of ICP-AES

Detection limits by PN-ICP-AES in pure solution, calculated as three times the standard deviation of the background signal (3σ) at the most sensitive emission line, are approximated as follows:

<1 ppb   Be, Mg, Ca, Sc, Mn, Sr, Y, Ba, Eu, Yb, Lu
<10 ppb   Li, B, Na, Si, Ti, V, Cr, Fe, Co, Ni, Cu, Zn, Zr, Mo, Rb,
           Ag, Cd, La, Hf, Re, Au, Hg, Nd, Sm, Gd, Tb, Dy, Ho,
           Er, Tm, Th
<100 ppb   Al, P, S, K, Ga, Ge, As, Se, Nb, Ru, Pd, In, Sn, Sb, Te, I,
           Ta, W, Os, Ir, Pt, Tl, Pb, Bi, Ce, Pr, U
<1 ppm   Rb

Practical detection limits are greater than these optimal instrumental limits, by factors generally from 2 to 5, depending on sample type (and interferences therein) and conditions of the analysis. Spectral interference from major elements in particular can preclude the use of a trace element's most sensitive line; hence, detection power can be further degraded. Realistic detection limits in ppm taken from Thompson and Walsh (1989) for sediments undergoing an $HF-HNO_3-HClO_4$ digestion (with a dilution factor of 120), are as follows: 0.1 to 0.5 for Ag, Be, Cu, Li, Sr, and V; 1 for Ba, Cd, Co, Cr, La, Mn, Ni, and Zn; and 5 for Mo, P, and Pb. Table 5-4 lists detection limits commonly reported by commercial laboratories employing ICP-AES following an *aqua regia* or $HF-HNO_3-HClO_4$ decomposition. It should be remembered that these values are ideal and that the presence of high concentrations of concomitant elements can increase these detection limits substantially. The impact of the new generation of ICP-AES instrumentation has not yet

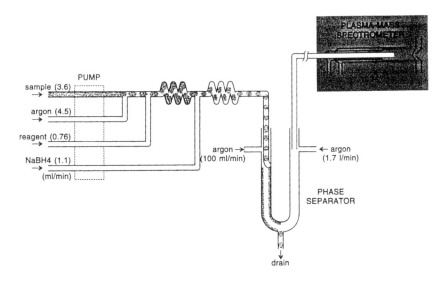

Figure 5-6 Hydride generation system used in the author's laboratory at the GSC.

Table 5-4 Detection Limits Quoted by Commercial Laboratories for Trace Elements Determined by ICP-AES Following an *Aqua Regia* or HF–HClO$_4$–HNO$_3$–HCl Decomposition

Element	Detection limit	Element	Detection limit	Element	Detection limit
Ag	0.2	Al*	0.01%	As	2–5
Ba*	2–10	Be*	0.2–0.5	Bi	2–5
Ca*	0.01%	Cd	0.2–0.5	Co	1
Cr*	1	Cu	1	Fe	0.01%
Ga	10	K*	0.01%	La*	2–10
Li	2	Mg	0.01%	Mn	1–5
Mo	1	Na	0.01%	Nb*	5
Ni	1	P	10	Pb	2
Sb*	2–5	Sc*	1–10	Sr*	1
Ta*	5	Th*	2	Ti*	0.01%
Tl*	10	U	5–10	V	2
W*	1–20	Zn	1	Zr*	1–5

Note: Values in ppm unless otherwise indicated. An asterisk denotes possible incomplete recovery.

been felt and will certainly lead to a significant lowering of these detection limits.

Although ICP-AES has been a workhorse in geoanalytical laboratories, certainly since the mid 1980s, the literature is not replete with descriptions of methods. A landmark paper published by Church (1981) describes the analysis of a diverse suite of 54 geological SRMs for 40 elements. The decomposition was based on HF–HClO$_4$–HNO$_3$–HCl with modifications to ensure solubilization of chromite-, phosphate- and carbonate-bearing samples. Fourteen standard solutions were designed to generate a matrix-matching calibration scheme to encompass the wide variety of materials analyzed, and an azeotropic acid mixture was employed for final dissolution to ensure constant acid strength. Overall, Church found that agreement with accepted values for trace elements lay in the range 5 to 10%. Of the 28 traces studied, 17 could be determined to crustal abundance levels. Precision, normally obtained in analysis by ICP-AES, is in the 1 to 2% range when measuring well above (>10 times) the detection limit. Using a similar digestion applied to SRMs MESS-1 and BCSS-1, McLaren et al. (1981) reported excellent accuracy and precision for Be, Co, Cu, Mn, Ni, Pb, V, and Zn, using simple aqueous standards for calibration. However, detection limits of 5 ppm for As, 0.2 ppm for Cd, and 1.5 ppm for Mo by their procedure proved inadequate for these nearshore marine sediments. Detection limits in the range 1 to 10 ppm by DCP-AES were achieved by Cantillo et al. (1984) for V, Cr, Mn, Fe, Co, Ni, Cu, Zn, Ga, Y, Zr, La, Ce, Dy, and Yb in estuarine sediments after a LiBO$_2$ fusion (dilution factor of 500). However, unpredictable interferences for some elements led to low results, a finding of numerous laboratories employing DCP-AES in the 1980s. Ten trace elements in sediment SRMs were accurately determined by Kanda and Taira (1988) by sequential ICP-AES without correction for spectral line overlap using the high resolving power of the echelle grating. Boron is an element not included in numerous multielement schemes because it is not compatible with the preparation procedure (e.g., use of $LiBO_2$ or loss by HF). Hall and Pelchat (1986) developed an ICP-AES method for geological materials to determine B to 1 ppm levels, together with Mo, V, and W to 4 ppm levels, based on an aqueous leach of a Na_2CO_3–KNO_3 melt.

Applications of HG-ICP-AES to the analysis of sediments for As, Sb, Bi, Se, Te, Sn, and Ge are well covered in a chapter of the book by Thompson and Walsh (1989). The average abundance of these elements in the lithosphere is as follows: As, 2 to 5 ppm; Bi, 0.1 ppm; Ge, 1 to 2 ppm; Sb, 0.1 to 0.3 ppm; Se, 0.01 to 0.05 ppm; Sn, 10 to 50 ppm; and Te, 0.001 ppm. This range in concentration would be reflected in sediments. Assuming a sample decomposition scheme producing a dilution factor of 20 to 50, the sensitivity of PN-ICP-ES for each element is such that only As and Sn could be determined to their abundance levels. Earlier work on HGAAS laid much of the foundation for detection by HG-ICP-AES in that optimum conditions of formation by reduction with $NaBH_4$ had been established and interferences classified. Two advantages over HGAAS are obvious: interference due to compound formation in the quartz tube would be eliminated in the hot plasma (particularly beneficial for Se and Te in the presence of excess As), and elements could be measured simultaneously. The efficiency with which the hydride is generated depends upon factors such as the acid concentration (normally HCl), the strength of the $NaBH_4$, the sample matrix, and the valency state of the analyte. The benefit of the simultaneous detection capability of ICP-AES is limited by incompatibility among the elements in their optimum conditions for hydride production. For example, As (V) is normally reduced to As (III) prior to generation of arsine by the addition of KI, but this is not compatible with the determination of Se. Arsenic, Bi, Sb, Se, and Te hydrides are best evolved from strong acid solutions, such as 4 to 6 M HCl, but hydrides of Ge and Sn require much lower acidities, at about 0.1 M. Attention must be paid to the valency state in which the element occurs after digestion. If Te and Se are in the (VI) state following a strongly oxidizing digestion, they must be prereduced to the reactive (IV) state by heating in a medium of 4 to 6 M HCl for ca. 30 min. Arsenic and Sb can both be reduced to state (III) by the addition of KI at room temperature. This modifying step can be carried out in-line prior to the automatic addition of sodium borohydride, as long as enough time (sufficient length of mixing coil) is allowed to ensure complete reduction. There are numerous inconsistencies in the literature concerning the effectiveness of prereduction procedures and degree of interference, probably resulting from the dependency on kinetics of these mechanisms. For example, Thompson and Walsh (1989) state that KBr can be used to reduce Se (VI), As (V), and Sb (V) prior to their simultaneous detection, whereas de Oliveira et al. (1983) found KBr under their conditions to be unsuitable for As and Sb. The latter group studied four decomposition methods (mixed acids and KOH and NaOH fusion) for the HG-ICP-AES determination of As, Sb, and Se in SRMs BCSS-1, MESS-1 and NIST certified reference material 1645, but found results for As only to be satisfactory across all methods. Thus, development of methods based on HG-ICP-AES must take into account compatibility of analytes through the digestion, prereduction, and hydride formation procedures, and often the different chemistries and mineralogies of analyzed samples can preclude simultaneous analysis for more than several elements (e.g., As, Sb and Bi as one package; Se and Te as another; and Sn and Ge as a third). Stability of analytes in solution must also be evaluated. For example, Sb can easily hydrolyze in low acid strength solutions but is stable as the chloro-complex. High concentrations of Bi in the system should be avoided, as this element can plate out on tubing and the plasma torch, leading to memory effects and adsorption of other hydrides. In spite of all these concerns, HG-ICP-AES offers a powerful tool by which to analyze sediments for these elements. A Canadian commercial laboratory offers the detection of As, Sb, Bi, Ge, Se, and Te by HG-ICP-AES to a determination limit of 0.1 ppm following HF–*aqua regia* decomposition with the caveat that the method is unsuitable for samples containing >0.5% Cu or Ni. It is probably fair to say that most commercial laboratories offer only As, Sb, and Bi by HGAAS or HG-ICP-AES at this time.

The sensitivity of detection by ICP-AES may be enhanced by using some form of preconcentration after digestion, such as ion exchange or solvent extraction. In seeking an appropriate medium or active reagent for solvent extraction or ion exchange, the goal is usually to include as many trace elements as possible in the package. Using Aliquat 336 and MIBK as solvent, detection levels can be lowered to 0.01 to 0.1 ppm for Cu, Pb, Zn, Co, Ni, Mn, Mo, As, Bi, Sb, Ag, Cd, Te, and Tl (Motooka, 1988). Recently, this procedure has gained popularity in commercial laboratories. Figure 5-7 illustrates the scheme adopted by Chemex Laboratories of Vancouver to provide ultrasensitive detection for trace elements. The oxidizing attack of HCl–$KClO_3$ is aggressive and not subject to volatilization loss of potentially volatile elements, such as As, Sb, or Hg. The leachate from a second attack using HF–$HClO_4$–HNO_3 is analyzed both directly by ICP-AES and following another extraction into MIBK with the complexing agent Aliquat 336. This procedure extends the usefulness of ICP-AES dramatically for the

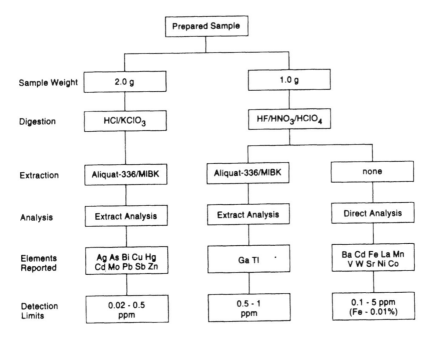

Figure 5-7 Scheme used by Chemex Laboratories of Vancouver for ultratrace analysis by ICP-AES.

analysis of sediments and, though it appears somewhat cumbersome, the extraction itself is fully automated using the concept now known as "flow injection" (FI).

Since its coining in 1975, to mean "the technique of injecting a small sample plug into an unsegmented carrier stream", there have been in excess of 4000 papers published on FI and its applications (Ruzicka, 1992). Two processes take place simultaneously: physical dispersion within the carrier stream and a chemical reaction between the analyte molecules and reagent molecules so supplied. In its simplest form, the reactor is a tube that has been tightly coiled to promote mixing of sample and reagent, but it can be designed to accommodate solvent extraction, gas diffusion, or ion exchange. As the process of FI can be repeated very reproducibly time after time, typical RSDs obtained can be better than 1%. Low consumption of sample (µL) and reagents and a high sampling rate (e.g., 2/min) are the hallmarks of FI, together with its computer compatibility and versatility. A microcolumn of the resin Chelex-100 has been used in FI systems to enhance ICP-AES measurement of elements such as Ba, Be, Cd, Co, Cu, Pb, Ni, and Zn (Christian and Ruzicka, 1987).

The recent commercial availability of true and flexible FI systems, together with more appropriate nebulizers such as the DIN to handle small volumes, is promoting rapid adoption of FI by high production laboratories. Such an approach would aid the determination of the REEs by ICP-AES as preconcentration via ion exchange is mandatory in order to reach adequate levels of detection (Thompson and Walsh, 1989). Most commercial laboratories offer INAA or ICP-MS detection of the REEs in sediments; ion exchange coupled with ICP-AES has been prohibitively expensive to date, but this may change with the implementation of FI strategies. Preconcentration by ion-exchange, with the added advantage of separating spectral interferences, is used extensively for the REEs to achieve adequate sensitivity (Jarvis and Jarvis, 1992). The sample is fused with $LiBO_2$ and dissolved in 0.5 M HCl. The REEs are then separated by cation exchange on BioRad AG 50W-X8 and preconcentrated by evaporation of the eluent to a volume of 5 mL. In this manner, the REEs can be determined down to chondrite levels with typical precisions of 3 to 5% RSD and accuracies of 5 to 10%. Designing a FI scheme to carry out both ion exchange and preconcentration of the REEs will be challenging, however, given the high volumes of acids used in the manual process.

The number of applications of ETV-ICP-AES and laser ablation ICP-AES to sediment analysis in nonresearch oriented laboratories is so small that these techniques do not warrant inclusion in this chapter.

5.5   Inductively coupled plasma mass spectrometry

The first commercial ICP-MS instrument was introduced by the Canadian manufacturer Sciex, in 1983, followed shortly by VG in the United Kingdom. Perhaps an indication of the acceptance of the technique is the fact that almost all major manufacturers of ICP emission spectrometers worldwide have now added the mass spectrometer to their product line, after considerable expenditures on research into optimum design. The instruments have become much more robust, and their operation is streamlined compared to the first generation. Nevertheless, only two major commercial geoanalytical laboratories in Canada use the technique routinely; this situation is poised to change with the proliferation of new models and applications. The attractive features of ICP-MS are simple spectra (much more so than in ICP-AES), wide linear dynamic range (10^4 to 10^5), flexibility, the ability to measure isotope ratios as well as elemental concentrations, and, perhaps above all, excellent detection limits in solution in the range 0.01 to 0.1 ng/mL (i.e., ppb) for most elements. The negative attribute of ICP-MS is its cost, about twice that of a simultaneous ICP emission spectrometer, but competition is forcing both these levels downward. Its high sensitivity makes it a very exciting technique for ultratrace analysis; it is not really suitable for major element determinations. International conferences devoted to ICP-MS theory and applications, such as those in the United

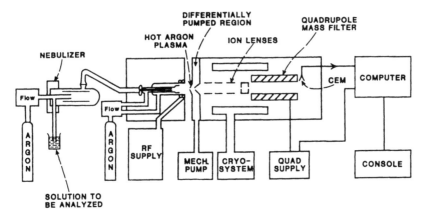

Figure 5-8 A conventional quadrupole ICP mass spectrometer.

Kingdom, known as the Surrey and Durham meetings, take place annually, and proceedings are published either in journals (e.g., *Chemical Geology* or *Journal of Analytical Atomic Spectrometry*) or as books (e.g., Holland and Eaton, 1993). The Plasma Winter Conference, held every 2 years in the United States and intended to cover all techniques related to excitation with a plasma, is becoming predominantly an ICP-MS meeting, as attested to by the papers given at the 1996 conference.

5.5.1 *Principles and instrumentation of ICP-MS*

The front end of an ICP mass spectrometer is identical to that of ICP-AES, but here the ions produced in the plasma are actively being sampled, coaxially to the horizontal torch, for measurement in the mass spectrometer (Figure 5-8). The ion beam passes through the sampler (made of Ni, Cu, or Pt) into a pressure region of 1 Torr and then through the skimmer to the ion lenses at about 10^{-5} Torr. Ions of selected m/z (mass/atomic number) range leave the mass analyzer and are deflected into a channel electron multiplier (CEM) for detection in a pulse-counting mode. Measurements are made by either scanning a chosen range of the mass spectrum for a fixed time or sitting on an analyte peak for a specified count time before advancing to the next peak.

Spectral interferences are far fewer than in ICP-AES and are relatively easy to predict. These mass overlaps are due to isotopes of another element, oxides (MO^+), doubly charged ions, hydroxides, and polyatomic ions formed from matrix elements and the plasma (e.g., ArO^+ on $^{56}Fe^+$). Fortunately, almost every element has an isotope that is free from mass overlap by another element. Oxide and hydroxide formation is minimized by judicious setting of operating conditions, particularly the plasma power and carrier gas flow rate. When the magnitude of this interference remains significant, mathematical corrections can be computed. Mixing of the argon carrier gas with nitrogen has been employed to reduce polyatomic interferences. Laborda et al. (1994) reported favorably on the reduction of Cl-based interferences on V, Cr, Zn, As, and Se isotopes by (a) adding 8% N_2 to the carrier gas and (b) using a high aerosol carrier gas flow-rate (0.95 L/min instead of 0.75 L/min). The $ArCl^+$ interference on As, by the presence of 0.05% Cl in solution, was effectively removed by both methods and, in addition, ClO^+ and ClO_2^+ were removed by N_2.

The unit mass resolution inherent in ICP-MS quadrupole systems is now thought inadequate for a wide range of applications. Two companies, VG and Finnigan Mat, have launched magnetic sector ICP mass spectrometers as an alternative to the quadrupole design; these are considerably more expensive but the much enhanced resolution obtained is producing far fewer spectral interferences and superior detection limits. A useful comparison of isobaric interferences evident in quadrupole- and sector-based instruments when nebulizing common acid mixtures has been documented by Reed et al. (1994). The performance characteristics of the Finnigan MAT double focusing magnetic sector mass analyzer, known as the ELEMENT, have been evaluated recently (Feldmann et al., 1994; Moens et al., 1995). This instrument can be operated in the low (R = 300) or high (R = 3000 and 7500) resolution mode. At low resolution, dramatically improved detection limits (compared to quadrupole-based instruments) in the order of fg/mL (ppq) were obtained. At high resolution, albeit at the sacrifice of sensitivity, most of the isobaric interferences are eliminated. Similar detection capability, in the ppq range, has been reported by Gunther et al. (1995) using the VG Model PQII+S, a quadrupole instrument pumped down to lower pressures, especially in the region between the sampler and skimmer. The resolution of this instrument, however, is typical of a quadrupole ICP-MS.

Nonspectral or matrix-induced interference is more problematic than spectral and, in simple terms, is manifested by a change in analyte signal due to other elements present in the sample solution. For example, salt deposition on the sampler and skimmer cause a reduction in sensitivity, which worsens with time. There is an ion sampling effect where the relative proportions of ions initially sampled do not remain constant through the skimmer and lens system. The lighter the analyte and the heavier the concomitant (interferent), the greater is the degree of interference. Hence, the effects of elements, such as Pb and U (matrix elements), on analytes, such as Li and B, can be severe. Elements below mass 80 tend to suffer the most interference, and care is needed when analyzing in this region. Xiao and Beauchemin (1994) investigated the effect of (a) plasma power and (b) mixing N_2 in the outer argon flow on the determination of 17 trace elements in the presence of 0.1 mol dm^{-3} Na (simulating seawater matrix). They found that boosting the power to 1.3 kW and using a mixed-gas plasma at 10% N$_2$ eliminated any effect of 0.01 mol dm^{-3} Na and led to a uniform effect across the mass range (m/z 27 to 208) at 0.1 mol dm^{-3} Na so that only one internal standard was required. There is still much activity into the elucidation of matrix effects, both in the plasma and in the spectrometer. Much of this research is carried out in the laboratories responsible for instrument design (Tanner, 1995), in the quest for minimal interference in the ultratrace analysis of high salt solutions by ICP-MS.

Naturally, calibration strategies are chosen with anticipation of the types of interferences that may arise. Selection of one or more internal standards that would behave in a similar fashion to the analytes in changing matrices is now a common strategy. Thus, an internal standard is close in mass and IP to the analyte and, of course, does not exist to any degree in the sample matrix. The best internal standard one can choose is an isotope of the analyte, but, other than in very precise work, calibration by isotope dilution is not yet widespread. Calibration by the method of standard addition is little used because it is time-consuming. The choice of acid in which the sample is presented to the nebulizer is not the popular HCl used in ICP-AES because Cl forms polyatomic ions which create spectral interference such as ClO (on V) and ArCl (on As); hence HNO$_3$, which forms species that already exist in the plasma, is now employed more frequently. However, it is the author's opinion that spectral interference by Cl species has been exaggerated and has had the effect of altering acid digestions without enough thought being given to the assurance that this non-trivial modification provides the same degree of dissolution.

In many cases the critical factor in computing the detection limit in a method based on ICP-MS is not the sensitivity of the analytical technique but the random contamination to the blank level created in the sample preparation procedure. The powerful sensitivity of ICP-MS benefits in two ways the determination of those elements of low natural abundance: detection limits below background can be achieved so that background itself can be estimated reliably and efficiently; therefore, cost can be improved so that multiple analytical techniques need not always be employed. For example, in the past, one would have chosen XRF to determine Zr and Nb and INAA to determine Hf and Ta. These now form part of a larger package (including Mo and W) based on LiBO$_2$ fusion and ICP-MS, where detection levels as low as 20 ppb in the sample can be reached (Hall and Plant, 1992).

5.5.2 *Alternatives to conventional nebulization*

Current research in ICP-MS centers on alternative sample introduction techniques to replace the inefficient nebulizer, to obviate interferences caused by concomitants in solution, and to increase the present upper limit tolerated of 0.1 to 0.2% in TDS. Different nebulizers, such as the DIN and HEN mentioned earlier in this chapter, are being used as alternatives to the conventional Meinhard pneumatic nebulizer in ICP-MS as they are in ICP-AES. Flow injection is being employed to mitigate interferences and increase tolerance to high salt solutions. The quadrupole mass spectrometer is not ideal for the measurement of transient signals such as those obtained in FI because it is essentially a single-channel detector. Thus, the multichannel mode must be employed, where the dwell time on each isotope is minimized and measurement cycles are repeated throughout the duration of the peak. The advantages of FI include shorter analysis time (ca. $\frac{1}{2}$ to $\frac{1}{4}$), the ability to use μL rather than mL sample solutions and, perhaps most importantly, reduction in the amount of solid material, which is introduced to the mass spectrometer. This is particularly beneficial in the analysis of high salt solutions. A report by Vickers et al. (1989) demonstrated that FI could dramatically reduce the severity of nonspectroscopic suppression with little sacrifice in sensitivity. Dean et al. (1988) described the capability of FI in analyzing solutions high in salt content, viscosity, and acid strength for elemental abundances and isotope ratios.

The use of FI in the automatic application of isotope dilution and on-line sample dilution has been studied by Barnes and colleagues at the University of Massachusetts (Lasztity et al., 1989; Viczian et al., 1990). Solutions of SRMs containing 7 to 90 ppb of Pb were measured by FI-isotope dilution with RSDs in the range 1 to 2% (6 replicates) and within 5% accuracy. Care must be taken in applying this technology to ensure that the spike isotope and analyte achieve chemical equilibrium. It is obvious that FI will greatly facilitate calibration by standard addition where the spike analyte can be merged with the sample stream, the amount of spike being controlled by the programmable time of injection.

Electrothermal vaporization (ETV) units are now offered commercially with ICP mass spectrometers, the advantages of the technique having first been pointed out by Gray and Date (1983) and demonstrated in the pioneering work of Park and colleagues (Park and Hall, 1986, 1987, 1988; Park et al., 1987). Eventually commercial manufacturers adopted Park's design, in that vapors generated during the drying, ashing, and cleaning steps are vented to waste while an auxiliary flow of argon bypasses the ETV chamber and enters the plasma. In this way, salt build-up on the sampler and skimmer is avoided. The advantages of ETV- over PN-ICP-MS are numerous: (1) transport efficiency to the plasma is typically 60 to 80% instead of 1 to 4% and hence detection limits are superior; (2) μL quantities rather than mL amounts are required; (3) oxide and hydroxide formation is drastically reduced as the sample vapor is introduced to a dry plasma; (4) potential interfer-

ences can be removed by thermal programming for selective volatilization of the analyte; (5) organic solutions can be easily handled; and
(6) the ashing step can be employed to remove salts, and thus solutions
of higher dissolved salts can be tolerated without build-up on the
sampler or skimmer. On the down side, analysis by ETV-ICP-MS (1)
requires several minutes; (2) is capable of determining only a limited
number of elements at once as transient rather than steady-state signals
are produced and rapid scans are necessary; and (3) can be subject to
variation in transport efficiency to the plasma. The last problem can be
overcome using calibration by ID, as both isotopes of the analyte will
be affected equally.

As is the case with sample introduction by ETV, direct sample
insertion devices initally designed for analysis by ICP-AES have been
applied to and modified for ICP-MS. Application reports in the literature are few to date, however, probably as commercial manufacturers
do not yet offer such an accessory. The advantage of DSI is that there
is no loss of analyte in transportation since the loop is inserted directly
into the plasma; however, unwanted matrix species are sampled by the
mass spectrometer while these can be vented to waste during thermal
programming of the ETV unit. Both techniques enjoy the advantages
over nebulization listed above, such as simpler and reduced background spectra. Blain et al. (1989) modified the DSI design to accommodate graphite and metallic probes, which could contain the dried
or powdered sample. Pelletizing the sample with graphite (e.g., 30 mg
sample + 200 mg graphite) led to a greater amount of sample inserted
into the plasma, and hence better detection limits, as well as improved
thermal transfer in the absence of container walls. The ability to handle
solid samples with ease is a major difference between DSI and ETV
introduction. The performance characteristics and improvements in
background spectra shown by a fully automated DSI device designed
and built at the University of Alberta, Canada, have been outlined in
a series of papers summarized by Karanassios and Horlick (1989).

Gray (1985) first reported the development and application of a
laser sampling system for the direct analysis of geological SRMs by
ICP-MS. In spite of the limitations inherent in the use of the low repetition rate and high powered ruby laser, Gray demonstrated the potential to determine, first, trace elements down to levels of about 10 to 100
ppb, and second, Pb isotope ratios. Since that time, numerous lasers
have been sold by ICP-MS manufacturers as an accessory to replace
the nebulization system when desired, and many of these are now
located in geoanalytical laboratories. Although it was realized early on
that coupling of the laser beam with geological materials was most
efficient in the IR spectrum (e.g., at 266 nm), the first commercial models
were based on the Nd-YAG laser operating at 1064 nm. The advantages
enjoyed by laser ablation over solution nebulization include (1) decom-

position and its associated potential problems of incompleteness, contamination, and volatilization losses are avoided; (2) the sample vapor is introduced to a dry plasma, thus minimizing the population of species such as MO^+, MOH^+, and $ArOH^+$; and (3) spatially resolved analysis can be accomplished to correlate composition with morphology. On the negative side, sample homogeneity is required for consistent results, and quantification still presents challenges. Sediments are much less homogeneous than, say, steels or glasses, which have been widely analyzed by direct solid techniques in the past. The amount of sample ablated per laser shot is unknown and varies with the matrix. This begs the question: to what extent are the signals measured at the detector and are they truly representative of the elemental composition of the sample? To overcome these problems, some kind of internal standardization is necessary. The element chosen should possess the same degree of homogeneity as the analyte(s) and cannot be of such concentration as to saturate the detector. If suitable standard samples are available, a common element of known concentration may be chosen. In spite of the hundreds of geological reference materials available, the wide range in composition of this type of sample makes matrix-matching difficult. Alternatively, if the sample is ground, mixed and pelletized, an internal standard can be added to the sample. Many laboratories are currently engaged in striving for values of precision and accuracy on a par with those obtained by nebulization by investigating various means of calibration and quantification. In the author's opinion, the strength of laser ablation-ICP-MS in geochemistry lies primarily in its ability to analyze material spatially (laterally and, to a certain extent, to depth) with resolution in the order of 10 to 50 μm, as well as in determining isotope ratios and in providing bulk abundance data rapidly with a semiquantitative quality. Extending our understanding of the processes taking place in ablation and transport will undoubtedly improve ability in bulk composition analysis, as well as promote many more new applications.

There are only a few reports in the literature of slurry nebulization ICP-MS. This method also offers the advantage of minimal sample preparation, but perhaps the limiting factor is that slurries of greater than 1 g of solid content per 100 mL cause deposition on the sampler and the consequent need to dilute impairs detection capability. Williams et al. (1987) published a feasibility study analyzing slurry concentrations of 0.05 g per 100 mL of tetrasodium pyrophosphate as the dispersing agent for the finely ground (<3 μm) sample particles. Mochizuki et al. (1989) found that the parameters of particle size and slurry concentration greatly influenced the sensitivity, precision, and accuracy obtained. Response increased markedly with hours of grinding, reaching a plateau after 8 to 10 h, the majority of particles being less than 3 μm at this stage. Slurry concentrations in 0.1% aqueous Triton X-100 solution of up to 0.05% could be tolerated without significant orifice blockage over a 100-min operational period. Using a known matrix element as the internal standard and calibrating with aqueous synthetic standards, precision in the range 1 to 6% RSD was obtained for the REEs (at 0.1 to 38 ppm) in geological SRMs. The time necessary for grinding and the possibility of contamination in that operation, together with suspected volatilization/atomization effects in the plasma for the more refractory involatile elements, are probably reasons why slurry nebulization lacks popularity.

The knowledge learned in the development of HGAAS and HG-ICP-AES has given rise to rapid growth in applications of HG-ICP-MS currently, particularly for the lower abundance elements such as Se and Te. Natural levels of As are high enough that the sensitivity of HG-ICP-MS is not required, which is fortunate since As is monoisotopic and suffers from spectral interference by ArCl at m/z 75. Unfortunately, the two major isotopes of Se, ^{80}Se (49.96% abundance) and ^{78}Se (23.61%) are overlapped by argon dimer species, Ar–Ar. Detection by ICP-MS is further degraded by its relatively high IP of 9.75 eV; consequently, its degree of ionization in the plasma is low, at about 33% (cf. Cu at 90%). However, these impediments are more than offset by the superior detection capability potential of HG-ICP-MS compared to its ICP-AES counterpart.

5.5.3   Applications of ICP-MS

It is not surprising that the impact of the first decade of ICP-MS in geoanalysis has been to provide methods of detection for the low abundance elements, hitherto not well served by more established techniques (see review by Hall, 1992). Much of the early development of ICP-MS methods for marine sediments was carried out at the National Research Council of Canada by McLaren and co-workers. External calibration (i.e., with synthetic solutions) was evaluated in the analysis of BCSS-1 following digestion in $HF–HClO_4–HNO_3$ in Teflon pressure vessels and subsequent dissolution in 0.1 M HNO_3 (McLaren et al., 1987a). Acceptable accuracy was found for As, Cd, Pb, and Zn, but the method of standard additions was necessary for Mo, V, Mn, Co, and Ni to obviate nonspectral interferences. A second paper demonstrated the power of isotope dilution ICP-MS in the analysis of BCSS-1 and MESS-1 for Sr, Mo, Sn, Tl, U, Pb, Ni, and Sb using the same decomposition (McLaren et al., 1987b). Precision in the range of 5 to 10% RSD was obtained for low abundance elements such as Sb and Tl at levels of 0.47 to 0.74 ppm. However, Cl (from residual $HClO_4$) molecular species created isobaric interference in the measurement of Cr and Zn. Later, a combination of external calibration and ID strategies was used to determine 16 elements in the marine sediment, PACS-

Table 5-5   Analysis of the Reference Marine Sediment,
PACS-1 by ICP-MS

| Element | External calibration | | ID | Certified |
	No IS	$^{40}Ar_2$ IS		
V	136 ± 3	139 ± 2	—	127 ± 5
Cr	105 ± 2	107 ± 2	—	113 ± 8
Co	20.6 ± 0.4	21.1 ± 0.4	—	17.5 ± 1.1
Ni	43.5 ± 1.9	44.5 ± 2.6	44.7 ± 1.6	44.1 ± 3.0
Cu	451 ± 15	446 ± 24	447 ± 15	452 ± 16
Zn	911 ± 23	932 ± 19	836 ± 16	824 ± 22
As	195 ± 12	204 ± 11	—	211 ± 11
Sr	250 ± 5	264 ± 5	275 ± 2	277 ± 11
Mo	11.8 ± 0.5	15.7 ± 0.7	12.7 ± 0.4	12.3 ± 0.9
Cd	2.42 ± 0.16	—	2.23 ± 0.14	2.38 ± 0.20
Sn	42 ± 2	—	41.0 ± 2.5	41.1 ± 3.1
Sb	195 ± 8	—	206 ± 9	171 ± 14
Hg	—	—	4.56 ± 0.09	4.57 ± 0.16
Tl	—	0.74 ± 0.05	0.77 ± 0.05	—
Pb	401 ± 18	—	402 ± 10	404 ± 20
U	—	2.9 ± 0.5	3.02 ± 0.04	—

Note: IS: Internal Standardization; ID: Isotope Dilution.

Modified from McLaren, J.W., Beauchemin, D., and Berman, S.S., *Spectrochim. Acta*, 43B, 413, 1988.

1; these results, together with the standard deviation of six replicate analyses, are shown in Table 5-5 (McLaren et al., 1988). Even though precision and, to a lesser extent, accuracy are improved by ID, it is not a strategy that has been widely adopted outside of certification laboratories. This may be due to several factors: isotopes are still not readily available at low cost; equilibration (e.g., through heating) with the analyte must be ensured; the second isotope of the element should be free of spectral interference; and the concentration of the anayte should be known roughly (within 1 to 2 orders of magnitude) so that the amount of isotope to add can be estimated. Boomer and Powell (1987) reported excellent results of another low abundance element, U, at 1.016 ± 0.004 ppm (certified at 1.1 ± 0.05 ppm) in NIST certified reference material 1645.

The early years of ICP-MS coincided with a high level of geochemical exploration for the precious metals; therefore, not surprisingly, this technique has found a niche in the analysis of sediments for these low abundance elements. While the initial applications focused on rocks, Hall and Bonham-Carter (1988) demonstrated the powerful combination of Pb fire assay and ICP-MS in reaching detection limits of 0.1 and 0.4 ppb, respectively, for Pt and Pd in 10-g samples of sediments and soils. This represented a 50-fold improvement over conventional methods (e.g., based on fire assay/ICP-AES) for Pt. Further homogeneity studies of candidate SRMs at the GSC clearly showed that detection limits for Au and Pd are dominated by the fluctuation of blank levels derived from the fusion itself, while the limit achievable for Pt was a reflection of the sensitivity of the ICP-MS measurement, not contamination (Hall and Pelchat, 1994). Applying this method to a regional geochemical orientation program in Jamaica, Simpson et al. (1990) discerned patterns reflecting underlying bedrock in results for Au, Pt, and Pd in stream sediments in the range 1 to 5 ppb.

Another area of impact of ICP-MS is in the determination of the REEs, the enormous advantage over ICP-AES being the elimination of the laborious separation and preconcentration step, thanks to the sensitivity of this technique (Hall and Plant, 1992). When the total concentration of the element is required, the $LiBO_2$ fusion is clearly the method of choice. Many trace elements can be determined from a dilution of the melt directly, or when detection limits in the low ppb range are required, some are still best measured after separation. Hall and Pelchat (1990) analyzed a large suite of sediment and rock SRMs in this manner using coprecipitation with cupferron to reach detection limits of 20 ppb for Zr, Hf, Nb, and Ta. Inductively coupled plasma mass spectrometry (ICP-MS) in the positive ion mode can also be employed to measure the halides, regardless of their low population as a positive ion. This potential was demonstrated by Date and Stuart (1988) in the analysis of urban particulate SRM, down to a detection limit of 0.7 ppm for I. Fortunately, the natural concentrations of these elements inversely follow their sensitivity (e.g., IP for I is 10.46 eV, cf. Cl at 13.02 eV). The research group at the University of Cincinnati has repeatedly demonstrated the application of the He microwave-induced plasma (MIP) to the determination of the halides, He having a higher IP than Ar; however, this alternative has yet to be adopted by ICP-MS manufacturers (Sheppard and Caruso, 1994).

The U.S. Environmental Protection Agency Method 6020 for contract laboratories is based on the ICP-MS detection of Al, Sb, As, Ba, Be, Ca, Cr, Co, Cu, Fe, Pb, Mg, Mn, Ni, K, Se, Ag, Na, Tl, V, and Zn (Method 6020, U.S. Environmental Protection Agency, 1993). A list of isobaric interferences to be corrected for is given, and nonspectral interferences are minimized using three internal standards (from a choice of Li, Sc, Y, Rh, In, Tb, Ho, Bi). The digestion of sediments (and other media) was altered from the *aqua regia* used with ICP-AES and is based on refluxing with HNO_3 and H_2O_2. While warnings are given concerning the lack of stability of Ag and Sb in this acid medium, the user should ensure that this digestion serves to dissolve the phases and mineralogy of the samples being analyzed, in line with the objectives of the program. The Community Bureau of Reference (BCR) of the European Community has been using ICP-MS for the certification of its SRMs since 1988. Moens et al. (1994) state that "while the method proved to be potentially accurate and precise, accuracy and precision cannot be taken for granted and that each analysis requires a thorough study in order to obtain results that meet BCR standards". The ICP-MS technique was shown to yield accurate results only if matrix effects and spectral interferences were rigorously studied and dealt with. Whereas in the period from 1982 to 1990, ICP-MS contributed to less than 1% of the certification work carried out on biological and environmental samples, its contribution has risen to 60% since then at the BCR. A concurrent paper by Beary et al. (1994) at NIST echoes the sentiments of BCR analysts. They state that attention to sampling, dissolution procedures, and instrumental analyte/matrix/plasma interactions are critically important to high accuracy. At NIST, the ID-ICP-MS method is now used extensively in certification, and the 0.1 to 0.2% ratio measurement precisions achievable under carefully controlled conditions result in analyses that rival thermal ionization MS for accuracy. In order to meet these stringent goals, a separation scheme is often necessary in order to negate spectral interferences. In fact, Beary and Paulsen (1993) reported that separation was mandatory in the ID-ICP-MS determination of trace Cd, Ag, Mo, and Ni in complex SRMs (e.g., soil, sediment) with an average bias uncertainty and imprecision of <1%.

Thermal ionization MS has been used traditionally to measure Pb isotope ratios in order to identify sources of Pb contamination. Hinners et al. (1987) evaluated ICP-MS for this application and found RSDs of 0.86 to 1.1% in the measurement of m/z 208, 207, and 204 in ratio to 206 in Pb ores. New instrumentation is now capable of providing precision in the region of 0.1 to 0.2% for these ratios where the concentration of Pb is sufficiently above background; doubtless this application will become more common. Mercury isotopes have recently been used as tracers in environmental studies. Hence good accuracy and precision are required in ratio measurement at low concentrations of Hg. Haraldsson et al. (1994) achieved a precision of 0.2% RSD in the ratio 202:199 Hg at 300 pg of Hg. This was accomplished by the concentration of Hg on gold traps with subsequent volatilization and addition of N_2 or H_2 to the central gas flow. Again, this use of ICP-MS for ultratrace measurement of ratios should grow in the environmental field.

Powell et al. (1986) reported a preliminary study of HG-ICP-MS, employing a commercial hydride generator (Plasma Therm) for automatically controlled mixing of acidified sample and borohydride and subsequent separation of gas from liquid prior to entry into the spray chamber of the ICP mass spectrometer. They reported detection limits (3σ) in solution from 4 ppt (Sb) to 100 ppt (Te), but were dissatisfied with the high volume of sample solution required using an uptake rate of 10 mL/min combined with over 2 min for washout and integration time. A much lower sample flow rate of 0.8 mL/min was used by Heitkemper and Caruso (1990) in their arrangement, where As, Se, Sb, Te, and Bi were measured as their hydrides simultaneously with elements in solution by conventional nebulization. One order of magnitude improvement over PN-ICP-MS was realized with RSDs of less than 5% throughout a 30-min period of measurement. Memory effects were observed in this work, appearing to be caused by adherence of analytes to the inner surfaces of the Teflon tubing (used to transport the hydrides), the nebulizer, and the spray chamber. In the author's laboratory at the GSC, the hydrides are passed directly into the central channel of the torch after the gas/liquid phase separation (Figure 5-6). This feature, combined with a high acid strength in the sample solution of 4 M HCl, produces short washout times and reduced memory. The detection limits obtained range from 1 ppt for Bi to 12 ppt for Se. Assuming an overall dilution factor of 100 following, for example, an *aqua regia* digestion, the determination limit (10σ) for Te would be 1 ppb in the original sample. Interferences from elements such as Cu and Ni can be mitigated, to a large degree, as the sample solution can be diluted considerably, thus avoiding precipitation when $NaBH_4$ is added. Mercury is also determined with this system, providing a detection limit in solution of 1 to 2 ppt. The benefits to be accrued from the marriage of this simple sample introduction equipment to the ICP mass spectrometer should be realized by more geoanalytical laboratories in the near future. As this takes place, analysis for Se, Te, and Bi, in particular in geological samples, will be far better served (INAA provides adequate detection limits for As and Sb in most matrices).

There have been relatively few reports in the literature of applications of ETV, laser, or slurry nebulization to the analysis of sediments. Using Park's custom ETV, Park and Hall (1987) determined Mo and W in geological materials down to levels of 30 and 60 ppb, respectively, and eliminated the need for separation of the analytes from the fusion leachate (high in salt content) by thermal programming. Similarly, Tl was determined using a W filament down to 9 ppb in rocks, soils, and sediments after an acid attack with an average RSD of 4% in the range 50 to 1400 ppb Tl (Park and Hall, 1988). Solutions containing 1% solids were easily tolerated. Spectral interference from Pb at concentrations greater than 500 ppm was negated by extraction of $TlCl_4^-$ into MIBK, a solvent easily analyzed by ETV-ICP-MS.

Hager (1990) demonstrated the performance of LA-ICP-MS in the semiquantitative analysis of the GXR series of reference materials, prepared by pressing 95% of the sample with 5% of paraffin wax at 20 tons. Calibration was carried out by measuring the signals for four well-characterized elements in GXR-6 and using these to derive the response table for the other elements of interest; Si was used as the internal standard. The Q-switched mode was chosen at a repetition rate of 10 pulses per sec and the sample rastered over an area of about 1 cm by 1 cm to create crater diameters of 100 μm. For the most part, results for Sc, Ti, V, Cr, Fe, Co, Zn, As, Zr, Sb, Ce, Sm, Eu, Ta, Pb, Th, and U were within 25% of the preferred value, but there were discrepancies such as 9.5 (obtained) vs. 31 ppm (preferred) for Sc in GXR-4, 35 vs. 10 ppm for Cr in GXR-1, 126 vs. 220 ppm for Zn in GXR-3, and 866 vs. 670 ppm for Pb in GXR-1. Similar work on these standard samples by Broadhead et al. (1990) extended the number of elements determined to 66 major, minor, and trace, with most yielding precision in the range 5 to 15% RSD. Imai (1990) at the Geological Survey of Japan demonstrated the importance of normalizing to an internal standard to improve accuracy in the Q-switched mode. Variation in the amount of material ablated for a series of SRMs ranged from 20 to 70 μg per analysis; normalization to Ba of known concentration dramatically reduced RSDs from 1 to 16% to 2 to 8%. More laboratories today are recognizing the real strength of the laser, in spatial analysis rather than absolute abundance measurement. For example, several geoanalytical laboratories are examining trace elements in Fe-oxide coatings in stream sediment material, taking care not to ablate down to the crystalline substrate.

In reviewing the ICP-MS literature over the past several years, it becomes clear that there is a dominance of applications where the technique is married to gas or liquid chromatography for speciation studies of elements, such as Cr, Hg, Sn, Se, Sb, and Pb. To date, little has been published concerning sediments, but rather this "hyphenated" technique has been developed as a speciation tool in the clinical and biological fields (Tomlinson and Caruso, 1995). In addition to high performance liquid chromatography (HPLC), other techniques, such as capillary zone electrophoresis (CZE) and supercritical fluid chromatography (SFC), are being married to ICP-MS. The current growth in studies of toxic trace and ultratrace elements present in various media in the surficial environment should lead to the adoption of these tremendously powerful tools in sediment analysis in the near future.

References

Ajilec, R., Cop, M., and Stupar, J., Interferences in the determination of chromium in plant materials and soil samples by flame atomic absorption spectrometry, *Analyst*, 113, 585, 1988.

Angino, E.E. and Billings, G.K., *Atomic Absorption Spectrometry in Geology*, Elsevier Science, Amsterdam, 1967, 144.

Barth, P., Krivan, V., and Hausbeck, R., Cross-interferences of hydride-forming elements in hydride-generation atomic absorption spectrometry, *Anal. Chim. Acta*, 263, 111, 1992.

Beary, E.S. and Paulsen, P.J., Selective application of chemical separations to isotope dilution inductively coupled plasma mass spectrometric analyses of standard reference materials, *Anal. Chem.*, 65, 1602, 1993.

Beary, E.S., Paulsen, P.J., and Fassett, J.D., Sample preparation approaches for isotopic dilution inductively coupled plasma mass spectrometric certification of reference materials, *J. Anal. At. Spectrom.*, 9, 1363, 1994.

Bendicho, C. and de Loos-Vollebregt, M.T.C., Solid sampling in electrothermal atomic absorption spectrometry using commercial atomizers, *J. Anal. At. Spectrom.*, 6, 353, 1991.

Bettinelli, M., Baroni, U., and Pastorelli, N., Microwave oven sample dissolution for the analysis of environmental and biological materials, *Anal. Chim. Acta*, 225, 159, 1989.

Bettinelli, M., Determination of trace metals in siliceous standard reference materials by electrothermal AAS after a Li tetraborate fusion, *Anal. Chim. Acta*, 148, 192, 1983.

Blain, L., Salin, E.D., and Boomer, D.W., Probe design for the direct insertion of solid samples in the inductively coupled plasma for analysis by atomic emission and mass spectrometry, *J. Anal. At. Spectrom.*, 4, 721, 1989.

Bock, R., *A Handbook of Decomposition Methods in Analytical Chemistry*, John Wiley & Sons, New York, 1979, 444.

Boomer, D.W. and Powell, M.J., Determination of uranium in environmental samples using inductively coupled plasma mass spectrometry, *Anal. Chem.*, 59, 2810, 1987.

Boumans, P.W.J.M., *Inductively Coupled Plasma Emission Spectroscopy*, Part 2, John Wiley & Sons, New York, 1987, 486.

Brenner, I.B., Bremier, P., and Lemarchand, A., Performance characteristics of an ultrasonic nebulizer coupled to a 40.68 MHz inductively coupled plasma atomic emission spectrometer, *J. Anal. At. Spectrom.*, 7, 819, 1992.

Brindle, I.D., Le, X., and Li, X., Convenient method for the determination of trace amounts of germanium by hydride generation direct current plasma atomic emission spectrometry: interference reduction by L-cystine and L-cysteine, *J. Anal. At. Spectrom.*, 4, 227, 1989.

Broadhead, M., Broadhead, R., and Hager, J.W., Laser sampling ICP-MS: semiquantitative determination of 66 elements in geological samples, *At. Spectrosc.*, 11, 205, 1990.

Cantillo, A.Y., Sinex, S.A., and Helz, G.R., Elemental analysis of estuarine sediments by lithium metaborate fusion and direct current plasma emission spectrometry, *Anal. Chem.*, 56, 33, 1984.

Chan, C.C.Y. and Sadana, R.S., Determination of arsenic and selenium in environmental samples by flow-injection hydride generation atomic absorption spectrometry, *Anal. Chim. Acta*, 270, 231, 1992.

Chao, T.T. and Sanzolone, R.F., Decomposition techniques, *J. Geochem. Explor.*, 44, 65, 1992.

Cheam, V., Chau, A.S.Y., and Horn, W., National Interlaboratory Quality Control Study No. 35: Trace Metals in Sediments, Report Series No. 76, Inland Waters Directorate, Environment Canada, Burlington, 1988.

Christian, G.D. and Ruzicka, J., Flow injection analysis: a novel tool for plasma spectroscopy, *Spectrochim. Acta*, 42B, 157, 1987.

Church, S.E., Multi-element analyses of fifty-four geochemical reference samples using inductively coupled plasma atomic-emission spectrometry, *Geostand. Newsl.*, 2, 133, 1981.

Clark, J.R., Electrothermal atomisation atomic absorption conditions and matrix modifications for determining antimony, arsenic, bismuth, cadmium, gallium, gold, indium, lead, molybdenum, palladium, platinum, selenium, silver, tellurium, thallium and tin following back-extraction of organic aminohalide extracts, *J. Anal. At. Spectrom.*, 1, 301, 1986.

Cresser, M.S., Armstrong, J., Cook, J., Dean, J.R., Watkins, P., and Cave, M., Atomic spectrometry update — environmental analysis, *J. Anal. At. Spectrom.*, 8, 1R, 1995.

Croudace, L.W., A possible error source in silicate wet-chemistry caused by insoluble fluoride, *Chem. Geol.*, 31, 153, 1980.

Date, A.R. and Gray, A.L., *Applications of Inductively Coupled Plasma Mass Spectrometry*, Blackie, Glasgow and London, 1989, 254.

Date, A.R. and Stuart, M.E., Application of inductively coupled plasma mass spectrometry in the simultaneous determination of chlorine, bromine and iodine in the National Bureau of Standards Reference Material 1648 urban particulate, *J. Anal. At. Spectrom.*, 3, 659, 1988.

De la Guardia, M., Carbonell, V., Morales-Rubio, A., and Salvador, A., On-line microwave assisted digestion of solid samples for their flame atomic spectrometric analysis, *Talanta*, 40, 1609, 1993.

de Oliveira, E., McLaren, J.W., and Berman, S.S., Simultaneous determination of arsenic, antimony, and selenium in marine sediments by inductively coupled plasma atomic emission spectrometry, *Anal. Chem.*, 55, 2047, 1983.

Dean, J.R., Ebdon, L., Crews, H.M., and Massey, R.C., Characteristics of flow injection inductively coupled plasma mass spectrometry for trace metal analysis, *J. Anal. At. Spectrom.*, 3, 349, 1988.

Dolezal, J., Povondra, P., and Sulcek, Z., *Decomposition Techniques in Inorganic Analyses*, Iliffe Books, London, 1968, 224.

Fassel, V.A. and Bear, B.R., Ultrasonic nebulization of liquid samples for analytical inductively coupled plasma-atomic spectroscopy: an update, *Spectrochim. Acta*, 41B, 1089, 1986.

Feldmann, I., Tittes, W., Jakubowski, N., Stuewer, D., and Giessmann, U., Performance characteristics of inductively coupled plasma mass spectrometry with high mass resolution, *J. Anal. At. Spectrom.*, 9, 1007, 1994.

Fuller, C.W., *Electrothermal Atomization for Atomic Absorption Spectrometry*, The Chemical Society, London, 1977, 127.

Gilman, L. and Grooms, W., Safety concerns associated with wet ashing samples under pressure heated by microwave energy, *Anal. Chem.*, 60, 1624, 1988.

Gorsuch, T.T., *The Destruction of Organic Matter*, Pergamon Press, Oxford, 1970, 152.

Govindaraju, K. and Melvelle, G., Fully automated dissolution and separation methods for inductively coupled plasma atomic emission spectrometry rock analysis, *J. Anal. At. Spectrom.*, 2, 615, 1987.

Govindaraju, K., Compilation of working values and sample description for 383 geostandards, *Geostand. Newsl.*, 18, 1994.

Gray, A.L. and Date, A.R., Inductively coupled plasma source mass spectrometry using continuous flow extraction, *Analyst*, 108, 1033, 1983.

Gray, A.L., Solid sample introduction by laser ablation for inductively coupled plasma source mass spectrometry, *Analyst*, 110, 551, 1985.

Greenfield, S., Jones, I.L., and Berry, C.T., High-pressure plasmas as spectroscopic emission sources, *Analyst*, 89, 713, 1964.

Gunther, D., Longerich, H.P., and Jackson, S.E., A new enhanced sensitivity quadrupole inductively coupled plasma mass spectrometer (ICP-MS), *Can. J. Appl. Spectrosc.*, 10, 111, 1995.

Hager, J.W., Laser sampling ICP-MS: geological exploration samples, *Perkin Elmer Laser Appl. Rep.*, 1, 1990.

Hale, M., Thompson, M., and Lovell, J., Large batch sealed tube decomposition of geological samples by means of a layered heating block, *Analyst*, 110, 225, 1985.

Hall, G.E.M., Inductively coupled plasma mass spectrometry in geoanalysis, *J. Geochem. Explor.*, 44, 201, 1992.

Hall, G.E.M. and Bonham-Carter, G.F., Review of methods to determine gold, platinum and palladium in production-oriented geochemical laboratories, with application of a statistical procedure to test for bias, *J. Geochem. Explor.*, 30, 255, 1988.

Hall, G.E.M. and Bouvier, J.L., A procedure to lower the limits of detection for silver, cadmium and lead in the analysis of geological materials by atomic absorption spectrometry, *Geol. Surv. Can. Pap.*, 87-27, 1988.

Hall, G.E.M. and Pelchat, J.C., Inductively coupled plasma emission spectrometric detemination of boron and other oxo-anion forming elements in geological materials, *Analyst*, 111, 1255, 1986.

Hall, G.E.M. and Pelchat, J.C., Analysis of standard reference materials for Zr, Nb, Hf and Ta by ICP-MS after lithium metaborate fusion and cupferron separation, *Geostand. Newsl.*, 14, 197, 1990.

Hall, G.E.M. and Pelchat, J.C., Analysis of geological materials for gold, platinum and palladium at low levels by fire assay — ICP mass spectrometry, *Chem. Geol.*, 115, 61, 1994.

Hall, G.E.M. and Plant, J.A., Analytical errors in the determination of high field strength elements and their implications in tectonic interpretation studies, *Chem. Geol.*, 95, 141, 1992.

Hall, G.E.M., de Silva, K.N., Pelchat, J.C., and Vaive, J.E., Advances in analytical methods based on atomic absorption spectrometry in the Geochemistry Laboratories of the Geological Survey of Canada, in Current Research, Part A, Geological Survey of Canada, Paper 87-1A, 1987, 477.

Hall, G.E.M., Vaive, J.E., Coope, J.A., and Weiland, E., Bias in the analysis of geological materials for gold using current methods, *J. Geochem. Explor.*, 34, 157, 1989.

Halls, D., Analytical minimalism applied to the determination of trace elements by atomic spectrometry, *J. Anal. At. Spectrom.*, 10, 169, 1995.

Haraldsson, C., Lyven, B. Ohman, P., and Munthe, J., Determination of mercury isotope ratios in samples containing sub-nanogram amounts of mercury using inductively coupled plasma mass spectrometry, *J. Anal. At. Spectrom.*, 9, 1229, 1994.

Heitkemper, D.T. and Caruso, J.A., Continuous hydride generation for simultaneous multielement detection with inductively coupled plasma mass spectrometry, *Appl. Spectrosc.*, 44, 228, 1990.

Hinners, T.A., Heithmar, E.M., Spittler, T.M., and Henshaw, J.M., Inductively coupled plasma mass spectrometric determination of lead isotopes, *Anal. Chem.*, 59, 2658, 1987.

Holland, G. and Eaton, A.N., Applications of Plasma Source Mass Spectrometry II, The Royal Society of Chemistry, Spec. Publ. 124, Cambridge, 1993.

Imai, N., Quantitative analysis of original and powdered rocks and mineral inclusions by laser ablation inductively coupled plasma mass spectrometry, *Anal. Chim. Acta*, 235, 381, 1990.

Jarvis, I. and Jarvis, K.E., Inductively coupled plasma atomic emission spectrometry in exploration geochemistry, *J. Geochem. Explor.*, 44, 139, 1992.

Kanda, Y. and Taira, M., Sequential multi-element analysis of sediments and soils by inductively-coupled plasma atomic emission spectrometry with a computer-controlled rapid-scanning echelle monochromator, *Anal. Chim. Acta*, 207, 269, 1988.

Karanassios, V. and Horlick, G., Elimination of some spectral interferences and matrix effects in inductively coupled plasma-mass spectrometry using direct sample insertion techniques, *Spectrochim. Acta*, 12, 1387, 1989.

Karanassios, V., Ren, J.M., and Salin, E.D., Electrothermal vaporization sample introduction system for the analysis of pelletized solids by inductively coupled plasma emission spectrometry, *J. Anal. At. Spectrom.*, 6, 527, 1991.

Kingston, H.M. and Jassie, L.B., Microwave energy for acid decomposition at elevated temperature and pressures using biological and botanical samples, *Anal. Chem.*, 58, 2534, 1986.

Kingston, H.M. and Jassie, L.B., *Introduction to Microwave Sample Preparation: Theory and Practice*, ACS Prof. Ref. Book, American Chemical Society, Washington, D.C., 1988, 263.

Laborda, F., Baxter, M.J., Crews, H.M., and Dennis, J., Reduction of polyatomic interferences in inductively coupled plasma mass spectrometry by selection of instrument parameters and using an argon-nitrogen plasma: effect on multi-element analyses, *J. Anal. At. Spectrom.*, 9, 727, 1994.

Lamothe, P.J., Fries, T.L., and Consul, J.J., Evaluation of a microwave oven system for the dissolution of geologic samples, *Anal. Chem.*, 58, 1881, 1986.

Landi, S. and Fagioli, F., The adaption of the dichromate digestion method for total mercury determination by cold-vapor atomic absorption spectrometry to the analysis of soils, sediments and sludges, *Anal. Chim. Acta*, 298, 363, 1994.

Langmyhr, F.J. and Sveen, S., Decomposability in hydrofluoric acid of the main and some minor and trace minerals of silicate rocks, *Anal. Chim. Acta*, 32, 1, 1965.

Lasztity, A., Viczian, M., Wang, X., and Barnes, R.M., Sample analysis by on-line isotope dilution inductively coupled plasma mass spectrometry, *J. Anal. At. Spectrom.*, 4, 761, 1989.

Legere, G. and Salin, E.D., Fast-clearing spray chamber for ICP-AES, *Appl. Spectrosc.*, 48, 761, 1994.

Legere, G. and Salin, E.D., Capsule-based microwave digestion, *Appl. Spectrosc.*, 49, 14A, 1995.

Littlejohn, D., Egila, J.N., Gosland, R.M., Kunwar, U.K., and Smith, C., Graphite furnace analysis — getting easier and achieving more?, *Anal. Chim. Acta*, 250, 71, 1991.

Liu, J., Sturgeon, R.E., and Willie, S.N., Open-focused microwave-assisted digestion for the preparation of large mass organic samples, *Analyst*, 120, 1905, 1995.

L'vov, B.V., Recent advances in absolute analysis by graphite furnace atomic absorption spectrometry, *Spectrochim. Acta*, 45B, 633, 1990.

Lynch, J.J., Provisional elemental values for eight new geochemical lake sediment and stream sediment reference materials LKSD-1, LKSD-2, LKSD-3, LKSD-4, STSD-1, STSD-2, STSD-3 and STSD-4, *Geostand. Newsl.*, 14, 153, 1990.

Macalalad, E., Bayoran, R., Ebarvia, B., and Rubeska, I., A concise analytical scheme for 16 trace elements in geochemical exploration samples using exclusively AAS, *J. Geochem. Explor.*, 30, 167, 1988.

Madrid, Y. and Camara, C., Lead hydride generation atomic absorption spectrometry: an alternative to electrothermal atomic absorption spectrometry: a review, *Analyst*, 119, 1647, 1994.

Matusiewicz, H., Thermal vaporisation for inductively coupled plasma optical emission spectrometry: a review, *J. Anal. At. Spectrom.*, 1, 171, 1986.

Matusiewicz, H. and Sturgeon, R.E., Present status of microwave sample dissolution for elemental analysis, *Prog. Analyt. Spectrosc.*, 12, 21, 1989.

McLaren, J.W., Berman, S.S., Boyko, V.J., and Russell, D.S., Simultaneous determination of major, minor and trace elements in marine sediments by inductively coupled plasma emission spectrometry, *Anal. Chem.*, 53, 1802, 1981.

McLaren, J.W., Beauchemin, D., and Berman, S.S., Determination of trace elements in marine sediments by inductively coupled plasma mass spectrometry, *J. Anal. At. Spectrom.*, 2, 277, 1987a.

McLaren, J.W., Beauchemin, D., and Berman, S.S., Application of isotope dilution inductively coupled plasma mass spectrometry to the analysis of marine sediments, *Anal. Chem.*, 59, 610, 1987b.

McLaren, J.W., Beauchemin, D., and Berman, S.S., Analysis of the marine reference material PACS-1 by inductively coupled plasma mass spectrometry, *Spectrochim. Acta*, 43B, 413, 1988.

Meier, A.L., Flameless atomic-absorption determination of gold in geological materials, *J. Geochem. Explor.*, 13, 77, 1980.

Mermet, J.M. and Poussel, E., ICP emission spectrometers: 1995 analytical figures of merit, *Appl. Spectrosc.*, 49, 12A, 1995.

Miller-Ihli, N.J., A systematic approach to ultrasonic slurry GFAAS, *At. Spectrosc.*, 13, 1, 1992.

Mochizuki, T., Sakashita, A., Iwata, H., Ishibashi, Y., and Gunji, N., Slurry nebulization technique for direct determination of rare earth elements in silicate rocks by inductively coupled plasma mass spectrometry, *Anal. Sci.*, 5, 311, 1989.

Moens, L., Vanhoe, H., Vanhaecke, F., Goosens, J., Campbell, M., and Dams, R., Application of inductively coupled plasma mass spectrometry to the certification of reference materials from the Community Bureau of Reference, *J. Anal. At. Spectrom.*, 9, 187, 1994.

Moens, L., Vanhaecke, F., Riondato, J., and Dams, R., Some figures of merit of a new double focusing inductively coupled plasma mass spectrometer, *J. Anal. At. Spectrom.*, 10, 569, 1995.

Montaser, A. and Golightly, D.W., *Inductively Coupled Plasmas in Analytical Atomic Spectrometry*, VCH Publishers, New York, 1987, 660.

Morales-Rubio, A., Mena, M.L., and McLeod, C.W., Rapid determination of mercury in environmental materials using on-line microwave digestion and atomic fluorescence spectrometry, *Anal. Chim. Acta*, 308, 364, 1995.

Motooka, J.M., An exploration geochemical technique for the determination of preconcentrated organometallic halides by ICP-AES, *Appl. Spectrosoc.*, 42, 1293, 1988.

Nakashima, S., Sturgeon, R.E., Willie, S.N., and Berman, S.S., Acid digestion of marine samples for trace element analysis using microwave heating, *Analyst*, 113, 159, 1988.

Nieuwenhuize, J., Poley-Vos, C.H., van den Akker, A.H., and van Delft, W., Comparison of microwave and conventional extraction techniques for the determination of metals in soil, sediment and sludge samples by atomic spectrometry, *Analyst*, 116, 347, 1991.

O'Leary, R.M. and Viets, J.G., Determination of antimony, arsenic, bismuth, cadmium, copper, lead, molybdenum, silver and zinc in geological materials by atomic absorption spectrometry using hydrochloric acid-hydrogen peroxide digestion, *At. Spectrosc.*, 7, 4, 1986.

Olesik, J.W., Kinzer, J.A., and Harkleroad, B., Inductively coupled plasma optical emission spectrometry using nebulizers with widely different sample consumption rates, *Anal. Chem.*, 66, 2022, 1994.

Park, C.J. and Hall, G.E.M., Electrothermal Vaporization as a Means of Sample Introduction into an Inductively Coupled Plasma Mass Spectrometer: a Preliminary Report of a New Analytical Technique, Geological Survey of Canada Paper, 86-1B, 767, 1986.

Park, C.J. and Hall, G.E.M., Analysis of geological materials by inductively coupled plasma mass spectrometry with sample introduction by electrothermal vaporisation, Part 1, Determination of molybdenum and tungsten, *J. Anal. At. Spectrom.*, 2, 473, 1987.

Park, C.J. and Hall, G.E.M., Analysis of geological materials by inductively coupled plasma mass spectrometry with sample introduction by electrothermal vaporisation, Part 2, Determination of thallium, *J. Anal. At. Spectrom.*, 3, 355, 1988.

Park, C.J., Van Loon, J.C., Arrowsmith, P., and French, J.B., Design and optimization of an electrothermal vaporizer for use in plasma source mass spectrometry, *Can. J. Spectrosc.*, 32, 29, 1987.

Potts, P.J., *A Handbook of Silicate Rock Analysis*, Blackie, Glasgow and London, 1987, 622.

Potts, P.J., Hawkesworth, C.J., van Calsteren, P., and Wright, I.P., Advances in analytical technology and its influence on the development of modern inorganic geochemistry: a historical perspective, in Magmatic Processes and Plate Tectonics, Pritchard, H.M., Alabaster, T., Harris, N.B.W., and Neary, C.R., Eds., Geological Society, Spec. Publ., 76, 501, 1993.

Powell, M.J., Boomer, D.W., and McVicars, R.J., Introduction of gaseous hydrides into an inductively coupled plasma mass spectrometer, *Anal. Chem.*, 58, 2864, 1986.

Puk, R. and Weber, J.H., Critical review of analytical methods for determination of inorganic mercury and methylmercury compounds, *Appl. Organomet. Chem.*, 8, 293, 1994.

Reed, N.M., Cairns, R.O., Hutton, R.C., and Takaku, Y., Characterization of polyatomic ion interferences in inductively coupled plasma mass spectrometry using a high resolution mass spectrometer, *J. Anal. At. Spectrom.*, 9, 881, 1994.

Rubeska, I., Multi-element pre-concentration by solvent extraction compatible with aqua regia digestion for geochemical exploration samples, *Analyst*, 112, 27, 1987.

Ruzicka, J., The second coming of flow-injection analysis, *Anal. Chim. Acta*, 261, 3, 1992.

Scifres, J., Cheema, V., Wasko, M., and McDaniel, W., Determination of ultra-trace level total mercury in sediment and tissue by microwave digestion and atomic fluorescence detection: Part 2, *Am. Environ. Lab.*, 6, 1, 1995.

Sen Gupta, J.G. and Bouvier, J.L., Direct determination of traces of Ag, Cd, Pb, Bi, Cr, Mn, Ni, Li, Be, Cu and Sb in environmental waters and geological materials by simultaneous multi-element graphite furnace atomic absorption spectrometry with Zeeman-effect background correction, *Talanta*, 42, 269, 1995.

Sheppard, B.S. and Caruso, J.A., Plasma mass spectrometry: consider the source, *J. Anal. At. Spectrom.*, 9, 145, 1994.

Simpson, P.R., Robotham, H., and Hall, G.E.M., Regional geochemical orientation studies for platinum in Jamaica, *Trans. Inst. Min. Metall.*, (Sect. B: Appl. Earth Sci.), 99, 183, 1990.

Slavin, W., Graphite Furnace AAS, A Source Book, The Perkin-Elmer Corporation, Ridgefield, CT, 1984.

Sturgeon, R.E., Desaulniers, J.A.H., Berman, S.S., and Russell, D.S., Determination of trace metals in estuarine sediments by graphite-furnace atomic absorption spectrometry, *Anal. Chim. Acta*, 134, 283, 1982.

Sturgeon, R.E., Willie, S.N., Sproule, G.I., and Berman, S.S., Sorption and atomisation of metallic hydrides in a graphite furnace, *J. Anal. At. Spectrom.*, 2, 719, 1987.

Sturgeon, R.E., Graphite furnace atomic absorption analysis of marine samples for trace metals, *Spectrochim. Acta*, 44B, 1209, 1989.

Sturgeon, R.E., Willie, S.N., Methven, B.A., Lam, J.W., and Matusiewicz, H., Continuous-flow microwave-assisted digestion of environmental samples using atomic spectrometric detection, *J. Anal. At. Spectrom.*, 10, 981, 1995.

Tanner, S., Characterization of ionization and matrix suppression in inductively coupled 'cold' plasma mass spectrometry, *J. Anal. At. Spectrom.*, 10, 905, 1995.

Tao, G. and Hansen, E.H., Determination of ultra-trace amounts of selenium (IV) by flow injection hydride generation atomic absorption spectrometry with on-line preconcentration by coprecipitation with lanthanum hydroxide, *Analyst*, 119, 333, 1994.

Thompson, M. and Walsh, J.N., *Handbook of Inductively Coupled Plasma Spectrometry*, Blackie, New York, 1989, 315.

Tomlinson, M.J. and Caruso, J.A., Plasma mass spectrometry as a detector for chemical speciation studies, *Analyst*, 120, 583, 1995.

Torres, P., Ballesteros, E., and Luque de Castro, M.D., Microwave-assisted robotic method for the determination of trace metals in soil, *Anal. Chim. Acta*, 308, 371, 1995.

U.S. Environmental Protection Agency (U.S. EPA), U.S. EPA Method 6020, *ICP Inf. Newsl.*, 18, 584, 1993.

van Berkel, W.W., Balke, J., and Maessen, F.J.M.J., Introduction of analyte-loaded poly(dithiocarbamate) into inductively coupled plasmas by electrothermal vaporization: spatial emission characteristics of the resulting dry plasmas, *Spectrochim. Acta*, 45B, 1265, 1990.

Van Loon, J.C., *Analytical Atomic Absorption Spectroscopy*, Academic Press, New York, 1980, 337.

Van Loon, J.C. and Barefoot, R.R., *Analytical Methods for Geochemical Exploration*, Academic Press, San Diego, 1989, 344.

Vickers, G.H., Ross, B.S., and Heiftje, G.M., Reduction of mass-dependent interferences in inductively coupled plasma-mass spectrometry by using flow-injection analysis, *Appl. Spectrosc.*, 8, 1330, 1989.

Viczian, M., Lasztity, A., Wang, X., and Barnes, R.M., On-line isotope dilution and sample dilution by flow injection and inductively coupled plasma mass spectrometry, *J. Anal. At. Spectrom.*, 5, 125, 1990.

Viets, J.G., Determination of silver, bismuth, cadmium, copper, lead and zinc in geological materials by atomic absorption spectrometry with tricaprylylmethylammonium chloride, *Anal. Chem.*, 50, 1097, 1978.

Viets, J.G. and O'Leary, R.M., The role of atomic absorption spectrometry in exploration geochemistry, *J. Geochem. Explor.*, 44, 107, 1992.

Wach, F., GFAAS, a mature technique for a modest investment, *Anal. Chem.*, 67, 51A, 1995.

Welz, B., Schlemmer, G., and Mudakavi, J.R., Palladium nitrate-magnesium nitrate modifier for electrothermal atomic absorption spectrometry, *J. Anal. At. Spectrom.*, 7, 1257, 1992.

Wendt, R.H. and Fassel, V.A., Inductively coupled plasma spectrometric excitation source, *Anal. Chem.*, 37, 920, 1965.

Wickstrom, T., Lund, W., and Bye, R., Determination of arsenic and tellurium by hydride generation atomic spectrometry: minimizing interferences from nickel, cobalt, and copper by using an alkaline sample solution, *Analyst*, 120, 2695, 1995a.

Wickstrom, T., Lund, W., and Bye, R., Determination of selenium by hydride generation atomic spectrometry: elimination of interferences from very high concentrations of nickel, cobalt, iron, and chromium by complexation, *J. Anal. At. Spectrom.*, 10, 803, 1995b.

Williams, J.G., Gray, A.L., Norman, P., and Ebdon, L., Feasibility of solid sample introduction by slurry nebulisation for inductively coupled plasma mass spectrometry, *J. Anal. At. Spectrom.*, 2, 469, 1987.

Willie, S.N., Sturgeon, R.E., and Berman, S.S., Hydride generation atomic absorption determination of selenium in marine sediments, tissues, and seawater with *in situ* concentration in a graphite furnace, *Anal. Chem.*, 58, 1140, 1986.

Winge, R.K., Fassel, V.A., Peterson, V.J., and Floyd, M.A., *Inductively Coupled Plasma-Atomic Emission Spectroscopy, An Atlas of Spectral Information*, Elsevier Science, Amsterdam, 1985, 584.

Xiao, G. and Beauchemin, D., Reduction of matrix effects and mass discrimination in inductively coupled plasma mass spectrometry with optimized argon-nitrogen plasmas, *J. Anal. At. Spectrom.*, 9, 509, 1994.

Yan, X.-P. and Ni, Z.M., Vapor generation atomic absorption spectrometry, *Anal. Chim. Acta*, 291, 89, 1994.

Zhou, C.Y., Wong, M.K., Koh, L.L., and Wee, Y.C., Orthogonal array design for the optimization of closed-vessel microwave digestion parameters for the determination of trace metals in sediments, *Anal. Chim. Acta*, 314, 121, 1995.

chapter six

Neutron activation analysis

José Marcus Godoy

6.1 Introduction

Methods of neutron activation analyses, described in many scientific books and papers, have been frequently used in different analyses of soils and rocks. In studies of aquatic ecosystems, neutron activation methods have been used in sediments and seawater (for example, Kamel et al., 1987; Fernandes et al., 1993; Cardoso and Godoy, 1995). The first part of this chapter, a general review of the principles of neutron activation analyses, is based on the concepts developed by Das et al. (1989) and Silva (1991). The literature review, carried out using the INIS data bank from 1978 to 1994, is focused on the application of the neutron activation methods in analyses of sediment samples. However, the application of the method to other matrices is also considered where necessary to provide complete information to the reader. The description of the methods is restricted to the instrumental neutron activation analysis (INAA), which is the one most commonly used. The chapter includes a general introduction to the nuclear reactions with neutrons, a description of the relative and absolute methods, the procedures and equipment involved in the method, and the application of the method in analyses of sediment samples.

6.2 Theory

The description of the theory is restricted to the instrumental neutron activation analysis (INAA); the theoretical fundament will be restricted to this. Information about other nuclear analytical techniques can be found in *Radioanalysis in Geochemistry* by Das et al. (1989). Information on the development and trends of neutron activation analysis can be found in the proceedings of the international conferences on Modern Trends in Activation Analysis (MTAA), published in the *Journal of*

1-56670-155-4/97/$0.00+$.50
© 1997 by CRC Press, Inc.

Nuclear and Radiochemical Analysis. The conference is held once every five years, with the last one held in Korea in 1995.

The neutron is an elementary particle without charge, weighing 1.0086654 units of atomic mass. It is found in every atomic nucleus, except ^1H, and decays spontaneously into one proton and one electron, with a half-life of 12.5 min:

$$n \rightarrow p^+ + e^- + \nu$$

Since the neutron has no charge, it is not exposed to a repulsive coulombianic force and can easily penetrate an atomic nucleus, generating nuclear reactions in it.

A neutron can be classified according its kinetic energy. This classification has arbitrary limits, changing from author to author. The classification described below follows that of Lyon (1964):

> *Thermal neutrons* are neutrons in thermic equilibrium with their surroundings. The thermal neutrons follow a Maxwell velocity distribution, and the most probable velocity is:

$$v = \sqrt{\frac{2kT}{M}} \tag{1}$$

> where, k = Boltzmann constant = 8.56×10^{-5} eV $\cdot$ K^{-1}; T = temperature in Kelvin grades; and M = neutron mass = 1.67470×10^{-24} g.
>
> *Epithermal neutrons* are neutrons with kinetic energy between 0.1 to 10^5 eV that are under the thermalization process.
>
> *Rapid neutrons* are the neutrons with kinetic energy greater than 10^5 eV that are classified as rapid neutrons.

A neutron can react with an atomic nucleus according to its kinetic energy. The different processes involved are

> *Elastic Scattering.* During this process only an elastic collision occurs, which means that there is momentum conservation, and the resulting nucleus is not excited.
>
> *Inelastic Scattering.* In this case there is no momentum or kinetic energy conservation.
>
> *Nuclear Reaction.* Nuclear reaction is the process involving the neutron absorption and further emission of another particle, x. Here, x could be a proton, an alpha particle, or more than one particle. The process is represented by (n, x), and, as a consequence, an element other than the original one is formed.

Neutron Capture. This is a particular case of nuclear reaction where
a gamma ray is emitted instead of a nuclear particle, a (n, γ)
reaction. The product is an isotope from the initial nucleus.

Nuclear Fission. After the neutron absorption, when the new nucleus formed decays through its fragmentation into two pieces, the
process is called nuclear fission. Two radioactive fission products
are built, and also two or three neutrons are emitted. This is
represented as a (n, f) reaction.

The neutron reaction depends on the nucleus target and the kinetic
energy of the incident neutron. For thermal neutrons, the most probable
reaction is the (n, γ). For higher energies the reactions (n, x) may occur.
Only heavy elements, starting with thorium, can undergo a fission
reaction after a nuclear reaction with neutrons. The INAA basically
utilizes the (n, γ) produced by thermal and epithermal neutrons.

The cross-section expresses the probability of occurrence of a determined nuclear reaction. For a (n, γ) reaction, the cross-section decreases
with the neutron kinetic energy. This can be explained by the fact that
by increasing the neutron velocity, its permanence time inside a nucleus
decreases; therefore, the probability of occurrence also decreases. This
is known as the 1/v law. For some neutron energies, the cross-section
increases sharply. These cross-section maximums are known as resonance energies. A resonance energy corresponds to the difference
between two nuclear energy levels.

Among the neutron sources (Table 6-1), the most useful for the
INAA is the nuclear reactor. It allows a greater analytical sensibility
(high intensity flux) and the activation of many isotopes (large neutron
spectrum). A typical neutron spectrum of a research swimming pool
reactor is shown in Figure 6-1. The figure expresses the neutron energy
in lethargy. Lethargy (U) is a dimensionless quantity, defined as
$\ln(E_o/E)$. E_o is usually assigned a value of 10 MeV, which means that
most neutrons in a reactor will possess a positive lethargy (International
Atomic Energy Agency, 1982). Plotting the neutron flux per lethargy
unit vs. the lethargy is a way to preserve the area representation of flux
densities and to give a clear indication of the relative contributions of

Table 6-1   Neutron Sources for Activation Analysis

Neutron source	Nuclear reaction	Neutron flux $(n \cdot cm^{-2} \cdot s^{-1})$	Characteristics
Laboratory sources	(α, n) (γ, n) (n, f)	10^6	Small flux
Particles accelerators	(p, n) (α, n) (d, n)	$10^{10} - 10^{11}$	Monoenergetic neutrons
Research reactors	(n, f)	$10^{12} - 10^{14}$	Large spectrum, high flux

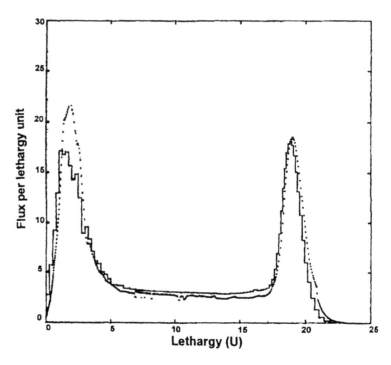

Figure 6-1 Neutron spectrum in a swimming pool reactor. (Adapted from Bitelli, U.d'U., Medida e Cálculo da Distribuição Espacial e Energética de Nêutrons no Núcleo do Reator IEA-R1, Masters Thesis, Instituto de Pesquisas Energéticas e Nucleares, São Paulo, Brazil, 1988.)

fast, epithermal and thermal neutrons (International Commission on Radiation Units and Measurements, 1969). In this case, the $1/E$ region is a horizontal line (International Atomic Energy Agency, 1982).

Such neutron spectrum depends on the reactor characteristics as moderator, nuclear fuel configuration, reactor potency, and burn-up. It depends also on the sample position inside the reactor. The thermal component of the neutron spectrum increases with the distance from the nuclear fuel, since the number of neutron interactions with the moderator also increases.

The activation analysis can be applied on the qualitative and quantitative determination of stable elements, despite their chemical form, in different matrix types in chemistry, geology, archeology, and many other fields. A simplified scheme of the activation analysis methodology is shown in Figure 6-2.

Consider the formation of a radioisotope (Y) based on the reaction of the stable isotope (X) with neutrons. The formation rate of Y is

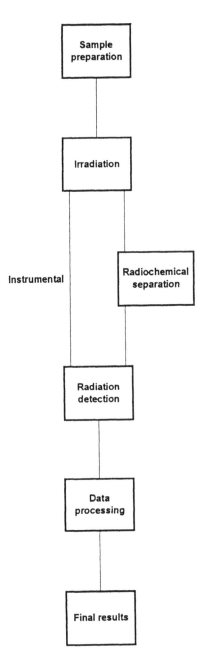

Figure 6-2 Typical Neutron Activation Analysis scheme, particularly the instrumental neutron activation analysis on the left.

proportional to the number of X atoms, to the neutron flux, and to the cross-section (the nuclear reaction probability):

$$F_Y = N_X \cdot \sigma \cdot \phi \tag{2}$$

where, F_Y = number of Y atoms formed per time unit (s^{-1}), N_X = number of X atoms present, σ = cross-section of X for neutron nuclear reaction producing Y ($1b = 10^{-24}$ cm^2), and ϕ = neutron flux ($cm^{-2} \cdot s^{-1}$).

But, since Y decays simultaneously with its formation, the formation rate of Y is

$$\frac{dN_Y}{dt} = (N_X \cdot \sigma \cdot \phi) - (\lambda \cdot N_Y) \tag{3}$$

where, λ = decay constant of the produced radionuclide Y (s^{-1}).

Solving equation (3) for an irradiation time of t_i, we obtain

$$N_Y = \frac{N_X \cdot \sigma \cdot \phi}{\lambda} (1 - e^{-\lambda t_i}) \tag{4}$$

The term $(1 - e^{-\lambda t_i})$ is called "saturation factor". However, since $\lambda \cdot N_Y$ is the activity of the radionuclide Y (A_Y)

$$A_Y = N_X \cdot \sigma \cdot \phi \cdot (1 - e^{-\lambda t_i}) \tag{5}$$

Taking into account that X is an isotope of a chemical element, say X', the number of atoms N_X is given by

$$N_X = \frac{N_0 \cdot f \cdot m}{M} \tag{6}$$

where, N_0 = Avogadro number = 6.02×10^{23} mol^{-1}, f = isotopic abundance of X, m = mass of X' present on the target (g), and M = X' atomic mass (g/mol).

As a consequence, equation (5) becomes

$$A_Y = \frac{N_0 \cdot f \cdot m}{M} \cdot \sigma \cdot \phi \cdot (1 - e^{-\lambda t_i}) \tag{7}$$

The produced activity has to be time corrected, taking into account decay that occurred between the irradiation end and the beginning of the Y activity measurement. This time period is frequently called "cooling time", t_c, and the factor $e^{-\lambda t_c}$ "cooling factor". Therefore

$$A_Y = \frac{N_0 \cdot f \cdot m}{M} \cdot \sigma \cdot \phi \cdot (1 - e^{-\lambda t_i}) \cdot e^{-\lambda t_c} \tag{8}$$

The activity A_Y is evaluated through the gamma spectrometry (see Section 6.4, this chapter) of the activated sample. During the gamma spectrometry, the area under the photopeak related to the radionuclide Y is measured. Using this peak area, A_Y is calculated

$$A_Y = \frac{\text{net peak area}}{\varepsilon \cdot b \cdot t_m} \tag{9}$$

where, ε = counting efficiency for the energy related with the photopeak of interest, b = emission probability of the photopeak of interest, and t_m = counting time (s).

For some radionuclides, it is necessary to correct A_Y due to the decay that occurred during the measurement. For these cases

$$(A_{obs})_Y = \frac{A_Y \cdot (1 - e^{-\lambda t_m})}{\lambda \cdot t_m} \tag{10}$$

where, $(A_{obs})_Y$ = observed A_Y activity without decay correction.

Therefore equation (9), including the decay correction, becomes

$$A_Y = \frac{\text{net peak area} \cdot \lambda}{\varepsilon \cdot b \cdot (1 - e^{-\lambda t_m})} \tag{11}$$

and, using this in equation (8) and solving m, we get

$$m = \frac{\text{net peak area} \cdot \lambda \cdot M}{N_0 \cdot f \cdot \varepsilon \cdot b \cdot \sigma \cdot \phi \cdot (1 - e^{-\lambda t_i}) \cdot e^{-\lambda t_c} \cdot (1 - e^{-\lambda t_m})} \tag{12}$$

In reality, we want to find the concentration of the element X' in the sample, c_s, knowing that $c_s = m/m_s$, with m_s the sample mass, and we finally obtain

$$c_s = \frac{\text{net peak area} \cdot \lambda \cdot M}{N_0 \cdot f \cdot \varepsilon \cdot b \cdot \sigma \cdot \phi \cdot m_s \cdot (1 - e^{-\lambda t_i}) \cdot e^{-\lambda t_c} \cdot (1 - e^{-\lambda t_m})} \tag{13}$$

6.3 Methods in instrumental neutron activation analysis

There are basically two methods of instrumental neutron activation analysis (INAA); the relative and the absolute. In the relative method, the sample is irradiated simultaneously with a standard containing one or more elements of interest. As the peak area is directly proportional to the mass of the element of interest present in the sample and in the standard (see equation 13), the ratio between the mass of the element of interest in the sample and in the standard is equal to the peak area ratio. The absolute method is slightly more complicated, since it implies that all members of the equation are well known and are under control as described below.

6.3.1 Relative method

The relative method is largely employed because it is simple and gives results with high accuracy. On the other hand, it is possible to quantify only the elements with a known concentration in the standard. The standard can be a reference material similar to the sample or an artificially prepared material.

If a similar amount of the sample and the standard are packed in the same way, irradiated simultaneously, and measured in the same detector using the same counting time, we obtain

$$c_s = c_{st} \cdot \frac{m_{st}}{m_s} \cdot \frac{(\text{net peak area})_s \cdot (e^{-\lambda t_c})_{st}}{(\text{net peak area})_{st} \cdot (e^{-\lambda t_c})_s} \quad (14)$$

where, c_{st} = concentration of the element of interest in the standard or in the reference material and m_{st} = mass of the standard or the reference material (g).

Using the relative method, it is fundamental to pay attention to the irradiation conditions in order to obtain the same neutron flux in the sample and in the standard.

6.3.2 Absolute method

Equation (2) represents an oversimplification of the real situation. The cross-section is a function of the incident neutron energy. Therefore, the product $\sigma \cdot \phi$ should be replaced by

$$\int \sigma(E) \cdot \phi(E) \cdot dE = \sigma_0 \cdot \phi_{th} + I \cdot \phi_{epi} \quad (15)$$

The integration limits go from thermal to rapid neutrons. However, due to the lower values of (n, γ) reaction cross-section for rapid neutrons and low rapid neutron flux when compared to those values on the thermal or epithermal region, it can be simplified as above. Here ϕ_{th} and ϕ_{epi} represent the thermal and epithermal neutron flux while σ_0 and I give the thermal cross-section and the resonance integral. The values of σ_0 and I are tabulated and can be found in the literature (e.g., Moens et al., 1979; Gryntakis and Kim, 1983).

Using stable reactors, only one flux monitor (see Section 6.2, this chapter) is necessary to evaluate the integral value, since the neutron spectrum is well defined for the different irradiation positions. Otherwise, two flux monitors with different I/σ_0 are used to estimate ϕ_{th} and ϕ_{epi}. Applying the values of ϕ_{th} and ϕ_{epi} obtained for each irradiation, the equation below can be applied to each radionuclide of interest produced in the sample:

$$c_s = \frac{\text{net peak area} \cdot \lambda \cdot M}{N_0 \cdot f \cdot \varepsilon \cdot b \cdot m_s \cdot (\sigma_0 \cdot \phi_{th} + I \cdot \phi_{epi}) \cdot (1 - e^{-\lambda \cdot t_i}) \cdot e^{-\lambda \cdot t_c} \cdot (1 - e^{-\lambda \cdot t_m})}$$

However, due to the lack of accurate nuclear data and because the neutron flux or the neutron spectrum is sometimes not well known for all irradiation positions, the absolute method lacks accuracy and is used more as a semiquantitative method (e.g., De Corte et al., 1982).

In reality, what is currently used is a way to combine the advantages of the INAA as a multielement method, without the limitations of the relative method and the lack of accuracy of the absolute method. There are two methods which are used: the k_0 method (Simonits et al., 1982) and the monostandard method (Kim and Stärk, 1970).

Using the k_0 method or the monostandard method, one of the two flux monitors above is also applied as a standard or a comparator for the determination of the elements of interest present in the sample. Our laboratory applies the monostandard method and flux monitors Al–Co and Al–Au wires. This application is based on our exchange program with the Radiochemistry Institute of the Technical University of Munich. The nuclear reactions involved are ^{59}Co (n, γ) ^{60}Co, $t_{1/2} = 7.27$ y, $\sigma_0 = 37.2$ barns and $I/\sigma_0 = 1.5$; ^{197}Au (n, γ) ^{198}Au, $t_{1/2} = 2.70$ d, $\sigma_0 = 98.8$ barns and $I/\sigma_0 = 15.8$. Co and Au have different I/σ_0 and are used to evaluate ϕ_{th} and ϕ_{epi}. Co is the standard for the quantification of the elements in the sample. The equation used in the monostandard method is (Kim, 1981)

$$c_s = \frac{m^*}{m_s} \cdot \frac{B^*}{B} \cdot \frac{S \cdot C}{C_1 \cdot C_2 \cdot C_3 \cdot C_4} \tag{16}$$

where, (*) is refereed to the standard:

$$\frac{B^*}{B} = \frac{\text{net peak area}^*}{\text{net peak area}} \cdot \frac{\lambda^*}{\lambda} \cdot \frac{(1-e^{-\lambda \cdot t_m})}{(1-e^{-\lambda \cdot t_m^*})}; \quad S = \frac{(1-e^{-\lambda \cdot t_i^*})}{(1-e^{-\lambda \cdot t_i})}$$

$$C_1 = \frac{M^* \cdot f}{M \cdot f^*}; \quad C_2 = \frac{b^*}{b}; \quad C_3 = \frac{\varepsilon^*}{\varepsilon}; \quad C_4 = \frac{\sigma^*}{\sigma}; \quad D = \frac{e^{-\lambda^* \cdot t_m^*}}{e^{-\lambda \cdot t_m}}$$

The cross-section (σ) here means the effective cross-section, which involves σ_0, I and ϕ_{th}/ϕ_{epi}.

It should be pointed out that the k_0 is more popular than the monostandard method, and other monitor combinations are utilized, mainly Al–Au wire and Zr foils (e.g., Simonits et al., 1980; Carmo Freitas and Martinho, 1989; Edtmann, 1990; Yuguang and Junlin, 1991).

6.4   Material and methods in activation analysis

A typical neutron activation analyses scheme is shown in Figure 6-2.

Dealing with the INAA, we have basically (n, γ) reactions. Therefore, the products are neutron-rich isotopes. This means they are β⁻ emitters. By properly adjusting the irradiation time, the cooling time, and the measuring or counting time, approximately fifty elements can be determined by INAA. The detection limits depend on many factors, as can be noted from equation (16). These factors will be discussed below. Table 6-2 shows which radionuclides are used in the INAA for the determination of elements in sediment samples. The table is based on studies by Nadkarni and Morrison (1978); Fong and Chatt (1987); Madaro and Moauro (1987); Song et al. (1987); Tian et al. (1987); Carmo Freitas and Martinho (1989); Guodong and Chunhan (1991); Ni and Tian (1991); Yuguang and Junlin (1991); and Smodis et al. (1993). The photopeak interferences experienced in the INAA of silicate samples are given by Potts (1983).

Many of the listed radionuclides emit gamma rays other than those shown in Table 6-2. It becomes clear that the high resolution gamma spectrometry is of primary importance in the application of the INAA.

By the end of the 1960s and at the beginning of the 1970s, the improvement of commercial manufacturing of large and long-term stable Ge(Li) detectors resulted in the replacement of the relatively poor resolution NaI(Tl). This allowed the development of the INAA and its success as a nondestructive multielemental instrumental analytical method. Ten years later, Ge(Li) was replaced by the high-purity Ge (HPGe) detector. At present, HPGe detectors covering an energy range from X-rays up to high-energy gamma rays and a relative efficiency of up to 100% are commercially available. A typical germanium detector

Table 6-2 Principal Characteristics of the Radionuclides Used in the Instrumental Neutron Activation Analysis of Sediment Samples

Element	Radionuclide	Half-life	E_γ (keV)	Element	Radionuclide	Half-life	E_γ (keV)
Na	^{24}Na	15 h	1368	Sb	^{122}Sb	2.72 d	564
Mg	^{27}Mg	9.5 min	1014	Sb	^{124}Sb	60.2 d	1691
Al	^{28}Al	2.3 min	1779	Ba	^{131}Ba	11.8 d	496
K	^{42}K	12.36 h	1525	Cs	^{134}Cs	2.1 y	796
Sc	^{46}Sc	84 d	889	La	^{140}La	40.3 h	487; 1596
Ca	^{47}Ca	4.54 d	1297	Ce	^{141}Ce	32.5 d	145
Ti	^{51}Ti	5.8 min	320	Nd	^{147}Nd	11.1 d	91; 531
Cr	^{51}Cr	27.7 d	320	Eu	^{152m}Eu	9.3 h	842
V	^{52}V	3.8 min	1434	Eu	^{152}Eu	12.7 h	1408; 122
Mn	^{56}Mn	2.58 h	847	Eu	^{154}Eu	16 y	725
Fe	^{59}Fe	45.1 d	1098	Sm	^{153}Sm	47 d	103
Co	^{60}Co	5.27 y	1332	Tb	^{160}Tb	72 d	299; 879
Zn	^{65}Zn	244 d	1115	Yb	^{175}Yb	4.2 d	395
Ga	^{72}Ga	14.1 h	834	Lu	^{177}Lu	6.8 d	208
Se	^{75}Se	119.8 d	264; 400	Lu	^{177m}Lu	161 d	208
As	^{76}As	1.1 d	559	Hf	^{181}Hf	42.2 d	482
Br	^{82}Br	36 h	776; 554	Ta	^{182}Ta	115 d	1222
Sr	^{85}Sr	64.8 d	514	Os	^{185}Os	94 d	646; 875
Rb	^{86}Rb	18.6 d	1077	W	^{187}W	23.9 h	686
Zr	^{95}Zr	65.5 d	724	Ir	^{192}Ir	74 d	317
Mo	^{99}Mo (^{99m}Tc)	66 h	140	Pt	^{193m}Pt	4.4 d	136
Ru	^{103}Ru	39.8 d	487; 610	Au	^{198}Au	2.7 d	412
Rh	^{104m}Rh	4.4 m	556; 1237	Hg	^{203}Hg	46.6 d	280
Pd	^{109}Pd	13.4 h	88	Th	^{233}Pa (^{233}Th)	27 d	312
Ag	^{110m}Ag	250 d	658; 885	U	^{239}Np (^{239}U)	2.36 d	228

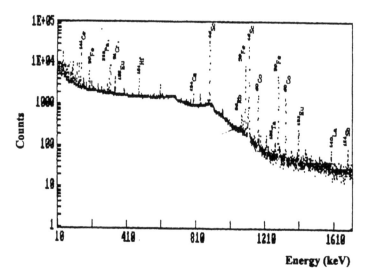

Figure 6-3   Gamma-ray spectrum obtained after the irradiation of a sediment sample.

has an energy resolution around 1.8 keV, for the 1332 keV ^{60}Co gamma line. For low energy gamma emitters, n-type coaxial germanium or planar detectors are necessary. The planar detectors are particularly useful for the low energy gamma emitters, since they reduce the spectral background from the high-energy photons (Debertin and Helmer, 1988). Depending on the crystal depth and diameter, planar detectors have 0.5 keV resolution for the 122 keV ^{57}Co gamma ray. There is much commercially available software available to analyze complex gamma spectrums and solve the great majority of peak overlapping. Their outputs serve as input data for activation analysis programs that give the concentration of the elements of interest in the sample as end products (e.g., Heft and Koszykowski, 1982).

Figure 6-3 shows a gamma ray spectrum of 250 mg sediment sample after 2 h irradiation at a neutron flux of 2×10^{12} n $\cdot$ cm^{-2} $\cdot$ s^{-1}, 2 weeks cooling time, and 4 h counting time, with a 15% relative efficiency HPGe detector. Figures 6-4 and 6-5 show different views of a counting room, including PC-based gamma spectrometers and germanium detectors inside lead shielding.

The term "relative efficiency" is a historical way to give an idea of the photopeak efficiency of a germanium detector. A 15% efficiency germanium detector has a counting efficiency for the 1332 keV ^{60}Co gamma ray, 15% of a 7.5 × 7.5 cm NaI(Tl) detector. Among the other types of interactions of a gamma ray with a germanium crystal, the photoelectric absorption is of principal interest, since it implies the total

Figure 6-4 Germanium detectors inside lead shielding, above, and a germanium detector with a sample holder utilized to fix the sample-detector distance, below.

Figure 6-5 PC-based counting room, above, and a monitor showing a spectrum in detail, below.

absorption of the gamma ray energy inside the detector crystal. As a consequence, the height of the produced electronic pulse is proportional to the gamma ray energy.

Since the counting efficiency ε is a function of the gamma ray energy and the sample geometry to the detector, it is necessary for each geometry to construct a counting efficiency curve as shown in Figure 6-6. Each curve is built with a mixture of primary calibration radionuclide standards (Debertin and Helmer, 1988). The use of primary calibration radionuclides is to avoid the so-called summing effect, which reduces the net peak area, giving a lower counting efficiency than the real one. The summing effect is an important factor to be considered in the activation analysis of sediment samples, using other methods than the relative one. The summing effect appears when counting geometries close to the detector are used, as they are utilized when the element of interest is present at a low concentration in the sample, and the radionuclide emits gamma rays in sequence. By increasing the

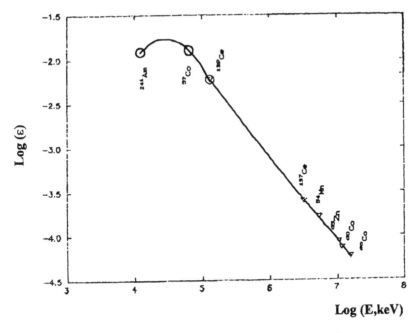

Log (E,keV)

Figure 6-6 Typical intrinsic germanium detector energy vs. efficiency curve.

distance of the sample-detector, the influence of the summing effect on the counting results is reduced. The problem of the summing effect and its consequence on the k_0 results are well discussed in the joint publications of the Activation Analytical Laboratories of the Central Research Institute for Physics, Budapest, Hungary, and the Institute for Nuclear Sciences, Gent, Belgium (Simonits et al., 1982; De Corte et al., 1987).

Following the scheme of Figure 6-2, the gamma spectrometry is the last step but one. The first step is the sample preparation for its irradiation, usually inside a research reactor. On the other hand, the sample preparation is connected with the choice of the conditions of the irradiation.

As indicated in Table 6-2, at least two different irradiation conditions are needed. The first one is a short irradiation using a few minutes of irradiation time and also a few minutes of cooling time. This short irradiation is usually used to determine elements such as Al and Mg, but depending on the concentration in the sample, it also can be used for Mn, Na, and Br. Using 2 min irradiation time and changing the cooling time from 2 to 60 min, Al, V, Cu, Ti, Mg, Mo, Rb, Sm, Th, U, Cl, Ba, Ni, Mn, Ba, Sr, K, and Na were determined in fossil fuels (Nadkarni, 1984). The second irradiation condition, a medium time irradiation, normally complements the first one. The irradiation time ranges from 1 h to a maximum of 4 h, and a variable cooling time is used, ranging from few days to one month, to avoid interferences. According to Nadkarni (1984), K, Na, Cu, As, Br, La, Sm, Hg, Mo, Sb, Ca, Yb, Lu, Nd, Ba, Rb, Th, Cr, Ce, Hf, Fe, Sr, Zr, Ni, Tb, Sc, Se, Ta, Zn, Ag, Cs, Co, and Eu can be determined by INAA under these conditions. Additionally, by epithermal neutron activation analysis, Ga, In, I, Ho, and W can also be determined by NAA.

The epithermal neutron activation analysis (ENAA) requires covering the sample with Cd foils (in some reactors there are Cd-lined irradiation sites) to cut off the thermal neutrons. For elements with high I/σ_0 ratio, such as U, Ba, Br, Cs, Rb, Sb, Sm, Ta, Tb, or Th (Alian and Sansoni, 1980; Song et al., 1987; Guodong and Chunhan, 1991), the ENAA appears to have more advantages than the conventional INAA, by providing lower detection limits and fewer spectral interferences. On the other hand, ENAA means higher costs and additional difficulties, due to the Cd-cover handling.

The preparation of the sediment sample for activation analysis is relatively simple. The sample is dried according to the elements of interest and fine ground. For the short and medium time irradiation, approximately 250 mg of sediment are sealed into high density polyethylene bags. Simultaneously, standards (for the relative method) and flux monitors are prepared by the same method as the samples. Figure 6-7 shows a set of samples and flux monitors ready for irradiation. The sample identification is arranged to be cut off after irradiation, decreasing the background signal. For higher irradiation time, high purity synthetic quartz (Suprasil) capsules should be used. The sealed

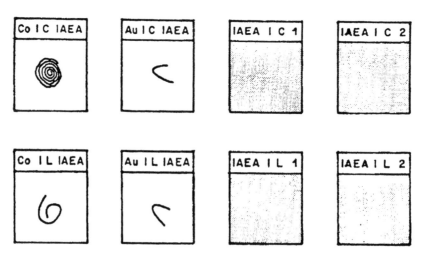

Figure 6-7 Sediment samples and gold and cobalt flux monitors ready for irradiation.

samples are then arranged on a polyethylene shuttle or Al-can and are sent to be irradiated under optimal conditions. If an epithermal irradiation is needed, the sample is covered with a Cd foil, taking care to refrigerate it.

After proper irradiation and cooling, the activated sample is placed over the detector; choosing the adequate counting time and geometry, the gamma spectrum is recorded and analyzed by a computer program. Special care is taken to avoid spectral interferences. Another computer program is used to calculate, from the obtained activity and the irradiation conditions, the concentration of the element of interest in the sample. Special care should be taken to avoid nuclear reaction interferences, as may be due to the uranium fission (e.g., Heft and Koszykowski, 1982; Tian et al., 1987; Das et al., 1989).

6.5 Analytical sensitivity, detection limits, and accuracy

In the INAA, sensitivity is defined as counts per µg of the element of interest. It is clear that all the factors involved in equation (16) can alter the sensitivity. However, factors such as cross-section, isotopic abundance, and half-life are fixed characteristics when choosing a specific nuclear reaction. The gamma ray emission probability is also a constant factor; however, a radionuclide can emit more than one gamma ray with different emission probability. Depending on the spectral interferences, the most adequate gamma ray is not always that of highest probability.

The higher the neutron flux, the higher is the sensitivity. But, in a nuclear reactor, the highest flux is obtained closer to the fuel elements, and here fast and epithermal components are also elevated. Therefore, when a more thermal flux is needed, the irradiation position chosen does not always have the highest possible neutron flux. Beside the neutron flux, the parameters that can really be changed to optimize the sensitivity are

t_i *(irradiation time)* — The saturation index, $(1 - e^{-\lambda t_i})$ shows that for long-lived produced radionuclide, the induced activity grows linearly with the irradiation time. For short-lived radionuclide, the produced activity does not change significantly for irradiation times higher than four half-lives. The irradiation time limits are determined by the sample recipient and by the reactor operation itself.

m_s *(sample amount)* — The sensitivity increases linearly with the sample mass. However, it is not possible to indefinitely increase the sample amount, since problems of flux gradient inside the sample occur, as well as self-absorption of low energy gamma rays. The typical sediment sample amount goes up to 1 g.

t_c *(cooling time)* — The cooling time is selected according to the half-life of the produced radionuclide, the gamma ray used for its measurement, the presence of interferences, and matrix effects. Normally, for one irradiation time, more than one cooling time is applied.

t_m *(counting time)* — The longer the counting time, the more impulses are accumulated by µg of one element, and the higher is the sensitivity. The detection limit, however, decreases only with the square root of the counting time. Therefore, by increasing the measurement time by a factor of four, the detection limit is improved by a factor of two. This, together with the number of detectors available, imposes limits to the counting time.

ε *(counting efficiency)* — The so-called counting geometry, together with the detector type, gives the counting efficiency. To improve counting efficiency, the sample must be measured closer to the detector, and it must have the highest possible relative efficiency. As was learned, the summing effect can impose restrictions to the counting close to the detector. Since the detector price is, among others, directly proportional to the relative efficiency, the detector size is a function of budget.

The detection limit defined as $\mu g_{\text{element of interest}} / g_{\text{sample}}$, will be, of course, a function of the parameters above. Beside those, the detection limit also depends on the background of the applied counting system in the spectrum region corresponding to the gamma ray to be analyzed. As shown in Figure 6-4, lead shielding is used to reduce this background. Special devices such as Anti-Compton shielding could also be utilized, mainly for low energy gamma rays. However, such systems can cost up to \$100,000 (U.S.), and, consequently, not all laboratories dealing with activation analyses can afford such a system.

Using the definition of Currie (1968), the detection limit, L_d, expressed in counts per time unit, is calculated according to equation (17). The result is applied to equation (16) to calculate it as $\mu g_{\text{element of interest}} / g_{\text{sample}}$. The background on the region of interest (s_b) is obtained from the irradiated blank sample.

$$L_d = \frac{2.73}{t_c} + 4.66 s_b \qquad (17)$$

The L_d definition above assumes that the blank and sample counting times are the same and assumes a confidence level of 95%. Figure

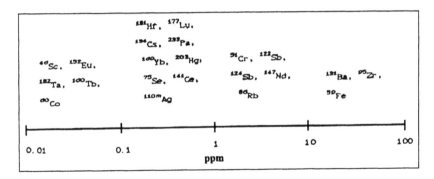

Figure 6-8   Detection limits range for several elements in a 250 mg sediment sample.

6-8 shows the detection limits range obtained for a 250 mg sediment sample analyzed as described in Section 6.4.

Another way to present the INAA detection limits and to demonstrate their applicability to the analysis of geological materials in special sediment samples is to show them in comparison to the average content of the elements in the earth's crust, as shown in Table 6-3. It is interesting to note here the influence of the irradiation and measurement conditions on the detection limits.

One can find much in the literature dealing with activation analysis accuracy when applied to environmental and geological samples (Kim and Born, 1973; Nadkarni and Morrison, 1978; Alian and Sansoni, 1980; Calmano and Lieser, 1981; Kim, 1981; Heft and Koszykowski, 1982; Xilei et al., 1984; Madaro and Moauro, 1985, 1987; De Corte et al., 1987; Fong and Chatt, 1987; Song et al., 1987; Tian et al., 1987; Carmo Freitas and Martinho, 1989a,b; Edtmann, 1990; Stegnar et al., 1991; Guodong and Chunhan, 1991; Ni and Tian, 1991; Obrusnik and Bode, 1992; Al-Jundi et al., 1993; Smodis et al., 1993; Cardoso and Godoy, 1995). This evaluation is normally undertaken using reference materials or through intercomparison studies.

As previously mentioned, the relative method offers, in general, better results than the k_0 or monostandard methods. The relative method accuracy is directly related to the standard accuracy. When using well established reference materials accuracy is high, if the conditions as previously outlined are followed. But the determination is limited by the certificated elements found in the standard. A synthetic standard may lead to preparation errors. An efficient way to prepare it seems to be the use of active carbon as described by Das et al. (1989).

The k_0 and the monostandard methods, which can be seen as absolute methods, mean that all parameters of equation (16) are under control. It is clear that it is not an easy task, and large errors can occur

Table 6-3   Instrumental Neutron Activation Analysis Detection
Limits in Relation to the Element's Average Concentration in the
Earth's Crust (values in µg/g, when not indicated)

Element	Crustal average*	Edtmann (1990)	Fong and Chatt (1987)	Guodong and Chunhan (1991)
Al	8.23%	—	130	0.02
As	1.8	0.05	9.7	0.08
Au	0.0004	0.0007	—	—
Ba	2.5	10	91	9
Br	425	0.04	4.9	0.09
Ca	4.15%	0.12%	0.3%	0.16%
Ce	60	0.3	0.26	0.3
Cl	130	—	276	80
Co	25	0.04	0.14	0.04
Cr	100	0.5	1.17	0.3
Cs	3	0.1	0.24	0.05
Dy	3	—	2.1	—
Eu	1.2	0.02	0.059	0.01
Fe	5.63%	0.004	105	30
Ga	15	0.6	1.2	—
Hf	3	0.08	0.33	0.03
K	2.09%	0.01%	0.09%	0.04
La	30	0.02	0.85	0.04
Lu	0.5	0.12	0.025	0.007
Mg	2.33%	—	0.08%	0.09%
Mn	950	—	16	6.5
Na	2.36%	1.1	0.036%	20
Rb	90	2.5	1.1	1
Sb	0.2	0.2	0.18	0.02
Sc	22	0.004	0.019	0.004
Se	0.05	0.7	4.3	—
Sm	6	—	0.15	0.009
Sr	375	50	100	30
Ta	2	0.06	0.077	0.02
Tb	0.9	0.035	0.10	0.01
Th	9.6	0.06	1.6	0.04
Ti	0.57%	—	431	220
U	2.7	0.00015	0.39	0.06
V	135	—	4.8	1
W	1.5	0.1	1.6	—
Yb	3	0.02	0.67	0.04
Zn	70	4.2	220	3

* Values from Taylor, S.R., *Geochem. Cosmochem. Acta*, 28, 1273, 1964.

(De Corte et al., 1987). Problems such as accuracy of the nuclear data, summing effects during the sample counting, cross-section deviation of the 1/v law, and non 1/E epithermal neutron flux distribution have to be under control. When it occurs, an accuracy around 5% can be achieved (De Corte et al., 1987).

6.6   Application to sediment samples

Previously, there was a tendency to believe that a new analytical technique would replace all others. Today, other multielemental methods, such as the ICP techniques or total-reflection X-ray fluorescence spectrometry (TXRF), are available. However, it is clear that each technique has its preferable application (Michaelis, 1986; Wennrich et al., 1988; Koopmann and Prange, 1991; Randle et al., 1993), and one is complementary to the other.

The INAA, when applied to a sediment sample, has the advantage of being a nondestructive, multielement method. Therefore, the uses of the INAA are derived from these characteristics. Among those we can find are environmental characterization of areas (Moore and Propheter, 1973; Nadkarni and Morrison, 1978; Calmano and Lieser, 1981; Sherief et al., 1981; Hedrich, 1983; Watterson et al., 1983; Li and Qian, 1986; Iovtchev et al., 1987; Zikovsky et al., 1987; Garcia Montero, 1988; James and Boothe, 1988; Van and Teherani, 1989; Guodong and Chunhan, 1991; Yuguang and Junlin, 1991; Crespi et al., 1993; Fernandes et al., 1994; Orlando et al., 1994), pollutant sources identification (Teherani et al., 1981; Drabaek et al., 1987; Schönburg, 1987; Golchert et al., 1991; Kalegeropoulos et al., 1993; Yusof and Wood, 1993), and rare earth elements distribution (Yamakoshi, 1984; Madaro and Moauro, 1987; Meloni et al., 1987; Peterson et al., 1987; Tian et al., 1987; Minai and Tominaga, 1989; Yuguang and Junlin, 1991; Crespi et al., 1993).

The environmental characterization of areas, or the pollutant source identification, based on the INAA of sediment samples, can be more than a simple survey. Its multielement character allows the application of multivariate statistics as discriminant, cluster, and principal factor analysis (Watterson et al., 1983; Crespi et al., 1993; Kalogeropoulos et al., 1993; Fernandes et al., 1994).

In both cases, it is also usual to combine the INAA with other techniques, such as sediment dating (Drabaek et al., 1987; Schönburg, 1987; Fernandes et al., 1993) or a complementary analytical method (Teherani et al., 1981; Schönburg, 1987; Fernandes et al., 1993; Yusof and Wood, 1993), especially to assess elements such as Pb and Cd.

Although up to 11 rare-earth elements (REE) can be determined by INAA in sediments (Song et al., 1987), usually between 6 and 9 REE are analyzed this way (Yamakoshi, 1984; Madaro and Moauro, 1987; Meloni et al., 1987; Tian et al., 1987; Peterson et al., 1987; Minai and

Tominaga, 1989; Yuguang and Junlin, 1991; Crespi et al., 1993). The REE are considered as a powerful tool to define the importance of remobilization mechanisms during diagenesis and weathering of sediments (Meloni et al., 1987). In order to understand these processes, the REE results are chondrite- or shale-normalized (Madaro and Moauro, 1987; Meloni et al., 1987; Minai and Tominaga, 1989; Yuguang and Junlin, 1991; Crespi et al., 1993).

Related to the sediment studies are other applications of the INAA technique. Some examples are

- REE phase distribution in a deep-sea carbonate sediment, where the INAA is associated with acid selective leaching (Peterson et al., 1987);
- Sorption and desorption water-sediment processes followed by INAA (Vertacnik and Biscan, 1993);
- Elements fractioning according to the suspended matter sedimentation rates (Niedergesäss et al., 1987).

6.7　Final remarks

The relative method can be used by many groups dealing with sediment studies, particularly for the determination of volatile elements such as Se, As, Sb, and Hg that produce intermediate or long-lived radionuclides. Those groups that already have germanium detectors can send their samples to be irradiated on a research nuclear reactor and perform the gamma spectrometry themselves. Germanium detectors are relatively popular and have been widely used for ^{210}Pb and ^{137}Cs dating.

For activation analysis, a growing difficulty is aging reactors. Since the great majority of research reactors are approximately 30 years old, many have been shut down. As a consequence, in the future, the main problem may be to find a reactor to irradiate the sample that is not too far away from the laboratory. Information about the nearest research reactor in operation can be found in the nuclear engineering departments of universities.

References

Alian, A. and Sansoni, B., Instrumental neutron activation analysis of geological and pedological samples: further investigation of epithermal neutron activation analysis using monostandard method, *J. Radioanal. Chem.*, 59, 511, 1980.

Al-Jundi, J., Randle, K., and Earwaker, L.G., Elemental analysis of the marine sediment reference materials MESS-1 and PACS-1 by instrumental neutron acivation analysis, *J. Radioanal. Nucl. Chem. Art.*, 174, 145, 1993.

Bitelli, U.d'U., Medida e Cálculo da Distribuição Espacial e Energética de Nêutrons no Núcleo do Reator IEA-R1, Masters Thesis, Instituto de Pesquisas Energéticas e Nucleares, São Paulo, Brazil, 1988.

Calmano, W. and Lieser, K.H., Trace element determination in suspended matter and in sediments by instrumental neutron analysis, *J. Radioanal. Chem.*, 63, 335, 1981.

Cardoso, K.M. and Godoy, J.M., Estudo do método monopadrão na análise por ativação neutrônica, *Química Nova*, 18, 435, 1995.

Carmo Freitas, M. and Martinho, E., Determination of trace elements in reference materials by the k_0-standardization method, *Talanta*, 36, 527, 1989a.

Carmo Freitas M. and Martinho, E., Neutron activation analysis of reference materials by the k_0-standardization and relative methods, *Anal. Chim. Acta*, 219, 317, 1989b.

Crespi, V.C., Genova, N., Tositti, L., Tubertini, G., Bettoli, G., Oddone, M., Meloni, S., and Berzero, A., Trace elements distribution in antarctic sediments by neutron activation analysis, *J. Radioanal. Nucl. Chem. Art.*, 168, 107, 1993.

Currie, L.A., Limits for qualitative detection and quantitative determination, *Anal. Chem.*, 40, 586, 1968.

Das, H.A., Faanhof, A., and Van der Sloot H.A., *Radioanalysis in Geochemistry*, Developments in Geochemistry 5, Elsevier Science, New York, 1989, 482.

Debertin, K. and Helmer, R.G., *Gamma- and X-ray Spectrometry with Semiconductor Detectors*, Elsevier, North Holland, 1988.

De Corte, F., Moens, L., Simonits, A., Sordo El-Hammani, K., De Wispelaere, A., and Hoste J., The effect of the epithermal neutron flux distribution on the accuracy of absolute and comparator standardization methods in (n, γ) activation analysis, *J. Radioanal. Chem.*, 72, 275, 1982.

De Corte, F., Simonits, A., De Wispelaere, A., and Hoste J., Accuracy and applicability of the k_0-standardization method, *J. Radioanal. Nucl. Chem. Art.*, 113, 145, 1987.

Drabaek, I., Eichner, P., and Rasmussen, L., Concentrations of rare earth elements in sediments, mussels and fish from a Danish marine environment, Lillebaelt, *J. Radioanal. Nucl. Chem. Art.*, 114, 29, 1987.

Edtmann, G., Applications of instrumental neutron activation analysis in the analytical division of the research center, Jülich (KFA), Juel-2561, Forschungszentrums Jülich GmbH, Zentralbibliotek, Postfach 1913, D-5170 Jülich, Federal Republic of Germany, 1990.

Fernandes, H.M., Cardoso, K.M., Godoy, J.M., and Patchineelan, S.R., Cultural impact on the geochemistry in Jacarepagua Lagoon, Rio de Janeiro, Brazil, *Environ. Technol.*, 14, 94, 1993.

Fernandes, E.A.N., Ferraz, E.S.B., and Oliveira, H., Trace elements distributions in the Amazon floodplain soils, *J. Radioanal. Nucl. Chem. Art.*, 179, 251, 1994.

Fong, B.B. and Chatt, A., Characterization of deep sea sediments by INAA for radioactive waste management purposes, *J. Radioanal. Nucl. Chem. Art.*, 110, 135, 1987.

Garcia Montero, G., Metodología para el análisis por activación neutrónica de sedimentos marinos de la plataforma de Cuba, *Nucleus*, 4, 12, 1988.

Golchert, B., Landsberger, S., and Hopke, P.K., Determination of heavy metals in the Rock River (Illinois) through the analysis of sediments, *J. Radioanal. Nucl. Chem. Art.*, 148, 319, 1991.

Gryntakis, E.M. and Kim, J.I., A compilation of resonance integral from hydrogen to fermium, *J. Radioanal. Chem.*, 76, 341, 1983.

Guodong, L. and Chunhan, T., Instrumental NAA of carbonate rocks and its application for the analysis of secondary sedimentary environments in shallow sea, *J. Radioanal. Nucl. Chem. Art.*, 147, 363, 1991.

Hedrich, E., Determination of layers in the sediment of Neusiedler Sea by neutron activation analysis, *Mikrochim. Acta*, I, 1, 1983.

Heft, R. and Koszykowski, R., Analysis of standard reference materials by absolute INAA, *J. Radioanal. Chem.*, 72, 245, 1982.

Iovtchev, M., Kinova, L., Nikolov, P., and Apostolov, D., Bestimmung von spurenelementen in proben maritiner Herkunft, *Isotopenpraxis*, 23, 346, 1987.

International Atomic Energy Agency, Dosimetry for Criticality Accidents: A Manual, Technical Report Series 211, 1982, 196.

International Commission on Radiation Units and Measurements, Neutron Fluence, Neutron Spectra and Kerma, ICRP Report 13, 1969.

James, W.D. and Boothe, P.N., Ocean-sediment analysis by neutron activation, *J. Radioanal. Nucl. Chem. Art.*, 123, 295, 1988.

Kalegeropoulos, N., Kilikoglou, V., Vassilaki, M., and Grimanis, A.P., Application of two INAA methods to pollution studies of sediments from Saranikus Gulf, Greece, *J. Radioanal. Nucl. Chem. Art.*, 167, 369, 1993.

Kamel, L., Godoy, J.M., and Nicolli, I., Accumulation of Cr, Cs, Fe, and Co in fishes and shrimps of Piraquara de Dentro Bay, Angra dos Reis, Brazil, in Second Rio Conference on the Chemistry of Tropical Marine Systems, Rio de Janeiro, Brazil, 1987.

Kim, J.I., Monostandard activation analysis: evaluation of the method and its accuracy, *J. Radioanal. Chem.*, 63, 121, 1981.

Kim, J.I. and Born, H.J., Monostandard activation analysis and its application: analysis of Kale Powder and NBS standard glass samples, *J. Radioanal. Chem.*, 13, 427, 1973.

Kim, J.I. and Stärk, H., Proceedings in Activation Analysis in Geochemistry and Cosmochemistry, NATO, Advanced Study Institute, Kjeller, Norway, 1970.

Koopmann, C. and Prange, A., Multi-element determination in sediments from the German Wadden Sea — investigation on sample preparation techniques, *Spectrochim. Acta*, 46B, 1395, 1991.

Li, X.X. and Qian, X.Z., Instrumental neutron activation analysis of suspended matters and sediments, in Papers Accepted for Presentation at 7th International Conference MTAA, Pt. 2, 1197, 1986.

Lyon, W.S., *Guide to Activation Analysis*, D. van Nostrand, New York, 1964.

Madaro, M. and Moauro, A., Instrumental neutron activation analysis results in an intercomparison campaign on lake and river sediments, *J. Radioanal. Nucl. Chem. Art.*, 90, 129, 1985.

Madaro, M. and Moauro, A., Trace elements determination in rocks and sediments by neutron activation analysis, *J. Radioanal. Nucl. Chem. Art.*, 114, 337, 1987.

Meloni, S., Genova, N., Oddone, M., Oliveri, F., and Vannuci, R., Rare-earth elements abundance and distribution in pelagic sediments by instrumental neutron activation analysis, *J. Radioanal. Nucl. Chem. Art.*, 112, 507, 1987.

Michaelis, W., Multielement analysis of environmental samples by total-reflection x-ray fluorescence spectrometry, neutron activation analysis and inductively coupled plasma optical emission spectrometry, *Fresenius Z. Anal. Chem.*, 342, 662, 1986.

Minai, Y. and Tominaga, T., Neutron activation analysis of rare earth elements in deep-sea sediments from the Pacific Ocean and the Japan Sea, *J. Radioanal. Nucl. Chem. Lett.*, 137, 351, 1989.

Moens, L., Simonits, A., De Corte, F., and Hoste, J., Comparative study of measured and critically evaluated resonance integral to thermal cross-section ratios, *J. Radioanal. Chem.*, 54, 377, 1979.

Moore, R.V. and Propheter, O.V., Neutron activation analysis of bottom sediments, in Environmental Protection Technology Series, Environmental Protection Agency, Athens, GA, 1973.

Nadkarni, R.A., Nuclear activation methods for the characterization of fossil fuels, *J. Radioanal. Nucl. Chem. Art.*, 84, 67, 1984.

Nadkarni, R.A. and Morrison G.H., Multielement analysis of lake sediments by neutron activation analysis, *Anal. Chim. Acta*, 99, 133, 1978.

Ni, B. and Tian, W., NAA of IAEA sediment RM SL-3 for 45 elements, *J. Radioanal. Nucl. Chem. Art.*, 151, 159, 1991.

Niedergesäss, R., Racky, B., and Schnier, C., Instrumental neutron activation analysis of Elbe River suspended particulate matter separated according to the settling velocities, *J. Radioanal. Nucl. Chem. Art.*, 114, 57, 1987.

Obrusnik, I. and Bode, P., Improved ease of operation in epithermal NAA by irradiation in plastic capsules and use of well-type Ge-spectrometry: A feasibility study, *J. Radioanal. Nucl. Chem. Art.*, 158, 343, 1992.

Orlando, R.C., Fernandes, E.A.N., and Szikszay, M., Trace elements in a profile of the unsaturated zone of the São Paulo Basin, *J. Radioanal. Nucl. Chem. Art.*, 179, 259, 1994.

Peterson, M.L., Tobler, L., and Wyttenbach, A., Rare earth element phase distributions in a deep-sea carbonate sediment, *J. Radioanal. Nucl. Chem. Art.*, 112, 515, 1987.

Potts, P.J., Gamma-ray photopeak interferences found in the instrumental neutron activation analysis, *J. Radioanal. Chem.*, 79, 363, 1983.

Randle, K., Al-Jundi, J., Mamas, C.J.V., Sokhi, R.S., and Earwaker, L.G., A comparative study of neutron activation analysis and proton-induced x-ray emission analysis for the determination of heavy metals in estuarine sediments, *Nucl. Instr. Meth. Phys. Res.*, B79, 568, 1993.

Schönburg, M., Radiometrische Datierung und Quantitative Elementbestimmung in Sediment-Tiefen-Profilen mit Hilfe Kernhysikalischer Sowie Röntgenfluoreszenz- und Atom-emissionsspektrometrischer Verfahren, Doctoral thesis, Institute of Physics, University of Hamburg, Federal Republic of Germany, 1987.

Sherief, M.K., Awadallah, R.M., and Grass, F., Trace elements in sediment samples of the Aswan High Dam Lake, *Chem. Erde*, 40, 178, 1981.

Silva, K.M.C., Estudo do Método Mono-padrão na Análise por Ativação Neutrônica, Masters Thesis, Departamento de Físico-Química, Universidade Federal do Rio de Janeiro, Rio de Janeiro, Brazil, 1991.

Simonits, A., Moens, L., De Corte, F., De Wispelaere, A., Elek, A., and Hoste, J., k_0 measurements and related nuclear data compilation for (n, γ) reactor neutron activation analysis, *J. Radioanal. Chem.*, 60, 461, 1980.

Simonits, A., De Corte, F., Moens, L., and Hoste, J., Status and recent development in the k_0-standardization method, *J. Radioanal. Chem.*, 72, 209, 1982.

Smodis, B., Jacimovic, R., Medin, G., and Jovanovic, S., Instrumental neutron activation analysis of sediment reference materials using the k_0-standardization method, *J. Radioanal. Nucl. Chem. Art.*, 169, 177, 1993.

Song, L., Macmahon, T.D., and Ward N.I., Determination of 34 elements in Chinese geochemical standard reference materials by instrumental neutron activation analysis, *J. Radioanal. Nucl. Chem. Art.*, 113, 285, 1987.

Stegnar, P., Smodis, B., and Jacimovic R., The k_0-standardization method of NAA as a tool for studying environmental pollution problems, in Coordinate Research Program on the Use of Nuclear and Nuclear-Related Techniques in the Study of Environmental Pollution Associated with Solid Wastes, International Atomic Energy Agency, Vienna, Austria, 1991.

Taylor, S.R., Abundance of chemical elements in the continental crust: a new table, *Geochem. Cosmochem. Acta*, 28, 1273, 1964.

Teherani, D.K., Wallish, G., Alltmann, H., and Putz, M., Untersuchungen von sedimenten aus stauräumen der drau unter besonderer berücksichtigung der schwermetalle, *Radiochem. Radioanal. Lett.*, 46, 57, 1981.

Tian, J., Chou, C., and Ehmann, W.D., INAA determination of major and trace elements in loess, paleosol and precipitation layers in a Pleistocene loess section, China, *J. Radioanal. Nucl. Chem. Art.*, 110, 261, 1987.

Van, L.T. and Teherani, D.K., Determination of trace elements and heavy metals in a lake sediment by neutron activation analysis, *J. Radioanal. Nucl. Chem. Lett.*, 135, 435, 1989.

Vertacnik, A. and Biscan, J., Behaviour of some macroelements, trace elements and RE in the water-sediment system, *J. Radioanal. Nucl. Chem. Lett.*, 175, 401, 1993.

Watterson, J.I.W., Sellschop, J.P.F., Erasmus, C.S., and Hart, R.J., The combination of multi-element neutron activation analysis and multivariate statistics for characterization in geochemistry, *Int. J. Appl. Radiat. Isot.*, 34, 407, 1983.

Wennrich, R., Niebergall, K., and Zwanziger, H., Geochemische untersuchungen an kupferschiefer mittels INAA und quantitative multielement-verteilungsanalyse am geologischen kompaktmaterial unter verwendung der Laser-ICP-AES, *Isotopenpraxis*, 24, 97, 1988.

Xilei, L., De Corte, F., Moens, L., Simonits, A., and Hoste, J., Computer-assisted reactor NAA of geological and other reference materials, using the k_0-standardization method; evaluation of the accuracy, *J. Radioanal. Chem.*, 81, 333, 1984.

Yamakoshi, K., Chemical composition of magnetic, stony spherules from deep-sea sediments determined by instrumental neutron activation analysis, *Geochem. J.*, 18, 147, 1984.

Yuguang, W. and Junlin, Z., Determination of rare earths and other trace elements in samples of Antarctica by neutron activation analysis, *J. Radioanal. Nucl. Chem. Art.*, 151, 345, 1991.

Yusof, A.M. and Wood, A.K.H., Environmental assessment of coastal sediments by the elemental rationing technique, *J. Radioanal. Nucl. Chem. Art.*, 167, 341, 1993.

Zikovsky, L., Kennedy, G., and Lasalle, P., The lognormal distribution of trace elements in glacial till, *J. Radioanal. Nucl. Chem. Art.*, 109, 275, 1987.

chapter seven

Determination of nutrients in aquatic sediments

Haig Agemian

7.1 Introduction

Carbon, nitrogen, and phosphorus are the most significant macronutrients studied in aquatic ecosystems. These elements move throughout the environment in their respective cycles and are distributed in the atmosphere, hydrosphere, biosphere, and geosphere in different organic and inorganic species, with concentrations that vary as a function of temporal and spacial conditions (Svensson and Soderlund, 1976; Bolin and Cook, 1983). In the aquatic environment, they are distributed between the water and sediment interface in both dissolved and particulate forms. Bioavailable nutrients are taken up and metabolized by aquatic organisms in the life cycle. In a specific aquatic ecosystem, nutrient dynamics partition these elements among water, sediment, and biota to attain a natural balance. This balance may change as nutrients are introduced from agricultural, industrial, and urban sources. The nutrients in soil are studied for their value to agricultural products. On the other hand, studies of nutrients in aquatic sediments are mainly related to the assessment of the effects of their increasing concentrations originating from different sources. Accumulation of nutrients in water and sediments can result in eutrophication of lakes, ponds, and rivers. The eutrophication is often accompanied by depletion of oxygen in the water and decrease of biodiversity in the affected water body (Vollenweider, 1968; Foehrenbach, 1973; Sutcliffe and Jones, 1992).

A comprehensive understanding of nutrient behavior in aquatic ecosystems requires their study in both the water and sediments. Bonanni et al. (1992) showed that sediments play an important role in the accumulation and regeneration of nutrients. Organic matter produced by phytoplankton in eutrophic shallow lakes settle to the sediment and decompose by aerobic and anaerobic processes, during which different carbon, nitrogen, and phosphorus compounds are produced (Anderson and Jensen, 1992). Furthermore, decomposing organic matter affects changes in oxygen concentrations and redox potential, which in turn affects nitrogen and phosphorus release from the sediments to the overlying water. Sediments are often thought of as nutrient sinks by their capability to adsorb and accumulate different forms of the nutrients (Bailey, 1968; Mulholland and Elwood, 1982). Bailey (1968) compared soils and geological strata to a three-dimensional anisotropic, chromatographic column, which absorbs and removes forms of nitrogen and phosphorus from the liquid phase. Relative to nitrogen, phosphorus is immobile and fixed to soil and sediment particles. Most phosphorus-bearing compounds reside in the sediments, and a relatively small fraction remains in the water. Mulholland and Elwood (1982) concluded that although the role of freshwater sediments as a sink in the global carbon cycle is small, compared to that of terrestrial systems and the oceans, it is, nevertheless, significant in the fixation of carbon. Furthermore, freshwater ecosystems have a large fraction of the sediment carbon fixed in organic forms compared to that in the oceans. The cycle of phosphorus is relatively slower than that of carbon and nitrogen. Unlike carbon and nitrogen, the phosphorus cycle is normally in one direction from rock deposits to the aquatic system, and would take millions of years to complete (Holtan et al., 1988).

Sediment particles in lakes and rivers are constantly resuspended and redeposited, depending on environmental conditions. The nutrient content in the particles is important in their transport from the bottom sediments into overlying waters. Suspended particles (also called suspended sediments) are sampled from the water column by filtration and analyzed for nutrients by analytical methods designed for sediments. The nutrients in the suspended particles are called particulate carbon, nitrogen, phosphorus, etc., and the knowledge of their concentrations is used to understand their movement within the aquatic ecosystem, particularly at the sediment–water interface. Analytical considerations for determination of nutrients in suspended particles are identical to those for bottom sediments. Differences in the methods of their determination are only in the sampling techniques (Keith, 1988; Mudroch and MacKnight, 1994; Mudroch and Azcue, 1995).

The main carbon species in suspended and bottom sediments include carbohydrates, fats, and proteins from decomposing biota, humic acids from decomposing humus, and carbonates from detritus. The ecological significance of carbon as a nutrient is through its organic forms. The concentration of total organic carbon is often used in correlation with other elements. For example, the carbon to nitrogen and carbon to phosphorus ratio is used to characterize the association of nitrogen and phosphorus in organic matter. Nitrogen species include organic nitrogen, ammonia (NH_3), nitrate (NO_3^-) and nitrite (NO_2^-). Biological activity in living and dead tissue produces reduced forms of organic nitrogen, ranging from simple amines to complex proteins. Ammonia is the most reduced form of inorganic nitrogen, and is the product of decomposition of organic matter. Bacterial oxidation in the nitrification cycle produces nitrite and nitrate. Phosphorus species in the environment include organic phosphorus compounds, inorganic phosphate, and mineralized inorganic complexes with iron, calcium, and aluminum. Phosphorus precipitates to form low solubility compounds and metallic complexes, and is relatively immobile compared to carbon and nitrogen. The natural abundance of the nutrients of interest is carbon > nitrogen > phosphorus. Due to this sequence, phosphorus is often considered to be the limiting nutrient in the ecological cycle.

Several analytical techniques, such as distillation/titration, colorimetry, ion chromatography, ion selective electrodes, microdiffusion, and combustion analyzers have been used to determine carbon, nitrogen, and phosphorus in sediments (Jackson, 1958; Black, 1965; Hesse, 1971; Page et al., 1982). Detection systems include ultraviolet/visible (UV/VIS) and infrared (IR) absorption, chemiluminescence, conductometry, potentiometry, and emission photometry (D'Elia, 1983; Dionex, 1987; Renmin et al., 1995). Determination of the nutrients in sediment samples requires a sample treatment by dry combustion or wet chemical digestion to either separate different nutrient phases or species from each other or to solubilize and oxidize the sample and provide an extract ready for quantitative determination by the appropriate instrumental technique. For example, the concentrations of total carbon and nitrogen in sediments can be determined by dry combustion using the carbon/hydrogen/nitrogen (CHN) analyzer. Furthermore, total carbon, nitrogen, and phosphorus, as well as their various forms, can be isolated for measurement by treating sediment samples with oxidants, reducing agents, acids, and salt solutions to dissolve the mineral phase in which the elements occur in the sediment. In this chapter, dry combustion and wet digestion methods are described for the determination of total carbon, nitrogen, and phosphorus, as well as extraction techniques for their forms/phases, such as NO_3^-, NO_2^-, NH_4^+, urea, organic nitrogen, loosely bound phosphorus, polyphosphates, organic phosphorus, phosphorus bound to aluminum, iron and calcium, and refractory phosphorus. The goal is not to present every possible analytical technique, since many such reviews are available (for example, Page et al., 1982). Rather, the recent state-of-the-art methods that use rapid multisample preparation techniques coupled with automated analytical systems with good sensitivity and precision are described in this chapter. The methods include efficient sample digestion systems, such as hot blocks, followed by segmented flow automated analytical systems using sensitive colorimetric techniques. These types of speedy and cost-effective systems have the significant advantage of allowing a large number of quality control samples to be analyzed concurrently with a batch of samples. As a result, statistically processed quality control can be implemented on an ongoing basis with control charts and established control limits and confidence intervals. Accuracy and precision can be optimized and maintained at the required levels without jeopardizing laboratory productivity and safety.

7.2 Chemical forms of nutrients in aquatic sediments: literature review

7.2.1 Nutrients in suspended particulate matter (i.e., suspended sediments)

Meybeck (1982) reviewed various forms of dissolved and particulate carbon, nitrogen, and phosphorus transported by world rivers. It was concluded that dissolved organic carbon (DOC) and particulate organic carbon (POC) are dependent on environmental conditions and level of suspended matter, respectively, and vary widely. Transported nitrogen forms include NH_4^+, NO_3^-, and NO_2^-. However, most river-borne nitrogen is related to particulate organic nitrogen (PON). The carbon to nitrogen ratio in river particulate matter was shown to be very constant and similar to surficial soils. Particulate phosphorus was found to represent 95% of the total phosphorus naturally carried by rivers, and approximately 40% of this phosphorus was in organic forms. Saunders and Lewis (1988) studied the transport of carbon, nitrogen, and phosphorus in tropical rivers and found that 68% of phosphorus, 54% of nitrogen, and 37% of carbon were in particulate forms. They also found that seasonal pattern in nutrient chemistry correlates highly with the hydrological cycle. During the high wet season, the contents of particulate carbon, nitrogen, and phosphorus increased sharply. The flood plain stored sediment, which was later remobilized. Andersen and Jensen (1992) studied sedimentation of phosphorus and nitrogen originating from decomposing seston, and observed that 47% phosphorus and 43% nitrogen in the particulate form were mineralized under constant rates. Furthermore, their comparison of carbon to nitrogen to phosphorus ratios in seston and sediments and calculated flux rates showed that nitrogen was mineralized faster than carbon and phosphorus.

7.2.2 Chemical forms of carbon, nitrogen, and phosphorus in sediments

7.2.2.1 Carbon species in sediments

The global carbon cycle is basic to all life processes (Trabalka and Reichle, 1986; Riemann and Sondergaard, 1986). In the sediment–water system, carbon cycles together with nitrogen and phosphorus by incorporation of inorganic forms into organic materials with subsequent mineralization of the organic material to specific inorganic species. In their review of the carbon cycle in sediment–water systems, Kerr et al. (1973) described the mechanism by which CO_2 and/or HCO_3^- were converted into organic carbon by autotrophic organisms. The organic matter gathered in pools of suspended particles, which were eventually incorporated into the sediment layer. Nitrogen, phosphorus, and other nutrients cycle simultaneously with carbon so that the decomposition of organic matter results in increased availability of all nutrients. The stoichiometric model of the mineralization of organic matter has been used to explain the regeneration, distribution, and interrelationship of carbon, nitrogen, and phosphorus (Redfield et al., 1963). The elemental composition of phytoplankton determines the relative abundance of the biophilic elements in the aquatic ecosystem. For example, the average chemical composition of marine plankton is $C_{106}(H_2O)_{106}(NH_3)_{16}PO_4$ (Wada and Hattori, 1991).

Kerr et al. (1973) concluded that organic-carbonate particles are sinks for both organic and inorganic carbon. Organic matter in surface sediments plays a major role in influencing community structure and metabolism of benthos (Graf et al., 1983). Types of organic carbon found in sediments include different compounds, such as proteins, carbohydrates, and lipids (Fabiano and Danovaro, 1994). Of these, simple sugars, fatty acids and proteins are labile compounds and rapidly mineralize. Other compounds, such as humic and fulvic acids and complex carbohydrates, are refractory and break down slowly (Fry, 1987). Hongve (1994) found that particulate nitrogen correlated well with POC. The sediment had a higher carbon to nitrogen ratio than seston, which indicated more effective mineralization of nitrogen than carbon. Hendricks and Silvey (1973) found a high degree of correlation between carbon and nitrogen but a very poor one between carbon and phosphorus. This indicated that nitrogen was prevalently in organic forms, and phosphorus was mainly associated with inorganic forms. Martinova (1993) confirmed this finding by showing that nitrogen accumulated in sediments was mostly in organic forms (over 90%), and depended on inputs of organic matter originating from phytoplankton or macrophytes. On the other hand, accumulated phosphorus in sediments was mostly in inorganic forms, such as apatite and residual phosphorus, originating from weathering of rocks.

7.2.2.2 Nitrogen species in sediments

The nitrogen cycle in sediment–water systems involves several biochemical processes, which include aminization, ammonification, nitrification, and mineralization (Bailey, 1968; Keeney, 1973; Delwiche, 1981; Wada and Hattori, 1991). Aminization of proteins by microbial enzymatic action produces complex amino compounds ($R–NH_2$). Urea is a common form of organic nitrogen and is normally produced by several biochemically important mechanisms, including hydrolysis of arginase, decay of dead fish, and excretion from zooplankton (Wada and Hattori, 1991). The information in scientific literature indicated that 45% of organic nitrogen in sediments consisted of amino acids in the form of proteins, peptides, and humic and fulvic compounds. Due to the diversity of organic forms of nitrogen, ecologists often use total organic nitrogen (TON) as a measure of the extent of organic nitrogen. Further, decomposition of organic matter is called ammonification, which produces NH_3, which is the most reduced form of nitrogen. In aqueous solution, NH_3 is in equilibrium with the ammonium ion (NH_4^+), which is its ionized form. Oxidation of NH_3 to NO_3^- and NO_2^- is the process called nitrification. Nitrite is the intermediate in the oxidation of NH_3 to NO_3^-. The conversion of organic to inorganic nitrogen is called nitrogen mineralization. Unlike phosphorus, nitrogen is not appreciably mineralized in forms that are associated with inorganic phases of metals in sediments.

7.2.2.3 Phosphorus species in sediments

Early limnologists (Einsele, 1936, 1938; Mortimer, 1941, 1942) studied iron–phosphorus interaction and the coupling of mobilization and redox conditions to explain iron–phosphorus binding in sediments and suspended matter. Gachter et al. (1988) and Gachter and Meyer (1993) discussed the role of bacteria and microorganisms in the release, mobilization, and fixation of phosphorus in sediments through the synthesis of polyphosphates in biological cells. They concluded that benthic bacteria do more than just mineralize organic phosphorus compounds, by regulating the flux of phosphorus across the sediment–water interface and contribute to fixation in refractory phosphorus compounds. This phenomenon is based on the classical Mortimer-Einsele concepts of phosphorus mobilization.

Boers et al. (1993) discussed the mineralization of organic phosphorus and emphasized that it is the first and driving step in benthic phosphorus cycling and sediment–water exchange processes. It was pointed out that the release and fixation of phosphorus are regulated by many factors and effects, which include complexation with iron, aluminum, calcium, and hydroxides. Syers et al. (1973) reviewed the phosphate chemistry in lake sediments, pointing out that inorganic species frequently form the major portion of phosphorus, and that evidence of discrete iron, aluminum, and calcium phases is based on solubility equilibria of variscite ($AlPO_4$), strengite ($FePO_4$), and hydroxyapatite ($Ca_{10}[PO_4]_6[OH]_2$) at the sediment–water interface. Furthermore, mixed ferric hydroxy-phosphates precipitate in most natural waters.

The most significant chemical form of phosphorus in nature is orthophosphate. However, there are three dissociated ionic forms of phosphoric acid (H_3PO_4) that are pH dependent. The most abundant species in the pH range of 4 to 6 is $H_2PO_4^-$. Further dissociation occurs at higher pH (i.e., 9 to 11) to form HPO_4^{2-}. The most highly dissociated form of phosphoric acid is PO_4^{3-}, which forms at an even higher pH. In soils and sediments, $H_2PO_4^-$ and HPO_4^{2-} are the most abundant forms, and may be complexed with metal ions (Bailey, 1968; Holtan et al., 1988). Research of phosphorus speciation in sediments includes studies of different phosphorus fractions, such as sorbed, reductant, soluble, organic, inert, and bound to iron, aluminum, manganese, and calcium (Broberg and Persson, 1988; Holtan et al., 1988). Organic phosphorus compounds in aquatic ecosystems consist of phosphoproteins, phospholipids, and nucleic acids.

7.3 Methods used in determination of carbon, nitrogen, and phosphorus in sediments: literature review

7.3.1 Determination of carbon in sediments

Conventional methods for the determination of carbon in sediments are adapted from techniques used in soil analysis (Jackson, 1958; Nelson and Sommers, 1982). These methods usually focus on total, organic, and inorganic carbon. The estimation of total organic matter is of great interest, while some investigators (Fabiano and Danovaro, 1994) have also determined the total lipids, proteins, and carbohydrates using test methods reported by Bligh and Dyer (1959), Rice (1982) and Gerchacov and Hatcher (1972), respectively.

Methods for the determination of the concentration of carbon include dry combustion or wet oxidation. In both techniques, carbon is converted to CO_2 followed by determination by volumetric, titrimetric, gravimetric, or conductometric methods (Nelson and Sommers, 1982). Dry combustion techniques use high temperature combustion systems to oxidize carbon to CO_2 in the presence of a catalyst, such as CuO. Several modern automated instrumental systems have recently been developed to analyze for carbon as well as other combustible volatile elements, such as nitrogen and sulfur. Some of the commonly used systems are the Carlo Erba Models NA-1500 (Verardo et al., 1990) and 1106, Leco CHN-1000 and Leco CNS-2000 (Leco Corporation, 3000 Lakeview Ave., St. Joseph, MO, 49085-2396, U.S.A.), and Perkin-Elmer CHN analyzers (Perkin-Elmer Corporation, 761 Main Ave., Norwalk, CT, 06859-0012, U.S.A.).

Wet digestion procedures use oxidation with acidic dichromate ($K_2Cr_2O_7$) followed by titration to determine the concentration of carbon. This procedure normally involves digesting the sediment in a mixture of $K_2Cr_2O_7$, H_2SO_4, and H_3PO_4 in a closed system, flushing the system with CO_2-free air in a weighing bulb filled with Ascarite (Allison, 1960; Nommik, 1971; Kotsch et al., 1976; Leventhal and Shaw, 1980). Wet digestion systems are less costly than instrumental automated systems but are often slow and require special and cumbersome glass apparatus.

The concentrations of organic carbon can be determined indirectly by the difference of total carbon and inorganic carbon or directly after removing inorganic carbon by dissolution of carbonates (Sandstrom et al., 1986). The concentration of organic carbon is determined directly by the simple loss on ignition (at 550°C) method (Parker, 1983) after removal of interfering carbonates by treatment of the sediment with 10% HCl (Buchanan, 1971).

7.3.2   Determination of nitrogen in sediments

Traditional methods for the determination of the concentration of nitrogen in soils and sediments usually employ a combination of wet or dry digestion/oxidation techniques to prepare the sample for analysis (Jackson, 1958; Hesse, 1971; Page et al., 1982). Analytical techniques include distillation/titration, specific ion electrodes, colorimetry, i.e., using UV/VIS, infrared absorption and emission photometry, chemiluminescence, conductometry, and ion and gas chromatography (Bremner and Mulvaney, 1982; Keeney and Nelson, 1982; D'Elia, 1983; Dionex, 1987; Renmin et al., 1995).

Total nitrogen is usually measured by dry combustion/oxidation, i.e., by Dumas method, or wet oxidation/extraction, i.e., by Kjeldahl method (Bremner and Mulvaney, 1982). The Dumas combustion/oxidation method was originally reported not to recover some refractory organics, such as hetrocyclic compounds (Bremner and Mulvaney, 1982). However, later modifications of commercially available systems, such as the Coleman Model 29 N analyzer (Stewart et al., 1964; Keeney and Bremner, 1967) and the LECO Model UO-14SP (Wong and Kemp, 1977), have resulted in good recoveries of nitrogen from sediments. More recent instruments of this type include the Antek 7000 NS (Antek Instruments Inc., 300 Bammel Westfield Rd., Houston, TX, 77090-3508,

U.S.A.), Carlo Erba NA-1500 (Verardo et al., 1990) and Carlo Erba 1106, Leco CHN-1000, and Leco CNS-2000, and Perkin-Elmer CHN analyzers, which are all equipped with state-of-the-art microprocessor control and automatic sampling systems. The usefulness of these analyzers has been discussed by Riley et al. (1994).

The classical Kjeldahl wet oxidation technique is used extensively by many laboratories to quantify the sum of complexed organic and free inorganic nitrogen (U.S. Environmental Protection Agency, 1983; American Public Health Association, 1992; Patton and Truitt, 1992; Fishman and Friedman, 1989; Fishman, 1993; National Laboratory for Environmental Testing, 1994), and has been discussed in several reviews (Bremner, 1965; Nelson and Sommers, 1980; Bremner and Mulvaney, 1982). In the Kjeldahl method, samples are digested in sulfuric acid (H_2SO_4), together with potassium sulfate (K_2SO_4), to raise the digestion temperature. Mercuric sulfate ($HgSO_4$) or copper sulfate ($CuSO_4 \cdot 5H_4O$) with traces of selenium are added to promote the oxidation (Nelson and Sommers, 1980). Organic compounds of nitrogen as well as free ammonia/ammonium in the sediment are converted to ammonium sulfate, which is then determined by different procedures. This extraction technique is called ammonia plus organic nitrogen, since NO_3^- and NO_2^- are not recovered by the digestion (Fishman and Friedman, 1989). However, treatment of samples with salicylic acid/thiosulfate has been reported to recover NO_3^- and NO_2^- in Kjeldahl extracts (Bremner and Mulvaney, 1982). Modern Kjeldahl digestion systems use hot block digestors with Pyrex tubes (Schuman et al., 1973; Gallaher et al., 1976; Jirka et al., 1976) to allow a large number of samples to be digested simultaneously. Hot blocks can be obtained commercially or can be constructed as described by Agemian et al. (1980). The use of a dialyzer unit has also been applied to automated analysis to deal with turbid or highly colored samples (Selmer-Olsen, 1971; Patton and Truitt, 1992). The popular color-forming reaction of the indophenol blue method is used in colorimetric determination of ammonium. In this reaction, phenate and NH_4^+ react in the presence of hypochloride and nitroprusside as a catalyst to produce the indophenol blue compound (Bremner and Mulvaney, 1982; Keeney and Nelson, 1982).

A common and informative technique for the extraction of the inorganic species of nitrogen, including NH_4^+, NO_3^-, and NO_2^-, is the extraction of the exchangeable fraction with KCl (2 N) (Keeney and Nelson, 1982). An appreciable amount of NH_4^+ is nonexchangeable and is associated with the crystal lattice of minerals in the sediment particles. The extraction of nonexchangeable NH_4^+ requires a stronger acid, such as HF (5 N) and HCl (1 N). Many different analytical systems have been used for the determination of NH_4^+, NO_3^-, and NO_2^- in sediment and soils (American Public Health Association, 1992). These include colorimetry, ion electrode, steam distillation/titration, microdiffusion, and ion chromatography. Keeney and Nelson (1982) found that the colorimetric techniques are the most sensitive for the determination of NH_4^+. They also reviewed the many different colorimetric techniques that have been used to determine NO_3^-. They indicated that the preferred method, based on sensitivity, reproducibility, and accuracy, is to reduce NO_3^- to NO_2^- by passing through a copperized cadmium column, followed by the determination of NO_2^- by the common colorimetric diazotization reaction with sulfanilamide. In this reaction, sulfanilamide in HCl converts NO_2^- to diazonium salt, which reacts with [N-(1-naphthyl)-ethylenediamide] to form an azo-compound with a reddish purple color, which can readily be analyzed by molecular absorption spectrophotometry.

As mentioned in section 7.2.2, this chapter, the significant organic forms of nitrogen in sediments are those derived from biological activity and the decomposition of biota and include various amino sugars, amino acids, and urea. Stevenson (1982) has described methods for the analysis of organic nitrogen in the form of amino sugars and amino acids, while Bremner (1982) has described methods for the determination of urea. Extraction methods using HCl (3 or 6 M) are used to extract acid soluble organic acids, which are separated into various classes of compounds. The KCl (2 M) extraction procedure mentioned above has also been used to extract urea from soils. These organic compounds are determined by either colorimetric or enzymatic techniques.

7.3.3 Determination of phosphorus in sediments

Extraction methods for the determination of the concentration of total phosphorus usually involve strong acid/oxidant digestion techniques, such as digestion by $HClO_4$, HF–HNO_3, H_2SO_4–$K_2S_2O_8$, dry ashing/HCl-digestion, or Na_2CO_3 fusion. Jackson (1958) has stated that, for most soils, $HClO_4^-$ extraction and $Na_2CO_3^-$ fusion techniques give similar results. Olsen and Dean (1965) recommended digestion by $HClO_4$, while Syers et al. (1968) preferred Na_2CO_3 fusion, since $HClO_4$ gave low results on strongly weathered samples. Furthermore, Sommers and Nelson (1972) found that the $HClO_4$ digestion underestimated the concentrations of total phosphorus by 1 to 6% compared to Na_2CO_3 fusion. Later both techniques were recommended, but it was suggested that the methods be compared on selected soil samples prior to the selection of the method to ascertain comparability (Olsen and Sommers, 1982). The $HClO_4$ digestion method has been successfully applied to particulate phosphorus in the aquatic environment (Kopacek and Hejzlar, 1993).

For geological samples, the HF–HNO_3 digestion method has been shown to be necessary to recover total phosphorus from silicate and acid-soluble rocks (Donaldson, 1982). Rao and Reddi (1990) used an

HF–*aqua regia* digestion mixture to determine the concentrations of phosphorus in rocks and minerals. Other dual acid combinations, such as the $HCl–H_2SO_4$ digestion mixture, have been used for soils (Flannery and Markus, 1980) and sediments (Upchurch et al., 1974).

The open digestion technique using $H_2SO_4–K_2S_2O_8$ has also been recommended as a good method for the determination of the concentrations of total phosphorus in aquatic sediments (Fishman and Friedman, 1989). Furthermore, Aspila et al. (1976) used $H_2SO_4–K_2S_2O_8$ in a Parr Teflon bomb to recover total phosphorus from aquatic sediments. They reported that when sediment extracts of the $H_2SO_4–K_2S_2O_8$ technique are further digested in $HClO_4$, the results are not significantly higher. A similar digestion mixture using $H_2SO_4–K_2SO_4$ has also been used with the total Kjeldahl nitrogen hot block digestion apparatus for multisample preparation, as well as simultaneous extraction with total Kjeldahl nitrogen (Jirka et al., 1976; Patton and Truitt, 1992). The method is relatively easy and safe to use, and the above authors have developed fast multisample automated colorimetric analysis systems adaptable to this digestion mixture. Recently, alkaline persulfate has been used as a digestion medium to recover sedimentary phosphorus in turbid waters (Lambert and Maher, 1995).

The safe and simple ignition or dry ashing method to destroy organic phosphorus followed by digestion in HCl (1 N) has been used to determine the concentration of total phosphorus in sediments (Anderson, 1976; Aspila et al., 1976). It was found that the dry ashing of sediment at 550°C followed by HCl (1 N) digestion revealed a close estimation of the concentration of total phosphorus in the sediments. When the ashing step was omitted, the HCl (1 N) extraction yielded the concentration of inorganic phosphorus. The concentration of organic phosphorus is calculated indirectly by the difference between total phosphorus (ashing + HCl [1 N]) and inorganic phosphorus (HCl [1 N]) (Aspila et al., 1976). Another approach to calculate the concentration of organic phosphorus is by the difference between the results obtained by extraction with $HClO_4$ for total phosphorus and consecutive HCl and NaOH extractions for organic phosphorus (Mehta et al., 1954; Sommers et al., 1970). An alternate method involves removing inorganic iron- and calcium-bound phosphates with Ca-NTA/dithinoite and Na-EDTA, with subsequent fractionating of the organic phosphorus into acidic-soluble organic phosphorus, residual organic phosphorus, and humic and fulvic acids by sequential extractions with HCl (0.5 M), NaOH (2 M), and concentrated H_2SO_4 (pH <2), respectively (De Groot, 1990; De Groot and Golterman, 1993).

Partial extraction techniques were originally applied to studies of phosphorus fractionation in soils by Chang and Jackson (1957, 1958). The procedure included sequential extraction steps using various dilute solutions of NH_4Cl, NH_4F, NaOH, HCl, citrate-dithionite-bicarbonate

(CDB), and hot NaOH (1 M at 85°C for 1 h) to recover labile phosphorus, aluminum-, iron-, and calcium-bound phosphorus, and reductant-soluble and refractory phosphorus, respectively. Modifications were later introduced, since it was revealed that the NH_4F-step extracted iron-bound phosphorus in addition to aluminum-bound phosphorus, for which it was intended (Fife, 1963; Williams et al., 1967, 1980). Furthermore, phosphate was found to resorb on CaF_2, which was a byproduct of the reaction. Hieltjes and Lijklema (1980) showed that NH_4Cl must be used to remove $CaCO_3$ to get satisfactory separation of iron- and aluminum-bound phosphorus from calcium-bound phosphorus. Otherwise, resorption of phosphate onto carbonate occurs in the NaOH extraction step. In these modified procedures, the following solutions are used: dilute NH_4Cl to extract labile phosphorus; NaOH–NaCl to extract nonoccluded phosphorus; CDB to extract reductant or nonapatite phosphorus; NaOH to extract iron- and aluminum-bound phosphorus; and HCl to extract calcium-bound or apatite phosphorus. These extractions are used in sequence to carry out the fractionation of phosphorus in sediments.

Van Eck (1982) compared eight extraction schemes for fractionation of metals and phosphorus in sediments. Seven different forms of phosphorus were assumed to be as follows: exchangeable, carbonate-bound, hydrolyzable, organic, iron-bound, aluminum-bound, and calcium-bound. A study was carried out to evaluate the efficiency of the extracting agents used to separate different combinations of phases, for example, NaCl (0.5 M) to extract exchangeable phosphorus; NH_4OAc (1 M) to extract exchangeable and part of the carbonate phosphorus; and NH_4Cl (1 M) or $NaHCO_3$ (0.5 M) to extract exchangeable, carbonate, and hydrolyzable phosphorus. Stronger agents, such as NaOH (1 M), extract iron-and aluminum-bound phosphorus, while NaOH + NaCl (1 M) extract iron-, aluminum- and carbonate-bound phosphorus. Even stronger agents, such as HCl (0.5 M), are needed to dissolve calcium-bound phosphorus. Psenner et al. (1984) introduced a five-step fractionation scheme to extract water-soluble phosphorus; dithionite-bicarbonate to extract reductant soluble phosphorus; NaOH to extract iron- and aluminum-bound phosphorus; HCl to extract calcium-bound phosphorus; and NaOH (1 M at 85°C) to extract refractory phosphorus. Pettersson et al. (1988) reviewed the speciation of phosphorus in sediments by comparing the study of the above authors. They concluded that the composition of sediments varies widely and that not a single fractionation scheme can be used generally. In their review of particulate and dissolved forms of phosphorus in freshwater, Broberg and Persson (1988) discussed similarities of techniques used in the determination of phosphorus in sediments.

Organic complexing agents, such as NTA and EDTA, have been used to selectively separate iron- and calcium-bound phosphorus (Golterman, 1982; Golterman and Booman, 1988; De Groot and Golterman, 1990). These methods are called "functionally defined" methods, since they isolate various components of the sample based on their chemical functions. This type of extraction has the advantage of being less harsh with respect to extremes in pH. Therefore, hydrolysis of organic phosphates or the degradation of clay structures is prevented. Furthermore, since the phosphorus removed from iron hydroxides is complexed with the extractant, the possibility of readsorption on $CaCO_3$ is avoided. In this approach, buffered solution of Ca-NTA/dithinoite (pH 8) was used to extract iron-bound phosphorus. This was followed by Na-EDTA (pH 8) extraction to isolate calcium-bound phosphorus. Since organic phosphorus compounds are not appreciably hydrolyzed, this technique can be used as an initial cleanup prior to the determination of the concentration of organic phosphorus (De Groot, 1990; De Groot and Golterman, 1993). Lucotte and D'Anglejan (1985) compared five functional extraction methods commonly used for the determination of iron hydroxides, including CDB, ammonium oxalate-oxalic acid, hydroxylamine hydrochloride-acetic acid, calcium nitrilotriacetic acid, and sodium acetate-sodium tartrate extractions. The CDB extraction was found to be the most reproducible. Barbanti and Sighinolfi (1988) studied the applicability of hydroxylamine hydrochloride/acetic acid and ammonium oxalate/oxalic acid to the extraction of heavy metals and phosphorus in sediments, and found that the latter technique was better suited to phosphorus analysis. Jensen and Thamdrup (1993) presented a modified five-step extraction scheme in which the iron-bound phosphorus was extracted with bicarbonate-buffered dithionite solution (BD) in preference to other chelating mixtures to ensure specificity of extraction of this phase. Ruttenberg (1992) presented a sequential extraction method where the two main categories of authigenic phosphate phases, namely ferric oxyhydroxide-phosphorus and authigenic carbonate fluorapatite, could be separated in marine sediments using CDB (pH 7.6) and acetate buffer (pH 4.0), respectively.

Broberg and Pettersson (1988) reviewed techniques used in the determination of the concentration of orthophosphate. The techniques were colorimetric, ion-association complexation, enzymatic, flame spectroscopic, fluorometric, gas chromatographic, ion exclusion chromatographic, inductively coupled plasma, and radiobiological bioassay. The most common and reliable technique was found to be the colorimetric technique using the molybdenum blue complex based on the Murphy and Riley (1962) method. The main interferences in the molybdate method are from the presence of arsenate, silicate, and germanium, which produce a similar color to that produced by phosphate. Van Schouwenburg and Walinga (1967) removed interference from arsenate by reducing it to arsenite by the addition of thiosulfate in acidic medium in an excess of metabisulfite. Downes (1978) developed an automated system incorporating the thiosulfate reagent to remove interferences from arsenate and silicate. Interferences may also be removed by isolating the phosphomolybdate complex from the aqueous phase by extraction into an organic solvent. Organic complexing and chelating agents used in fractionation studies may also interfere with the reduction step if their excess is not removed by extraction into an organic phase. Watanabe and Olsen (1962) used isobutyl alcohol to eliminate interferences from organic matter. Others have used different solvents, such as butyl acetate (Won, 1964), chloroform and n-butanol (Sugavara and Kanamori, 1961), butanol (Namiki, 1961), and n-hexanol (Golterman and Wurtz, 1961). The disadvantage of these techniques is that they require an additional manual step, which renders the analytical procedure tedious and laborious.

7.4   Considerations for the selection of methods for determination of nutrients in sediments

The determination of the concentrations of nutrients in solid materials typically requires thermal and chemical treatment prior to analysis. Dry combustion techniques, such as those using a carbon/hydrogen/nitrogen analyzer, can accommodate sediment samples directly in the instrumental system with no special preparation. This is only applicable to the determination of total concentrations of volatile elements, such as carbon and nitrogen. However, other forms of these elements are labile, and the use of special chemical extractants is required to prevent thermal or chemical degradation of the analyte prior to instrumental analysis. Phosphorus is relatively nonvolatile and is incorporated, to a large extent, into mineral phases of sediments. Therefore, extensive extraction techniques are required to separate the different forms of phosphorus from the matrix without degradation in the preparation of chemically-comparable extracts for instrumental analysis.

Generally, there are two types of wet chemical sample preparation techniques. In the first type of technique, strong oxidizing agents and/or strong mineral acids are used to oxidize organic matter and degrade organic species of the nutrients. These techniques also dissolve and decompose refractory compounds and complexes of the nutrients. The reaction products are stable chemical forms, such as NH_4^+, NO_3^-, and PO_4^-, which can be determined readily by standard analytical techniques. In dry combustion methods where carbon and nitrogen are oxidized with the addition of a catalyst, the chemical forms that are measured are CO_2 and N_2. The above sample preparation techniques are designed to extract total as well as organic and inorganic forms of nutrients from the sediment. These types of methods, presented in the description of the methods in this chapter, include the dry combustion technique for the determination of carbon and nitrogen, the Kjeldahl (TKN) digestion method for nitrogen, the H_2SO_4–$K_2S_2O_8$ digestion method for total concentrations of phosphorus, and the HCl (1 N) extraction/ignition method for the determination of the concentrations of inorganic and organic phosphorus. In the second type of techniques, mild extractants, such as dilute mineral acids, salt solutions, reducing agents, and organic complexing agents, are used. The objective is either to extract species of nutrients that are labile/exchangeable or dissolve specific phases of minerals that contain the nutrient of interest. The efficiency and validity of these methods often depend on the geochemistry of the sediments and are called "operationally defined" methods. Methods of this type, presented under the description of methods in this chapter, include the KCl (2 N) extraction method for the determination of exchangeable NH_4^+, NO_3^-, NO_2^-, and urea, the HCl (1 N) extraction method for the determination of the concentrations of inorganic phosphorus, and the partial extraction methods for the fractionation of phosphorus.

7.4.1 Carbon

For the determination of the concentration of carbon in sediments, wet oxidation with $K_2Cr_2O_7$ is simple and economical to carry out. However, it is slow and suffers from several interferences, which must be removed. The chemical oxidation is affected by other reducing agents that may be present in the sample, such as ferrous and manganous ions and sulfides. Furthermore, in this method, chloride interferes by the formation of chromyl chloride (CrO_2Cl_2). This interference can be removed by employing an additional step of leaching or precipitation of chloride as AgCl. Wet oxidation methods usually employ either a combustion, which is time consuming, or the Van Slyke-Neil apparatus, which requires a great deal of skill by the operator and is expensive (Nelson and Sommers, 1982). In contrast, the automated carbon/hydrogen/nitrogen dry combustion technique is simple, rapid, and has good precision. In modern environmental laboratories where the requirement is fast, multisample analysis, the choice is definitely the automated dry combustion technique.

Organic matter is often of interest in interelement correlation studies. Since the predominant element of organic matter in the sediments is carbon, which occurs quite uniformly in a given type of sediment, a factor is often used to calculate organic matter indirectly from the concentration of organic carbon. The factor is a function of the composition of organic matter and is assumed to be 1.74 in aquatic sediments. However, it may vary with the composition of organic matter. According to Nelson and Sommers (1982), it can range from 1.724 to 2.0. Therefore, a good estimate of the factor must be made before the indirect method is used to calculate the concentration of organic matter in sediments.

7.4.2 Nitrogen

Traditionally, total Kjeldahl nitrogen (TKN) has been found to be a useful determination of the concentration of total available nitrogen in waters and sediments. The sum of the concentrations of organic nitrogen and ammonia is determined by this method. The conversion of all organic forms of nitrogen in sediment to NH_4^+ is achieved by wet oxidation. The Kjeldahl mixture for the oxidation of nitrogen in sediments requires a catalyst (i.e., mercury or copper/selenium) in addition to H_2SO_4 and K_2SO_4 to ensure the breakdown of organic nitrogen. Mercury in the form of mercuric sulfate, or copper with traces of selenium (10% selenium w/w of copper) have both been used (Nelson and Sommers, 1980). The copper/selenium catalyst is preferred, since the use of mercuric sulfate often results in complexes that form between mercury and NH_4^+ and require treatment with $S_2O_3^{2-}$ to liberate NH_4^+ prior to its determination (Nelson and Sommers, 1980; Bremner and Mulvaney, 1982).

The TKN oxidation method is carried out most efficiently when many samples are digested simultaneously using hot block digestion apparatus described below in the methods for the determination of nitrogen in this chapter. The digestion mixture of H_2SO_4–$K_2S_2O_8$, also described in the methods for the determination of total phosphorus in this chapter, closely resembles the TKN digestion mixture. This has led investigators to demonstrate that the TKN extraction mixture, utilizing hot block apparatus, can recover total phosphorus from wastewaters (Jirka et al., 1976; Patton and Truitt, 1992). However, it was stated that applicability to sediments must be tested. Sediment geochemistry will determine whether this technique could be applied to a simultaneous extraction of nitrogen and phosphorus from sediments. Some sediments may require stronger oxidants, such as $K_2S_2O_8$, or a dry oxidation method.

A deficiency of the TKN digestion method is its inability to recover nitrate and nitrite. This can be achieved by treatment of the sediment prior to the TKN digestion with salicylic acid/sulfuric acid to convert NO_3^- to other nitrogen compounds, followed by treatment of the sample by thiosulfate for conversion of different forms of nitrogen to amino compounds. The same result could also be achieved by oxidizing NO_2^- to NO_3^- using permanganate in sulfuric acid followed by reduction to ammonia with iron (Bremner and Mulvaney, 1982). However, these pretreatment techniques have been applied to soils and must be further tested to demonstrate their suitability to sediments. Alternatively, if direct extraction of NO_3^- and NO_2^- is required, the KCl (2 N) extraction method could be used. This technique has the advantage of extracting simultaneously other parameters of interest, such as NH_4^+ and urea. The extraction by KCl (2 N) is called exchangeable nitrogen extraction because KCl (2 N) is a mild extractant. Relatively large amounts of some nitrogen compounds, such as NH_4^+, bound to sediment particles are nonexchangeable. Ammonium ion associated with the crystal lattice of minerals in sediments can be recovered using HF (5 N)/HCl (1 N) extraction (Keeney and Nelson, 1982).

Keeney and Nelson's (1982) comparison of sensitivities and inter-ferences of detection methods for nitrogen showed that colorimetric techniques are by far the most sensitive. In addition, these techniques either do not suffer from any interference from cations or the interfer-ence may be overcome by distillation or addition of complexing agents, such as EDTA or potassium-sodium tartrate/sodium citrate. Further-more, colorimetric methods are fast and simple, and several investiga-tors have adapted them to automated flow-through systems. This per-mits continuous nonstop analysis of a large number of samples with the advantages of high productivity, improved data quality, and unat-tended operation (Kamphake et al., 1967; Henriksen and Selmer-Olsen, 1970; Selmer-Olsen, 1971).

7.4.3 Phosphorus

Analytical methods for the determination of the concentration of phos-phorus in soils and sediments have been of interest to geologists, agron-omists, and environmental scientists. The requirements for methods are often very similar in these disciplines and typically include the determination of the concentrations of total, organic, and inorganic phosphorus, in addition to partial extractions of different mineral phases of phosphorus. However, in the determination of total phos-phorus, differences in extraction techniques are evident. For example, extraction by HF is often used for geological samples, while extraction by $HClO_4$ or fusion are common for agricultural soils. For environmen-tal samples, such as aquatic sediments, methods for the determination of total phosphorus include dry ashing (550°C) followed by extraction with HCl or H_2SO_4–$K_2S_2O_8$. This is due to the basic differences in the geochemistry of phosphorus in rocks, soils, and aquatic sediments. In the latter, large amounts of decaying organic matter and various adsorption mechanisms at the sediment–water interface generate chemical forms of phosphorus, which, in general, do not require rig-orous extraction provided by HF, $HClO_4$, and extraction techniques by fusion.

The dry ashing (550°C)/HCl-digestion technique has been shown to be a close estimation of the concentration of total phosphorus. How-ever, Aspila et al. (1976) noticed that the method is not as rigorous as the fusion technique and yields up to 8% lower recoveries of phosphorus in some sediments. However, the method has good applicability to environmental samples (Wildung et al., 1974), and can be used in the determination of the concentration of phosphorus in soils (Hesse, 1971). Furthermore, in bioassessment studies, phosphorus associated with silicate crystal lattice in sediment particles is not considered of great significance. Unlike the safety hazards of $HClO_4$ and the slow, lengthy, and tedious fusion procedure, the ashing of the sediment with subsequent digestion by HCl (1 M) or $H_2SO_4–K_2S_2O_8$ is safe, simple, and directly amenable to automated multisample applications.

Results of phosphorus fractionation provide soil scientists with information in studies of extraction of phosphorus from the soil by plants (Olsen and Sommers, 1982). In ecological studies, the results of phosphorus fractionation provide information on bioavailability of phosphorus (Bostrom et al., 1988). Chemical fractionation procedures do not necessarily extract the same fraction of phosphorus in sediments as algae and other aquatic biota (Hegemann et al., 1983). However, many extracting agents have been used to quantify discrete chemical forms to better understand the behavior of phosphorus in sediments. In these types of studies, a series of chemical solutions is used in sequence to extract different chemical species of phosphorus. Inorganic phosphorus extracted by HCl (1 N) contains several inorganic sub-phases, which can be further fractionated. The objective is to extract chemical species not associated with the crystal lattice of clays in sediments, to obtain more environmentally significant information. Agemian and Chau (1976, 1977) showed that more dilute extractants, such as HCl (0.5 N), do not appreciably degrade the structure of clay particles and are useful in extracting bioavailable species of many elements.

In the fractionation scheme presented in the description of methods in this chapter, the commonly used CDB extraction is not recommended, since the citrate component dissolves substantial amounts of calcium-bound phosphorus (Hieltjes and Lijklema, 1980). The extraction by CDB was originally designed to extract only aluminum- and iron-bound phosphorus (Williams et al., 1967). In contrast, the use of bicarbonate-dithionite successfully dissolves aluminum- and iron-bound phosphorus without the above-mentioned problem. Another extractant often used to isolate labile phosphorus from aluminum- and iron-bound phosphorus is NH_4Cl (1 N). However, it is not recommended because it dissolves small amounts of aluminum- and iron-bound phosphorus prior to their extraction in the fractionation scheme (Pettersson et al., 1988). The use of NaCl (0.46 M) in the fractionation scheme has the effect of reducing the extraction of calcium-bound phosphorus. In addition, it prevents reabsorption of phosphorus on sediment particles and is used to wash the sediment sample after every step of the sequential fractionation (Jensen and Thamdrup, 1993).

In the commonly used molybdenum blue method, orthophosphate reacts with acidic molybdate in the presence of potassium antimonyl tartrate as catalyst to form a yellow molybdophosphoric acid, also called phosphomolybdate complex. The complex is subsequently reduced by either stannous chloride or ascorbic acid to a blue compound called the molybdenum blue complex. The analysis is performed by molecular absorption spectrophotometry, either in the aqueous phase or after extraction of the complex into an organic phase to remove interferences. The ascorbic acid reductant was considered to be superior to stannous chloride due to the following features: more stable molybdenum blue color, reduced salt error, independence of color development on temperature, and improved reproducibility. However, the ascorbic acid method has been reported to give a higher result due to possible hydrolysis of organic phosphates. This problem only exists in fractionation studies of materials in which organic phosphorus is still intact. As a result, Broberg and Pettersson (1988) suggested that it be called phosphorus determination, based on the phosphomolybdate complex, molybdate-reactive phosphorus.

The review given by Broberg and Pettersson (1988) provides a good comparison of different methods for the formation of the molybdenum blue complex. They summarized experimental conditions used by different authors, such as final concentration of reagents, types of reductants, organic solvents for preanalysis extraction, and removal of interferences.

7.5   Description of selected methods for determination of nutrients in sediments

In this section, six analytical procedures are described for the unattended multisample extraction of nutrients from sediments, followed by accurate, precise, and cost-effective automated analytical techniques. The procedures include dry ashing or wet extraction followed by the automated analytical techniques. Procedures for manual techniques are also described for laboratories not equipped with automated apparatus. The described procedures are for the determination of the following parameters in sediments:

1. Carbon and nitrogen by dry combustion
2. Nitrogen by TKN digestion
3. Ammonia, nitrate, nitrite, and urea by KCl (2 N) extraction
4. Total, inorganic, and organic phosphorus by HCl (1 N) extraction–dry combustion

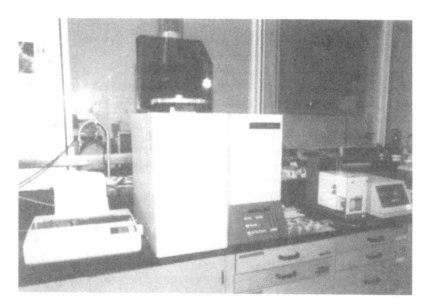

Figure 7-1   The Perkin-Elmer Model 2400 CHN elemental analyzer for the dry combustion analysis of carbon and nitrogen in aquatic sediments.

    5. Total phosphorus by H_2SO_4–$K_2S_2O_8$ digestion
    6. Phosphorus in chemical fractions

With the exception of the fully automated CHN dry combustion analyzer (Figure 7-1), the described automated procedures employ colorimetric flow manifolds using traditional Autoanalyzer II systems (Figure 7-2). Other commercially available automated flow analyzer systems may also be used with minor modification of the flow manifolds to achieve the specified flow rates. For example, analytical methods for some parameters described in the Manual of Analytical

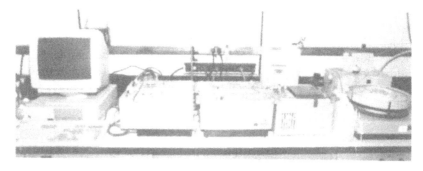

Figure 7-2   Technicon (Bran and Luebbe) autoanalyzer Model AA II system for the automated segmented flow analysis of ammonia, nitrate, nitrite, or phosphorus.

Figure 7-3 Technicon (Bran and Luebbe) autoanalyzer Model TRAACS 800 system for the automated segmented flow analysis of ammonia, nitrate, nitrite, or phosphorus.

Methods of the National Laboratory for Environmental Testing (1994) have been developed using the recent state-of-the-art Technicon (now Bran and Luebbe, Inc., 1025 Busch Parkway, Buffalo Grove, IL 60089-4516, U.S.A.) Model TRAACS 800 analyzer (Figure 7-3). The automated flow manifolds described in this chapter are typical examples and may be modified to suit specific needs. For high concentrations of PO_4^{-3}, NH_4^+, NO_3^-, and NO_2^-, dilution of sediment digests may be required and could be automatically performed using the flow manifold shown in Figure 7-4.

Prior to the determination of total concentrations of nutrients, dried sediment samples should be ground to pass a 0.15-mm or 100-mesh screen. In nutrient fractionation studies, sediment samples should be homogenized by mixing. However, the particle size should not be altered by grinding, since this will affect extractability of chemical species of nutrients from different geochemical components of the sediment. Descriptions of methods of aquatic sediment sampling, sample handling, and preparation are beyond the scope of this book, and are described elsewhere (Keith, 1988; Mudroch and MacKnight, 1994; Mudroch and Azcue, 1995).

7.5.1 Determination of carbon and nitrogen by dry combustion

7.5.1.1 Principle

The dry combustion method is based on the thermal decomposition and oxidation of carbon and nitrogen compounds in sediments or suspended matter obtained by filtration of water samples. Modern instrumentation to perform this type of analysis is available from several manufacturers, such as Perkin-Elmer, Leco, Carlo Erba, Antek, etc. The solid sample is first oxidized in a pure oxygen environment at

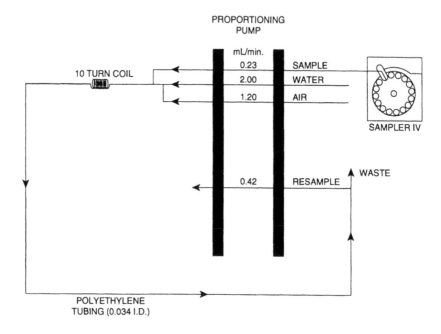

Figure 7-4 Dilution manifold for the automated dilution of sediment extracts for the analysis of nitrogen and phosphorus species.

elevated temperatures. The product gases are separated under steady-state conditions and are measured by thermal conductivity or infrared detectors. Operating gases include oxygen for combustion of the sample and helium or argon as a carrier gas. Several catalysts are employed to ensure complete conversion of carbon and nitrogen components to CO_2 and N_2, respectively. For instance, when the Perkin-Elmer Model 2400 CHN analyzer is used, product gases are passed through a combustion tube filled with three layers of oxidizing catalysts, including chromium/nickel oxide, silver tungstate coated on magnesium oxide, and silver vanadate to completely oxidize species, such as carbon monoxide to carbon dioxide. A reduction tube filled with copper is also used in this system to reduce oxides of nitrogen to N_2 (Nelson and Sommers, 1982; National Laboratory for Environmental Testing, 1994).

The dry combustion technique can be used to determine the concentrations of both total and organic carbon in suspended and bottom sediments. The concentration of organic carbon can be determined by two methods. In the first method, inorganic carbon is removed by washing the suspended sediment collected by filtration of water with 0.3% (v/v) sulfurous acid. The washing removes the inorganic carbon, which is usually present as calcite in the sediment. The sediment is then analyzed by combustion to determine the remaining organic carbon. In the second method, differential combustion at two temperatures is used to first oxidize organic carbon at 500 to 550°C. This is followed by a complete oxidation of inorganic carbon at 1000 to 1500°C. Exact temperature conditions depend on the design and manufacturer of the equipment used in the method.

7.5.1.2   Apparatus

- Perkin-Elmer Model 2400 CHN elemental analyzer (Figure 7-1), equipped with printer and PE AD-6 Ultramicro-balance or equivalent (Leco and Carlo Erba analyzers are also popular equivalents.)
- Tin capsule (6 mm × 4 mm) for solid samples and tin or aluminum disk for water samples
- Whatman GF/C filter papers

7.5.1.3   Reagents

- Deionized-distilled water: prepare by passing distilled water through an ion-exchange cartridge
- Sulfurous acid, 8% (v/v)
- Standards to calibrate CHN analyzer: weigh between 1.000 and 10.000 mg of microanalytical standard cyclohexanone 2,4-dinitrophenylhydrazone, $C_6H_{10}NNHC_6H_3(NO_2)_2$, or microanalytical standard acetanilide, $C_6H_5NOCH_3$, and analyze in the same manner as the samples. The amount of carbon and nitrogen in these materials is obtained from the certificate of analysis.
- Chromium/nickel oxide, silver tungstate on magnesium oxide, and silver vanadate for the combustion tube
- Copper and copper oxide for the reduction tube

7.5.1.4   Procedure

7.5.1.4.1   Sample preparation.   Organic carbon analysis in sediment: Place approximately 0.5 to 1.0 g of a representative subsample in a clean porcelain crucible with a lid. With the crucible in the fume hood, add dropwise approximately 1 to 5 mL of sulfurous acid to the sample and stir gently with a clean glass rod. Allow the sample to stand in the fume hood for approximately 3 to 5 h with the lid placed loosely over the crucible to decompose carbonates in the sediment. Wash the sample mixture with a few mL of carbon-free distilled water and transfer into a clean 15-mL centrifuge tube. Centrifuge for approximately 5 min at 5000 rpm and discard the supernatant. Wash the sample in the centrifuge tube three to five times with carbon-free distilled water to completely remove the acid. After completion of the washing, add 1 to 2 mL of carbon-free distilled water to the sample in the centrifuge tube, mix well, and transfer the sample from the tube into a clean petri dish. Repeat until the entire sample is in the petri dish. After the particles have settled to the bottom of the dish, remove the excess water with a pipette, and leave the petri dish overnight in the fume hood to dry the sample. When the sample is completely dry, cover the dish with an airtight cover, and store in a desiccator until analysis.

Inorganic and total carbon: To determine the concentrations of inorganic and total carbon, analyze the sediment before and after acid treatment. The difference between total and organic carbon equals the inorganic carbon concentration.

7.5.1.4.2 Instrumental analysis. Calibrate, optimize, and operate instrument as per manufacturer's instructions.

7.5.1.4.3 Analysis of suspended matter from filtered water samples. Cut out a rectangular sample portion of the filter paper containing the suspended matter sample using clean scissors and tweezers, and fold the filter paper twice. Place the folded filter on a 2 cm × 2 cm piece of precombusted tin foil and roll it into a ball (2 to 3 mm in diameter). Place the ball into the sampler tray according to the required tray pattern. Start the analysis and calculate the results as per manufacturer's instructions.

7.5.1.4.4 Analysis of sediments. Place a tin capsule on the microanalytical balance and tare. Remove the capsule and add 1 to 10 mg of homogenized sediment sample. Fold the capsule into a small ball (2 to 3 mm in diameter) with tweezers and weigh. Record the actual weight. Prepare all samples in this manner, and place them in the CHN analyzer sampler tray. Start the analysis and calculate the results according to manufacturer's instructions.

7.5.1.5 Comments

This method has been shown to be useful for the determination of carbon and nitrogen concentrations in suspended and bottom sediments over a range of 0.005 to 50.0 mg carbon per liter and 0.002 to 10.0 mg nitrogen per liter in suspended sediments in natural waters, and 0.01 to 30% carbon and nitrogen in sediments (National Laboratory for Environmental Testing, 1994).

7.5.2 Determination of nitrogen by Kjeldahl digestion

7.5.2.1 Principle

In the Kjeldahl method, sediment samples are digested in sulfuric acid in the presence of potassium sulfate and copper sulfate/selenium catalyst in glass test tubes set in a hot block (Schuman et al., 1973; Jirka et al., 1976; Nelson and Sommers, 1980; Patton and Truitt, 1992). Sulfuric acid provides a strong acid environment. Potassium sulfate increases the temperature of the oxidation, while the copper/selenium act as a catalyst to increase the rate of oxidation. Organic compounds of nitrogen, as well as free inorganic forms, are converted to ammonium sulfate. The NH_4^+ is determined colorimetrically using the indophenol blue method in which it reacts with phenate in the presence of hypochlorite and nitroprusside as catalyst.

The Kjeldahl extraction technique determines the concentrations of organic nitrogen plus ammonium nitrogen. Nitrate and nitrite nitrogen are not recovered by this technique. Organic nitrogen can be calculated indirectly from the total Kjeldahl nitrogen by subtracting the ammonium nitrogen.

7.5.2.2 *Apparatus*

- Autoanalyzer II system (Bran and Luebbe), consisting of sampler IV, (30/h, 2:1) and manifolds (Figure 7-5), or equivalent (Alpkem instrumentation is a good alternative) (Alpkem Corporation, P.O. Box 1260, Clackamas, OR, 97015, U.S.A.). Note: The dilution manifold (Figure 7-4) is used as part of the system to dilute sediment extracts automatically.
- Dialyzer — with H type membrane Alpkem part No. 303-0807P00 or equivalent (optional depending on turbidity of extracts)
- Colorimeter (*Automated analysis*): 15 mm F/C with 2.0 mm I.D., 630-nm filters
- Spectrophotometer (*Manual analysis*): simple colorimeter with 630-nm filter or variable UV/VIS spectrophotometer
- Hot block digestion tubes
- Hot block digestion apparatus — Technicon (Figure 7-6), Tecator (Schuman et al., 1973) or custom-built hot block (Agemian et al., 1980; see Figure 7-7 for dimensions)
- Centrifuge, model IEC Centra -7 or equivalent

7.5.2.3 *Reagents*

7.5.2.3.1 *Reagents for sediment digestion*

- Potassium sulfate (K_2SO_4)
- Concentrated sulfuric acid (H_2SO_4), specific gravity 1.84
- Copper sulfate ($CuSO_4 \cdot 5H_4O$)
- Selenium
- Potassium sulfate-catalyst mixture: Mix 200 g K_2SO_4, 20 g of $CuSO_4 \cdot 5H_4O$, and 2 g of selenium using a mortar and pestle

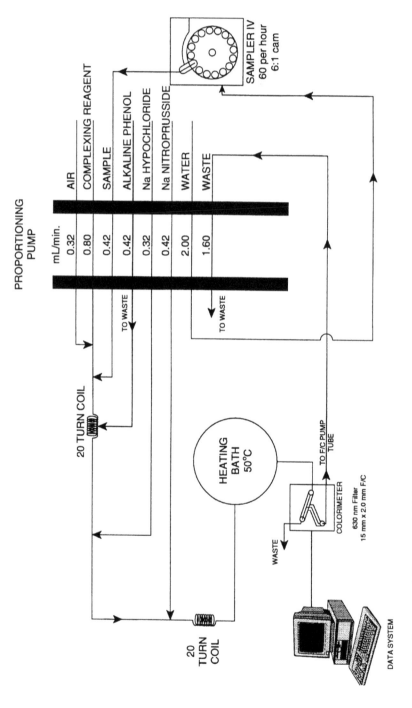

Figure 7-5   Automated flow manifold for the determination of ammonia by AA II Technicon (Bran and Luebbe) autoanalyzer.

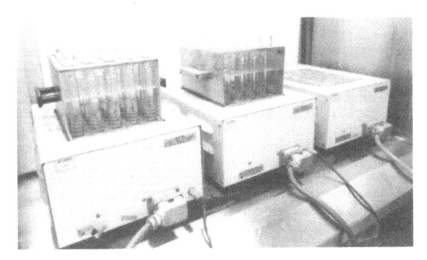

Figure 7-6   Commercially available Technicon hot block digestor for the acid digestion of sediments.

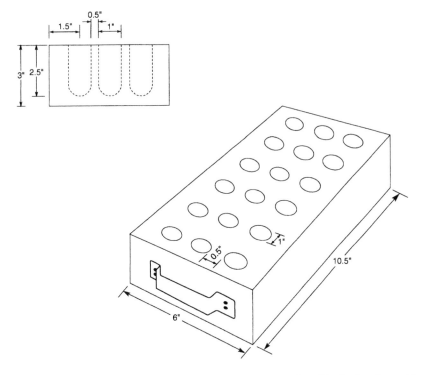

Figure 7-7   Schematic of a custom built aluminum hot block for the acid digestion of sediments. (From Agemian et al., *Analyst*, 105, 126, 1980. With permission.)

7.5.2.3.2   Reagents for ammonia analysis

- Complexing reagent: Mix 350 g of sodium hydroxide (NaOH) and 100 g of sodium potassium tartrate (NaKC$_4$H$_4$O$_6$ · 24H$_2$O) in distilled water and make up to 1 L volume
- Alkaline phenol solution: Mix 200 g of NaOH and 276 mL 88% liquefied phenol dissolved in distilled water and make up to 1 L volume
- Sodium hypochlorite solution: Commercially available bleach such as Chlorox® (5% available Cl$_2$)
- Sodium nitroprusside solution: Mix 0.5 g of sodium nitroprusside in distilled water and make up to 1 L volume
- Calibration standards: Standards can be made from any one of several different analytical grade nitrogen compounds, including glutamic acid, urea, ammonium oxalate, ammonium sulfate, and ammonium chloride. For example, a stock nitrogen standard of 100 µg/mL can be made by dissolving 1.050 g of glutamic acid and 2 mL of H$_2$SO$_4$ in 1 L of distilled water (Note: the glutamic acid should be kept dry by heating at 105°C for 1 h).

7.5.2.4   Procedure

7.5.2.4.1   Sediment digestion.   Weigh an accurate quantity of dry sediment (i.e., 0.5 g) and transfer to a hot block digestion tube. Add 3.5 g of K$_2$SO$_4$-catalyst mixture to the test tube. Accurately dispense 10 mL of concentrated H$_2$SO$_4$ to the test tube. Stir the resulting mixture with a brisk circulation. Allow a few minutes for any unusual effervescence to subside. Add Teflon boiling chips, and heat samples for 30 min at 200°C. Continue the digestion for 2 h at 370°C. To speed the digestion, samples could be transferred between two hot blocks sets, with one set at 200°C and the second at 370°C. After completion of the digestion, cool samples, and dilute to 100 mL.

7.5.2.4.2   Manual analysis.   Add 5 mL of sediment extract and 1 mL of complexing reagent to a 25 mL volumetric flask and mix well. After 1 min, add 1 mL of alkaline phenol solution, 2 mL of nitroprusside solution, and 1 mL of hypochlorite solution. Place flask in a water bath set at 40°C for 30 min. Remove flask from the bath and let cool to room temperature. Measure absorbance at 630 nm and compare to standards prepared in the same manner.

7.5.2.4.3   Automated analysis.   Analyze samples using an automated manifold, as shown in Figure 7-5. Use the dilution manifold (Figure 7-4) to dilute extracts in an automated manner prior to use of the ammonia manifold (Figure 7-5).

7.5.2.5 Comments

The use of a dialyzer unit in the automated system is recommended in cases where particulate matter causes a serious interference problem. However, its use may reduce analytical sensitivity, and therefore it is often excluded. If the latter option is chosen, care must be taken to ensure that the centrifugation step removes the particulate matter.

7.5.3 Determination of exchangeable ammonia, nitrate, nitrite, and urea by KCl (2 N) extraction

7.5.3.1 Principle

The analysis of exchangeable ammonia, nitrate, nitrite, and urea can be used in bioassessment studies of aquatic sediments, since bioavailable species often occur in the exchangeable phase. The KCl (2 N) equilibrium extraction method involves the shaking of a sediment sample with a solution of KCl (2 N) at room temperature for a specified period of time (i.e., 1 h). The method of Keeney and Nelson (1982), applied to sediment samples for the extraction of exchangeable NH_4^+, NO_3^-, NO_2^-, and urea, provides a simple and rapid method to assess bioavailable nutrients in sediments. The following four parameters are determined by the method:

NH_4^+ — colorimetrically, using the indophenol blue method, in which it reacts with phenate in the presence of hypochlorite and nitroprusside as catalyst

NO_2^- — colorimetrically, by reacting with sulfanilamide ($NH_2C_6H_4SO_2NH_2$) under acidic conditions, to form a diazo compound that couples with N-1(Napthyl)-ethylenediamine dihydrochloride ($C_{12}H_{16}Cl_2H_2$) to form a reddish-purple azo dye (National Laboratory for Environmental Testing, 1994)

NO_3^- — by reduction to NO_2^- by passing the sample through a cadmium coil in the automated system, and then determining the sum of NO_3^- and NO_2^-, using the NO_2^- system described above. The concentration of NO_3^- is calculated by the difference between (NO_3^- + NO_2^-) and NO_2^- (National Laboratory for Environmental Testing, 1994)

Urea — using the colorimetric analysis of the red color formed when urea reacts with diacetyl-monoxime in the presence of thiosemicarbazide (Nakas and Litchfield, 1977; Bremner, 1982)

7.5.3.2 Apparatus

- Autoanalyzer II system (Bran and Luebbe) consisting of sampler IV, (30/h, 2:1) and manifolds (Figure 7-5 for NH_4^+ and Figure 7-8

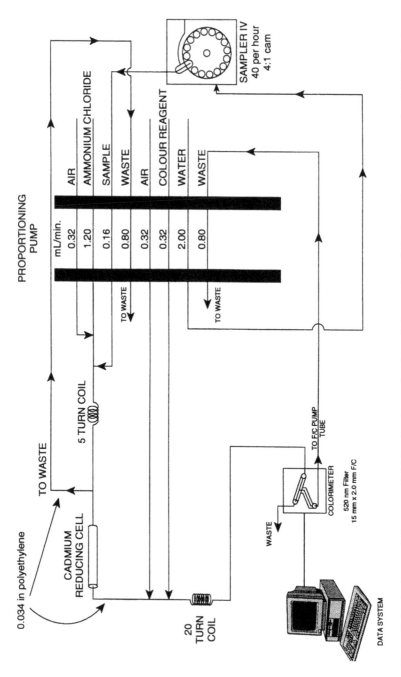

Figure 7-8 Automated flow manifold for the determination of nitrate plus nitrite in sediment extracts by AA II Technicon (Bran and Luebbe) autoanalyzer.

for NO_3^- + NO_2^-) or equivalent. Note: The dilution manifold (Figure 7-4) is used as part of the system to dilute sediment extracts automatically.
- Dialyzer — with H-type membrane Alpkem or equivalent (optional, depending on turbidity of extracts)
- Colorimeter (*Automated analysis*): 15 mm F/C with 2.0 mm I.D., 630 nm filters for NH_4^+ and 520 nm for NO_2^-
- Colorimeter or spectrophotometer (*Manual analysis*): any simple colorimeter or variable UV/VIS spectrophotometer (e.g., Bausch & Lomb® Model 20) with absorbance detection capability at 630 nm for NH_4^+, 520 nm for NO_2^-, and 527 nm for urea. Note: The manual method is the only technique presented for urea.
- Mechanical shaker
- Filtration apparatus
- Boiling water bath
- Cooling water bath set at 5°C

7.5.3.3   Reagents

- Potassium chloride, KCl (2 M): dissolve 150 g of reagent grade KCl in distilled water and dilute to 1 L.

7.5.3.3.1   Reagents for NH_4^+

- As described in the method for the determination of nitrogen by Kjeldahl digestion.

7.5.3.3.2   Reagents for NO_3^- and NO_2^-

- Copper (II) sulfate solution for buffer solution ($CuSO_4 \cdot 5H_2O$): dissolve 1.5 g of copper sulfate in 500 mL of distilled water.
- Ammonium chloride buffer solution, NH_4Cl: adjust the pH of 3 L of distilled water to approximately 8.5 (between 8.3 and 8.5) with a freshly prepared 1.5% v/v ammonium hydroxide solution. To this solution, add 30 g of ammonium chloride and mix thoroughly until all the salt has dissolved. The pH of the resulting solution should be approximately 5.5.
- Ammonium chloride working buffer solution, NH_4Cl (for daily use): add 1 mL of the copper (II) sulfate solution and 0.1 mL of the wetting agent Brij-35 (30%) solution to 200 mL of the ammonium chloride buffer solution.
- Color reagent stock solution: mix 300 mL of 85% phosphoric acid, 30 g of sulfanilamide ($NH_2C_6H_4SO_2NH_2$), and 1.5 g of N-1-(-Naphthyl)-ethylenediamine dihydrochloride ($C_{12}H_{16}Cl_2N_2$) and 1.5 mL of the wetting agent Brij-35 (30%) solution in distilled

water, and make up to 3 L volume. Store solution in an amber polyethylene bottle.

- Copper (II) sulfate activating solution ($CuSO_4 \cdot 5H_2O$): dissolve 0.6 g of copper (II) sulfate in 500 mL of distilled water. This solution is used to activate the Cd reduction cell (see Comments in this section describing the determination of urea).
- Hydrochloric acid 2 N (HCl): add 84 mL of concentrated HCl (36.5 to 38.0%) to approximately 400 mL of distilled water, mix thoroughly, and dilute to 500 mL. This solution is used to activate the Cd reduction cell.
- Cadmium reduction cell: loosely pack one end of a 10 cm long (0.17 cm I.D.) polyethylene tubing with glass wool. The tube is then filled by scooping the Cd chips from the bottle through the other end of the polyethylene tubing. Gently tap the tubing while filling to set the Cd chips as compactly as possible. The Cd-filled portion of the tubing should be about 8 cm long. This Cd reduction cell is good for approximately 65 h of running time.
- Nitrate stock standard solution (1000 mg/L NO_3, as N): dissolve 7.218 g of potassium nitrate, previously oven-dried at 105°C for 1 h (KNO_3) in distilled water and make up to 1 L volume. Store this stock standard solution at 4°C.
- Nitrite stock standard solution (1000 mg/L NO_2, as N): dissolve 6.072 g of potassium nitrite, previously oven-dried at 105°C for 1 h (KNO_2) in distilled water and make up to 1 L volume. Store this stock standard solution at 4°C.

7.5.3.3.3   Reagents for urea analysis

- Potassium chloride (2 N): phenylmercuric acetate (0.005%) solution: dissolve 1500 g of KCl and 50 mg of phenylmercuric acetate in 10 L of distilled water.
- Diacetyl monoxime solution (2.5%): dissolve 2.5 g of diacetyl monoxime in 100 mL of distilled water.
- Thiosemicarbazide solution (0.25%): dissolve 0.25 g of thiosemicarbazide in 100 mL of distilled water.
- Acid reagent: mix 300 mL of 85% phosphoric acid and 10 mL of sulfuric acid, and make up to 500 mL volume with distilled water.
- Color reagent: mix 25 mL of 2.5% diacetyl monoxime solution, 10 mL of 0.25% thiosemicarbazide solution, and 500 mL of acid reagent. Note: This solution is prepared just before use.
- Urea nitrogen stock standard solution (100 mg/L urea as N): dissolve 0.2144 g of previously dried analytical grade urea in 1 L of potassium chloride (2 N)-phenylmercuric acetate (0.005%) solution.

7.5.3.4   Procedure

7.5.3.4.1   KCl (2 N) extraction of NH_4^+, NO_3^-, and NO_2^-.   Weigh an accurate amount (i.e., 10 g) of dry sediment. Transfer to a 250 mL widemouth plastic bottle. Add 100 mL of KCl (2 N) to the sediment in the bottle and replace the cap. Shake the sediment in the bottle on a mechanical shaker for 1 h. Filter solution using standard filtration equipment and Whatman #42 filter paper. Store the filtrate at 4°C until analysis.

7.5.3.4.2   Automated or manual analysis of NH_4^+.   Determine the concentrations of NH_4^+ in the samples by the method described in the section on determination of nitrogen by Kjeldahl digestion.

7.5.3.4.3   Automated analysis of NO_3^- and NO_2^-.   Determine the concentrations of $NO_3^- + NO_2^-$ using the flow manifold shown in Figure 7-8. Determine the concentration of NO_2^- using the same manifold, without the Cd-reducing cell. Calculate the concentration of NO_3^- from the difference between the concentrations of $(NO_3^- + NO_2^-)$ and NO_2^-.

7.5.3.4.4   Manual analysis of NO_3^- and NO_2^-.   Place 2 to 5 mL of the sediment extract into a 50 mL volumetric flask. Add 1 mL of sulfanilamide 5%-HCl (2.4 N) solution and mix well. After 5 min add 1 mL of N-1(-Naphthyl)-ethylenediamine dihydrochloride 3%-HCl (0.12 N) solution. Mix, make up to volume, and allow to stand for 20 min. Measure absorbance at 540 nm, and compare to standards prepared in the same manner. This gives the concentration of NO_2^-. To measure $(NO_3^- + NO_2^-)$, pass the sample through a Cd column to convert NO_3^- to NO_2^-, and then repeat the above procedure. Calculate the concentration of NO_3^- from the difference between the concentrations of $(NO_3^- + NO_2^-)$ and NO_2^-.

7.5.3.4.5   KCl (2 N) extraction and determination of urea.   Weigh an accurate amount (i.e., 10 g) of previously frozen, thawed sediment, and transfer to a 250 mL widemouth plastic bottle. Add 100 mL of KCl (2 N)-phenylmercuric acetate (0.005%) solution to the bottle. Replace the cap and shake the bottle on a mechanical shaker for 1 h. Filter the solution using standard filtration equipment and Whatman #42 filter paper. Store the filtrate at 4°C until analysis. For the determination of urea, add 10 mL of extract and 30 mL of color reagent to a volumetric flask and mix. Heat the sample in a boiling water bath for 27 min and then immediately insert flasks for 15 min into a cooling bath, set at 5°C. Analyze samples by measuring the absorbance of the red complex at 527 nm.

7.5.3.5 Comments

To avoid biodegradation, sediment samples for the determination of the concentration of urea should be frozen immediately upon retrieval of the sediment and thawed just prior to analysis. During the KCl (2 N) extraction, phenylmercuric acetate is used together with the extracting agent to act as a urease inhibitor to prevent enzymatic degradation of the urea. The cadmium-reduction cell must be activated every day prior to use. This is accomplished by pumping HCl (2 N) through the ammonium chloride line for 1 min, followed by the copper (II) activating solution for 2 min, and HCl (2 N) for another 5 min.

The concentration of urea in the sediment is based on sediment wet weight. However, after determination of the water content in the sediment, the concentration of urea can be calculated for the sediment dry weight.

7.5.4 Determination of total, inorganic, and organic phosphorus by dry combustion followed by digestion by HCl (1 N)

7.5.4.1 Principle

For the determination of the concentration of total phosphorus, dry sediments are ignited at 550°C for 2 h. After ignition, the residue is digested for 16 h with HCl (1 N). The orthophosphate concentration determined in the acid digest is a measure of the total phosphorus content in the sediment (Aspila et al., 1976). The inorganic phosphorus content of the sediment is determined by measuring the orthophosphate released when the dry, unignited sediment is digested by HCl (1 N). The concentration of organic phosphorus is calculated from the difference between total and inorganic phosphorus.

The determination of phosphate in the dilute acid extract is based on the reaction of ammonium molybdate with orthophosphate to form molybdophosphoric acid, which is then reduced by ascorbic acid to an intensely colored heteropoly molybdophosphoric acid complex. The intensity of produced color is proportional to the concentration of orthophosphate and is measured colorimetrically at 660 nm wavelength (Figure 7-9 — detection method I). The addition of antimony potassium tartrate accelerates the development of the color and produces a more sensitive system with optimum sensitivity at 885 nm (Figure 7-9 — detection method II). The choice of the method depends on the concentration of phosphorus in the samples.

7.5.4.2 Apparatus

- Constant temperature shaker bath
- Constant temperature bath, 32°C

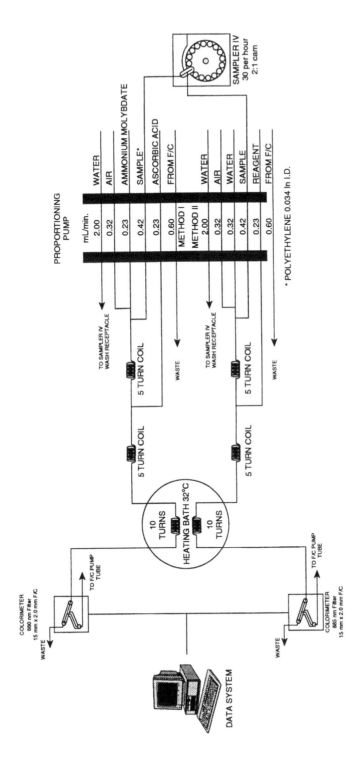

Figure 7-9   Automated flow manifold for the determination of orthophosphate in sediment extracts by AA II Technicon (Bran and Luebbe) autoanalyzer.

- Autoanalyzer II system (Bran and Luebbe), consisting of sampler IV (30/h, 2:1) and manifolds (Figure 7-9) or equivalent. Note: The dilution manifold (Figure 7-4) may be used as part of the system to automatically dilute sediment extracts with high concentrations of phosphorus.
- Dialyzer — with H type membrane, such as Alpkem (optional, based on turbidity of extracts)
- Colorimeter (*Automated analysis* — see Figure 7-9):
    Method I: 15 mm F/C with 2.0 mm I.D., 660-nm filters.
    Method II: 15 mm F/C with 2.0 mm I.D., 885-nm filters.
- Colorimeter or spectrophotometer (*Manual analysis*): any simple colorimeter with 660-nm and 885-nm filters or UV/VIS variable wavelength spectrophotometer
- Zirconium or platinum crucibles
- Muffle furnace
- Centrifuge (model IEC Centra -7 or equivalent)

7.5.4.3   Reagents

- Hydrochloric acid, HCl (1 N): dilute 89 mL of concentrated HCl to 1 L using distilled water

7.5.4.3.1   Phosphate standards

- Stock phosphate standard: dissolve 0.4394 g of anhydrous potassium dihydrogen phosphate (KH_2PO_4) in 800 mL of distilled water. Add 1 mL of concentrated sulfuric acid (H_2SO_4), and then dilute to 1000 mL with distilled water. This solution will contain 100 pg/mL phosphorus.
- Working phosphate standards: prepare working phosphate standards from the stock orthophosphate standard (100 µg/mL P) by adding 0.25 mL, 0.50 mL, 0.75 mL, 1.0 mL, 1.25 mL, 1.50 mL, 1.75 mL, and 2.00 mL to a series of 100-mL volumetric flasks. To each volumetric flask, dispense with a suitable pipettor 10.0 mL of 10% (v/v) H_2SO_4. Dilute this mixture to 100.0 mL using distilled water. These standards will be 1% (v/v) H_2SO_4, and the concentration range is suitable for most sediments when 0.5 g is employed.

7.5.4.3.2   Reagents required for Method I (Figure 7-9)

- Ascorbic acid reagent: dissolve 18 g of ascorbic acid ($C_6H_8O_6$) in a 1 L volumetric flask containing a small amount of distilled water. Add 50.0 mL of acetone (CH_3COCH_3) and dilute to ap-

proximately 999 mL using distilled water. Add 0.50 mL of Levor IV and then adjust the volume to 1 L using distilled water. Shake flask to ensure complete mixing.
- Ammonium molybdate reagent: add 10.0 g of ammonium molybdate [$(NH_4)_6Mo_7O_{24} \cdot 4H_2O$] to a 1-L volumetric flask containing about 600 mL of distilled water. While swirling this mixture, add 63 mL of concentrated H_2SO_4. Shake contents of the flask vigorously to effect dissolution. Dilute the solution to 1 L. Store flask in an amber bottle. It may be necessary to filter undissolved material.

7.5.4.3.3   *Reagents required for Method II (Figure 7-9)*

- H_2SO_4 (4.9 N): add 136 mL of concentrated H_2SO_4 (specific gravity 1.84) to 800 mL of distilled water while cooling. After the solution has cooled, dilute to 1 L with distilled water.
- Ammonium molybdate reagent: dissolve 40 g of ammonium molybdate [$(NH_4)_6Mo_7O_{24} \cdot 4H_2O$] in 800 mL of distilled water, and then dilute to 1 L with distilled water.
- Ascorbic acid: dissolve 18 g of $C_6H_8O_6$, in 800 mL of distilled water. Dilute to 1 L with distilled water.
- Antimony potassium tartrate [$K(SbO)C_4H_4O_6 \cdot 1/2H_2O$]: dissolve 3.0 g of $K(SbO)C_4H_4O_6 \cdot 1/2H_2O$, in 800 mL of distilled water. Dilute to 1 L with distilled water.
- Combined working reagent: combine the prepared reagent solutions in the following order:
    (a) 50 mL of sulfuric acid (4.9 N);
    (b) 15 mL of ammonium molybdate;
    (c) 30 mL of ascorbic acid solution; and
    (d) 5.0 mL of the antimonyl tartrate solution.
    The mixed reagent is stable for approximately 8 h.
- Water diluent: to 1 L of distilled water add 0.50 mL of levor IV wetting agent.

7.5.4.3.4   *Reagents required for manual analysis*

- Same as above combined working reagent.

7.5.4.4   *Procedure*

7.5.4.4.1   Ignition of sediment.   Weigh an accurate amount (0.1 to 0.5 g) of a previously dried sediment. Transfer to a dry crucible (approximately 10 to 30 mL capacity). Ignite the sample in the crucible at 550°C for approximately 2 h.

7.5.4.4.2   Extraction of orthophosphate from sediments. F o r   t h e
determination of the concentration of inorganic phosphorus, transfer
0.1 to 0.5 g of dry sediment to a 100-mL volumetric flask. For the
determination of the concentration of total phosphorus, transfer the
dry, ignited sediment sample to a dry 100-mL volumetric flask. Add
50 mL of HCl (1 N) to the volumetric flask and place on a shaker
assembly set at room temperature. Shake the sediment-acid mixture
with a carefully controlled and consistent motion for a period of 16 h.
Caution should be taken to prevent excessive agitation, which makes
sediment particles move up the sides of the flask. After extraction,
decant a portion of the mixture from the 100-mL volumetric flask to a
15-mL centrifuge tube and clarify the solution by centrifuging at 2000
rpm for approximately 5 min. Remove 1 mL of the supernatant with a
pipette and dilute with distilled water to a 10-mL volume.

7.5.4.5   Analysis

7.5.4.5.1   Automated analysis.   Decant a portion (i.e., 2 to 3 mL)
of the dilute extract into the sample vials of the Technicon sampler and
determine the concentration of orthophosphate using the manifolds
shown in Figure 7-4.

7.5.4.5.2   Manual analysis.   Analyze manually using a standard
laboratory colorimeter within 10 min of the addition of the color-forming reagents. Add 20 mL of "combined working reagent" (see Reagents
for Method II) to 30 mL of sample placed in a 50-mL volumetric flask,
and determine the intensity of the color at 885 nm wave length (Harwood et al., 1969). Since the intensity of the blue color is time-depen-
dent, ensure that the time of analysis after the addition of the reagent
is the same for all samples and standards.

All standards should contain the same amounts of reagents used
in samples in either automated or manual methods. This is critical since
acid concentrations affect the sensitivity of the reaction.

7.5.4.6   Comments

This method is applicable to the determination of the concentrations
of total, organic, and inorganic phosphorus in sediments (Aspila et al.,
1976). The analytical range is from 100 to 2000 µg/g phosphorus. The
range of sample size for analysis is 0.1 to 0.5 g. When the concentration
of total phosphorus exceeds 2000 µg/g, the range may be extended by
appropriate dilution of the sample solutions, either manually or by
using the automated flow manifold shown in Figure 7-4. Application
of very sensitive phosphate detection systems should be considered
with caution, since contamination from sediment dust or from dirty
glassware can be a problem.

Of the two systems described for the determination of the concentration of orthophosphate, the first method is recommended for high concentrations of phosphorus (using ascorbic acid only). The alternative system (i.e., the second method) is recommended for low concentrations of phosphorus with the use of a dialyzer to remove fine interfering particles (Patton and Truitt, 1992). However, its use may reduce analytical sensitivity; therefore, it is often excluded. If the latter option is chosen, care must be exercised to ensure that the centrifugation step removes such particles. The dual phosphorus manifold permits more flexibility when a large number of samples with very low and very high levels of phosphorus are being analyzed. The anticipated range of phosphorus in sediments is 200 to 2000 μg/g.

7.5.5 *Determination of total phosphorus by H_2SO_4–$K_2S_2O_8$ digestion*

7.5.5.1 *Principle*

Dry sediments are digested at 135° ± 5°C in a sealed Teflon bomb (Aspila et al., 1976) or test tubes in a hot block at 200°C/370°C (Jirka et al., 1976; Patton and Truitt, 1992) with concentrated H_2SO_4 and potassium persulfate. The persulfate is employed as an oxidizing agent to convert organic phosphorus to orthophosphate. Sulfuric acid is used as an agent in assisting the oxidation as well as the dissolution of phosphate minerals. The determination of the concentration of orthophosphate in the acid extract is carried out as described in the above method for the determination of orthophosphate.

7.5.5.2 *Apparatus*

- Constant temperature shaker bath and constant temperature bath, 32°C
- Autoanalyzer, colorimeter systems, and dialyzer (see Apparatus for Section 7.5.4)
- PARR 4745 acid digestion bomb (25 mL capacity) (PARR Instrument Company, Moline, IL, 61265, U.S.A.)
- Hot block digestion tubes
- Hot block digestion apparatus — Technicon (Figure 7-6), Tecator (Schuman et al., 1973), or custom built hot block (Agemian et al., 1980; see Figure 7-7 for dimensions)

7.5.5.3 *Reagents*

- Stock phosphate standard: see Reagents for Phosphate Standards in Section 7.5.4.

- *Working phosphate standards*: see Reagents for Phosphate Standards in Section 7.5.4.
- Ascorbic acid reagent: see Reagents for Method I in Section 7.5.4.
- Ammonium molybdate reagent: see Reagents for Method I in Section 7.5.4.
- Potassium persulfate ($K_2S_2O_8$)
- Concentrated sulfuric acid (H_2SO_4), specific gravity 1.84
- Antimony potassium tartrate: see Reagents for Method II in Section 7.5.4.

7.5.5.4 Procedure

7.5.5.4.1 Bomb digestion.   Weigh an accurate amount of dry sediment (0.1 to 0.5 g) and transfer to a dry Teflon bomb. Add 2.0 g of $K_2S_2O_8$ to the bomb. Accurately dispense 5.0 mL of concentrated H_2SO_4 to the bomb. Stir the resulting mixture with a brisk circulation of the bomb. Allow a few minutes for any unusual effervescence to subside. Carefully seal the Teflon bomb and place in a preheated oven set at 135 to 140°C for 2 h. After 2 h remove the bomb and allow to cool to the room temperature (i.e., approximately 1 to 2 h). Quantitatively transfer the contents of the bomb into a 500-mL volumetric flask. A large quantity of distilled water should be used to complete the transfer and reduce the increase in temperature when the water and acid are mixed. Dilute the sample extract to 500 mL using distilled water, and then shake briskly to ensure homogeneity of the solution. Transfer an aliquot (approximately 12 mL) of the sediment extract to a 15-mL centrifuge tube and clarify the solution by centrifuging at 2000 rpm.

7.5.5.4.2 Hot block digestion.   Weigh an accurate amount of dry sediment (0.5 g) and transfer to a hot block digestion tube. Add 3.5 g of $K_2S_2O_8$ to the test tube. Accurately dispense 10.0 mL of concentrated H_2SO_4 to the test tube. Stir the resulting mixture with a brisk circulation. Allow a few minutes for any unusual effervescence to subside. Add Teflon boiling chips and heat samples for 30 min at 200°C. Continue digestion for 2 h at 370°C. To speed the digestion, samples could be transferred between two sets of hot blocks; one set at 200°C and the second set at 370°C. Cool samples to room temperature and dilute to 100 mL.

7.5.5.5 Analysis

Determine the concentrations of orthophosphate in sediment extract by the automated system (Figure 7-9), or manually as described above in Sections 7.5.4.5.1 and 7.5.4.5.2, respectively. Note: The dilution manifold (Figure 7-4) is used as part of the system to dilute sediment extracts automatically.

7.5.5.6   Comments

This method is applicable for the determination of the concentrations of total phosphorus in sediments. The analytical range is from 100 to 2000 $\mu g/g$ phosphorus. The range of sample size for analysis is 0.1 to 0.5 g. When the concentration of phosphorus exceeds 2000 $\mu g/g$, the analytical range may be extended by appropriate dilution of the sample solution.

Some samples containing high concentrations of organic matter may exhibit a grey color after the contents of the bomb are diluted to 500 mL (1% v/v H_2SO_4). However, no significant spectral interference at 660 nm has been observed. The operator should check for samples with unusual color by measuring the sample response in the absence of the molybdate salt in the molybdate reagent line.

7.5.6   Determination of phosphorus in chemical fractions

7.5.6.1   Principle

Chemical fractionation experiments are designed to separate and quantify specific forms of phosphorus using operationally defined extraction schemes. Since the composition of sediments varies widely, none of the many suggested fractionation schemes can apply to all situations (see Section 7.2). However, there are several partial extraction techniques that have been recommended repeatedly by many investigators and that can provide useful information on the different forms of phosphorus in sediments. In addition to the techniques for the determination of the concentration of organic and inorganic phosphorus in sediments, the inorganic phosphorus can be further fractionated by the application of the following scheme, which includes an operationally defined extraction sequence incorporating universally used extraction techniques (Van Eck, 1982; Psenner et al., 1985; Jensen and Thamdrup, 1993), (see Figure 7-10).

Extraction step	Form of P extracted
A) NaCl (0.5 M)	Loosely sorbed exchangeable phosphorus
B) Bicarbonate-Dithionate (BD)	Iron-aluminum-bound phosphorus
C) NaOH (0.1 M)	Polyphoshates (Note: if the BD step is not performed, NaOH (0.1 M) extracts iron- and aluminum-bound phosphorus plus polyphosphates)
D) HCl (0.5 M)	Calcium-bound phosphorus
E) Hot (85°C) NaOH (0.1 M)	Refractory phosphorus

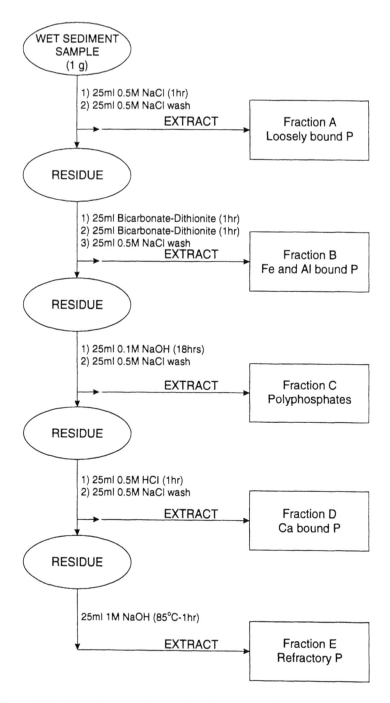

Figure 7-10   Flow chart for the fractionation of phosphorus species in aquatic sediments.

7.5.6.2 Apparatus

- Test tube shaker
- Autoanalyzer, colorimeter systems, and dialyzer (see Apparatus for Section 7.5.4)
- Centrifuge (see Apparatus for Section 7.5.4)

7.5.6.3 Reagents

- Stock phosphate standard: see Reagents for Phosphate Standards in Section 7.5.4
- Working phosphate standard: see Reagents for Phosphate Standards in Section 7.5.4
- Ascorbic acid, ammonium molybdate, antimony potassium tartrate, and combined working reagent: see Reagents for Method II in Section 7.5.4
- Concentrated sulfuric acid (H_2SO_4), specific gravity 1.84
- NaCl (0.5 M)
- Bicarbonate-Dithionite (BD): $Na_2S_2O_4$ (0.11 M) and $NaHCO_3$ (0.11 M) buffered to a pH of 7.0
- NaOH (0.1 M)
- HCl (0.5 M)

7.5.6.4 Procedure

Weigh 1 g of wet sample into 100 mL centrifuge tube with a lid.

7.5.6.4.1 Fraction A. Shake the sediment in the centrifuge tube for 1 h with 25 mL of NaCl (0.5 M). Centrifuge for 10 min at 5000 rpm and decant the supernatant into a 100-mL volumetric flask (Fraction A). Repeat the extraction with a second aliquot of NaCl (0.5 M), centrifuge, and combine the second supernatant with Fraction A. Acidify Fraction A in the volumetric flask by adding 1 mL of H_2SO_4 (1 M) and make up to volume. This Fraction A contains loosely bound or exchangeable phosphorus.

7.5.6.4.2 Fraction B. Extract sample left in centrifuge tube from the previous extraction step by shaking with 25 mL of BD reagent for 1 h. Centrifuge for 5 min at 5000 rpm and transfer the supernatant to a 200-mL volumetric flask (Fraction B). Repeat with a second 25 mL of BD reagent. Centrifuge, and add the supernatant to Fraction B in the volumetric flask. Wash the sediment in the centrifuge tube with 25 mL of NaCl (0.5 M), centrifuge, and add the supernatant to Fraction B in the volumetric flask. Acidify Fraction B in the volumetric flask with 8 mL of H_2SO_4 (1 M) and make up to volume. This Fraction B contains iron- and aluminum-bound phosphorus.

7.5.6.4.3   Fraction C.   Extract sample left in the centrifuge tube from the the previous extraction step by shaking with 25 mL of NaOH (0.1 M) for 18 h. Centrifuge for 5 min at 5000 rpm and transfer the supernatant to a 100-mL volumetric flask (Fraction C). Wash the sediment in the centrifuge tube with 25 mL of NaCl (0.5 M), centrifuge, and add the supernatant to Fraction C in the volumetric flask. Acidify Fraction C in the volumetric flask with 3 mL of H_2SO_4 (1 M) and make up to volume. This Fraction C contains polyphosphates.

7.5.6.4.4   Fraction D.   Extract the sediment left in the centrifuge tube after the previous extraction by shaking with 25 mL of HCl (0.5 M) for 1 h. Centrifuge for 5 min at 5000 rpm and transfer the supernatant to a 100-mL volumetric flask (Fraction D). Wash the sediment in the centrifuge tube with 25 mL of NaCl (0.5 M), centrifuge, and add the supernatant to Fraction D in the volumetric flask. Make up to the volume. This Fraction D contains the calcium-bound phosphorus.

7.5.6.4.5   Fraction E.   Extract the sediment left in the centrifuge tube after previous extraction step with 25 mL of NaOH (1 M) at 85°C for 1 h. Cool, centrifuge for 5 min at 5000 rpm, and transfer to a 100-mL volumetric flask. Acidify with 3 mL of H_2SO_4 (1 M) and make up to volume. This is Fraction E, and it contains residual or refractory phosphorus.

The determination of the concentrations of phosphorus in the extracts can be performed as described above, except for Fraction B. For this fraction, the solution must be extracted into an organic phase to remove interference from the reagents. This can be carried out by extracting the phosphomolybdate complex into hexanol/isopropanol with subsequent manual colorimetric determination. The other fractions provide extracts in the aqueous phase and could be readily analyzed for phosphorus by either the automated or manual methods given above.

7.5.6.5 Comments

It is very important to ensure that the standards contain the same concentrations of reagents used in the samples. The pH of the extracts must be compatible with the used detection system, whether it is automated or manual. This is particularly important when using the more sensitive but also more pH-dependent, antimony tartrate version of the ascorbic acid/molybdate technique. If the BD step is eliminated due to analytical complications, the NaOH (1 M) extract (Fraction C) would contain iron- and aluminum-bound phosphorus together with the polyphosphates. This option may be considered if speed and analytical productivity are of concern.

7.6   Data quality control

The International Standards Organization (ISO) has developed guidelines by which high standards of laboratory conduct can be achieved by the implementation of an effective quality management system. The International Standards Organization Guide 25 (International Standards Organization, 1990) is focused on quality management systems for analytical laboratories. In addition, the Canadian Association of Environmental Analytical Laboratories (CAEAL), in cooperation with the Standards Council of Canada (SCC), has developed the guide CAN/CSA-Z753-94 (Canadian Standards Association, 1994) as a basis for laboratory accreditation programs. Furthermore, the Association of Official Analytical Chemists, International (AOAC) peer-verified methods program provides a comprehensive set of guidelines on the validation of analytical methods (Association of Official Analytical Chemists, International, 1993). Laboratory quality control and good laboratory practices are the cornerstones of acceptable laboratory quality control systems. These activities include the use of properly validated analytical methods with good accuracy and precision. The ongoing use of control samples, duplicates, spikes, and blanks ensure that accuracy, precision, analyte recovery, and level of contamination are being monitored. The importance of the use of certified reference materials on the quality assessment of analytical data has been emphasized by Chau and Aspila (1991). Control and verification standards that are prepared from a different source than the calibration standards and that are analyzed frequently within a given analytical run ensure that an accurate and stable calibration method is maintained. Although quality assurance activities are very time consuming and expensive to implement, they have become the only acceptable way of performing analytical tests. This has made time- and cost-saving measures, such as laboratory automation, a necessity. Most tedious and manual methods for the sequential processing of a single sample at a time are typically rejected for economic and quality reasons. Investigators have sought multisample preparation techniques that save time, while ensuring that the whole batch of samples is treated exactly the same way. In addition, modern analytical systems are equipped with automated sample introduction and sample management procedures. Furthermore, robotic sample preparation systems have been introduced to improve productivity, quality, and level of safety. Figure 7-11 shows an application of a robot arm to the automated preparation of samples for total Kjeldahl nitrogen hot block digestion (Agemian et al., 1992).

Figure 7-11   Application of a robot arm to the automated preparation of TKN samples for hot block digestion. (Agemian, H., Ford, J.S., and Madsen, N.K., *Chemometr. Intelligent Lab. Syst.*, 17, 145, 1992. With permission.)

References

Agemian, H. and Chau, A.S.Y., Evaluation of extraction techniques for the determination of metals in aquatic sediments, *Analyst*, 101, 761, 1976.

Agemian, H. and Chau, A.S.Y., A study of different analytical extraction methods for nondetrital heavy metals in aquatic sediments, *Arch. Environ. Contam. Toxicol.*, 69, 6, 1977.

Agemian, H., Ford, J.S., and Madsen, N.K., A robotic system for the preparation of environmental samples for the determination of nutrients, *Chemometr. Intelligent Lab. Syst.*, 17, 145, 1992.

Agemian, H., Sturtevant, D.P., and Austen, K.D., Simultaneous acid extraction of six trace metals from fish tissue by hot-block digestion and determination by atomic absorption spectrometry, *Analyst*, 105, 125, 1980.

Allison, L.E., Wet-combustion apparatus and procedure for organic and inorganic carbon in soil, *Soil Sci. Soc. Am. Proc.*, 24, 36, 1960.

American Public Health Association, Standard Methods for the Examination of Water and Wastewater, 18th ed., American Public Health Association, Washington, D.C., 1992.

Anderson, F.O. and Jensen, H.S., Regeneration of inorganic phosphorus and nitrogen from seston in a freshwater sediment, *Hydrobiologia*, 228, 71, 1992.

Anderson, J.M., An ignition method for determination of total phosphorus in lake sediments, *Water Res.*, 10, 329, 1976.

Aspila, K.I., Agemian, H., and Chau, A.S.Y., A semiautomated method for the determination of inorganic, organic and total phosphate in sediments, *Analyst*, 101, 187, 1976.

Association of Official Analytical Chemists, International, AOAC peer-reviewed methods program: Manual on Policies and Procedures, AOAC International, Arlington, VA, 1993.

Bailey, G.W., Role of Soils and Sediment in Water Pollution Control — Part I: Reactions of Nitrogen and Phosphorus Compounds with Soils and Geological Strata, U.S. Department of the Interior, Federal Water Pollution Control Administration, Southeast Water Laboratory, 1968, 90.

Barbanti, A. and Sighinolfi, G.P., Sequential extraction of phosphorus and heavy metals from sediments: methodology considerations, *Environ. Tech. Lett.*, 9, 127, 1988.

Black, C.A., Methods of Soil Analysis — Part 2 — Chemical and Microbiological Properties, 1st ed., Agronomy Series No. 9, American Society of Agronomy, Inc. and Soil Science of America, Inc., Madison, WI, 1965.

Bligh, E.G. and Dyer, W., A rapid method for total lipid extraction and purification, *Can. J. Biochem. Physiol.*, 37, 911, 1959.

Boers, P.C.M., Cappenberg, T.E., and Raaphorst, W. van, The third international workshop on phosphorus in sediments — summary and synthesis, *Hydrobiologia*, 253, XI, 1993.

Bolin, B. and Cook, R.B., *The Major Biogeochemical Cycles and Their Interactions*, SCOPE 21 (Scientific Committee on Problems of the Environment), John Wiley & Sons, New York, 1983, 532.

Bonanni, P., Caprioli, R., Ghiara, E., Mignuzzi, C., Orlandi, C., Paganin, G., and Monti, A., Sediment interstitial water chemistry of the Orbetello lagoon (Grosseto, Italy); nutrient diffusion across the water-sediment interface, *Hydrobiologia*, 235, 553, 1992.

Bostrom, B., Persson, G., and Broberg, B., Bioavailability of different phosphorus forms in freshwater systems, *Hydrobiologia*, 170, 133, 1988.

Bremner, J.M., Organic nitrogen in soil, in Soil Nitrogen, Bartholomew, W.V. and Clark, F.E., Eds., Agronomy Series No. 10, American Society of Agronomy, Madison, WI, 1965.

Bremner, J.M., Nitrogen-Urea, in Methods of Soil Analysis — Part 2 — Chemical and Microbiological Properties, Page, A.L., Miller, R.H., and Keeney, D.R., Eds., 2nd ed., Agronomy Series No. 9 (Part 2), American Society of Agronomy, Inc. and Soil Science of America, Inc., Madison, WI, 1982, 699.

Bremner, J.M. and Mulvaney, C.S., Nitrogen-Total, in Methods of Soil Analysis — Part 2 — Chemical and Microbiological Properties, Page, A.L., Miller, R.H., and Keeney, D.R., Eds., 2nd ed., Agronomy Series No. 9 (Part 2), American Society of Agronomy, Inc. and Soil Science of America, Inc., Madison, WI, 1982, 595.

Broberg, O. and Persson, G., Particulate and dissolved phosphorus forms in freshwater: composition and analysis, *Hydrobiologia*, 170, 61, 1988.

Broberg, O. and Pettersson, K., Analytical determination of orthophosphate in water, *Hydrobiologia*, 170, 45, 1988.

Buchanan, J.B., Sediment, in *Methods for the Study Benthos*, Holme, N.M. and McIntyre, A.D., Eds., Blackwell, Oxford, 1971, 30.

Canadian Standards Association, Requirement for the Competence of Environmental Analytical Laboratories, CSA, Consolidated mailing list, 178 Rexdale Blvd., Rexdale (Toronto), Ontario, Canada, M9W-1R3, 1994.

Chang, S.C. and Jackson, M.L., Fractionation of soil phosphorus, *Soil Sci.*, 84, 133, 1957.

Chang, S.C. and Jackson, M.L., Soil phosphorus fractions in some representative soils, *J. Soil Sci.*, 9, 109, 1958.

Chau, A.S.Y. and Aspila, K.I., Reference materials and certified reference materials: an essential ingredient in environmental research and management, *Accountability Res.*, 1, 207, 1991.

De Groot, C.J., Some remarks on the presence of organic phosphates in sediments, *Hydrobiologia*, 207, 303, 1990.

De Groot, C.J. and Golterman, H.L., Sequential fractionation of sediment phosphate, *Hydrobiologia*, 192, 143, 1990.

De Groot, C.J. and Golterman, H.L., On the presence of organic phosphate in some Camargue sediments: evidence for the importance of phytate, *Hydrobiologia*, 252, 117, 1993.

D'Elia, C.F., Nitrogen determination in seawater, in *Nitrogen in the Marine Environment*, Carpenter, E.J. and Capone, D.G., Eds., Academic Press, New York, 1983, 731.

Delwiche, C.C., *Denitrification, Nitrification and Atmospheric Nitrous Oxide*, John Wiley & Sons, New York, 1981, 286.

Dionex, Determination of nitrite, nitrate and ammonia in KCl soil extracts, Application Update Au 118, Dionex Corporation, P.O. Box 3603, Sunnyvale, CA, 94088-3603, 1987.

Donaldson, E.S., Methods for the Analysis of Ores, Rocks and Related Materials, Monograph 881, 2nd ed., Canadian Centre for Mineral and Energy Technology, 555 Booth St., Ottawa, Ontario, Canada, K1A-0G1, 1982.

Downes, M.T., An automated determination of low reactive phosphorus concentrations in natural waters in the presence of arsenic, silicon and mercuric chloride, *Water Res.*, 12, 743, 1978.

Einsele, W.G., Uber die beziehungen des eisenkreislaufs zum phosphatkreislauf im eutrophen, *Arch. Hydrobiol.*, 29, 664, 1936.

Einsele, W.G., Uber chemische und kolloidchemishe vorgange in eisen-phosphat-systemen unter limnochremischen und limnogeologischen gesichtspunkten, *Arch. Hydrobiol.*, 33, 361, 1938.

Fabiano, M. and Danovaro, R., Composition of organic matter in sediments facing a river estuary (Tyrrhenian Sea): relationships with bacteria and microphytobenthic biomass, *Hydrobiologia*, 277, 71, 1994.

Fife, C.V., An evaluation of ammonium fluoride as a selective extractant for aluminum-bound soil phosphate: IV, *Soil Sci.*, 96, 112, 1963.

Fishman, M.J., Methods of analysis by the U.S. Geological Survey National Water Quality Laboratory — Determination of Inorganic and Organic Constituents in Water and Fluvial Sediments, U.S. Geological Survey, Open-file report 93-125, Denver, CO, 1993.

Fishman, M.J., and Friedman, L.C., Techniques of water-resources investigations of the United States Geological Survey, in *Methods for Determination of Inorganic Substances in Water and Fluvial Sediments*, Book 5, chapter A1, U.S. Department of Interior, Geological Survey, U.S. Government Printing Office, Washington, D.C., 1989.

Flannery, R.L. and Markus, D.K., Automated analysis of soil extracts for phosphorus, potassium, calcium, and magnesium, *J. Assoc. Off. Anal. Chem.*, 63, 779, 1980.

Foehrenbach, J., Eutrophication, *J. Water Poll. Contr. Feder.*, 45, 1237, 1973.

Fry, J.C., Detritus and microbial ecology in aquaculture, in *ICLARM Conf. Proc.*, Moriarty, D.J.W. and Pullin, R.S.V., Eds., 14, 83, 1987.

Gachter, R. and Meyer, J.S., The role of microorganisms in mobilization and fixation of phosphorus in sediments, *Hydrobiologia*, 253, 103, 1993.

Gachter, R., Meyer, J.S., and Mares, A., Contribution of bacteria to release and fixation of phosphorus in lake sediments, *Limnol. Oceanogr.*, 33, 1542, 1988.

Gallaher, R.N., Weldon, C.O. and Boswell, F.C., A semiautomated procedure for total nitrogen in plant and soil samples, *Soil Sci. Soc. Am. J.*, 40, 887, 1976.

Gerchacov, S.M. and Hatcher, P.G., Improved technique for analysis of carbohydrates in sediments, *Limnol. Oceanogr.*, 17, 938, 1972.

Golterman, H.L., Differential extraction of sediment phosphates with NTA solutions, *Hydrobiologia*, 92, 683, 1982.

Golterman, H.L. and Booman, A., The sequential extraction of Ca and Fe-bound phosphates, *Int. Assoc. Theor. Appl. Limnol.*, 23, 904, 1988.

Golterman, H.L. and Wurtz, I.M., A sensitive rapid determination of inorganic phosphate in presence of labile phosphate esters, *Anal. Chim. Acta.*, 25, 295, 1961.

Graf, G., Schulz, R., Peinert, R., and Meyer-Reil, L.A., Benthic response to sedimentation events during autumn to spring at a shallow water station in Western Kiel Bay I. Analysis of processes on a community level, *Mar. Biol.*, 77, 235, 1983.

Harwood, J.E., Van Steenderen, R.A., and Kuhn, A.L., A rapid method for orthophosphate analysis at high concentrations in water, *Water Res.*, 3, 417, 1969.

Hegemann, D.A., Johnson, A.H. and Keenan, J.J., Determination of algal-available phosphorus on soil and sediment: a review and analysis, *J. Environ. Qual.*, 12, 12, 1983.

Hendricks, A.C. and Silvey, J.K.G., Nutrient ratio variation in reservoir sediments, *J. Water Poll. Contr. Feder.*, 45, 490, 1973.

Henriksen, A. and Selmer-Olsen, A.R., Automated methods for determining nitrate and nitrite in water and soil extracts, *Analyst*, 95, 514, 1970.

Hesse, P.R., *A Textbook of Soil Chemical Analysis*, Chemical Publishing Company Inc., New York, 1971, 520.

Hieltjes, A.H.M. and Lijklema, L., Fractionation of inorganic phosphates in calcareous sediments, *J. Environ. Qual.*, 9, 405, 1980.

Holtan, H., Kamp-Nielson, L., and Stuanes, A.O., Phosphorus in soil, water and sediment: an overview, *Hydrobiologia*, 170, 19, 1988.

Hongve, D., Nutrient metabolism (C, N, P, Si) in trophogenic zone of a meromicide lake, *Hydrobiologia*, 277, 17, 1994.

International Standards Organization, General requirements for the competence of calibration and testing laboratories, Guide 25, 1990, ISO, Case postale 56, CH-1211, Geneva 20, Switzerland.

Jackson, M.L., *Soil Chemical Analysis*, Prentice-Hall, Englewood Cliffs, NJ, 1958, 498.

Jensen, H.S. and Thamdrup, B., Iron-bound phosphorus in marine sediments as measured by bicarbonate-dithionate extraction, *Hydrobiologia*, 253, 47, 1993.

Jirka, A.M., Carter, M.J., May, D., and Fuller, F.D., Ultramicro semiautomated method for simultaneous determination of total phosphorus and total Kjeldahl nitrogen in wastewaters, *Environ. Sci. Technol.*, 10, 1038, 1976.

Kamphake, L.J., Hannah, S.A., and Cohen J.M., Automated analysis for nitrite by hydrazine reduction, *Water Res.*, 1, 205, 1967.

Keeney, D.R., The nitrogen cycle in sediment–water systems, *J. Environ. Qual.*, 2, 15, 1973.

Keeney, D.R. and Bremner, J.M., Use of the Coleman Model 29A analyser for total nitrogen analysis of soils, *Soil Sci.*, 104, 358, 1967.

Keeney, D.R. and Nelson, D.W., Nitrogen-inorganic forms, in Methods of Soil Analysis — Part 2 — Chemical and Microbiological Properties, Page, A.L., Miller, R.H. and Keeney, D.R., Eds., 2nd ed., Agronomy Series No. 9 (Part 2), American Society of Agronomy, Inc. and Soil Science of America, Inc., Madison, WI, 1982, 643.

Keith, L.H., Principles of Environmental Sampling, American Chemical Society, 1988, 458.

Kerr, P.C., Brockway, D.L., Paris, D.F., and Craven, S.E., Carbon cycle in sediment–water systems, *J. Environ. Qual.*, 2, 46, 1973.

Kopacek, J. and Hejzlar, J., Direct determination of particulate phosphorus in water with perchloric acid digestion of whole membrane filters, *Int. J. Environ. Anal. Chem.*, 54, 27, 1993.

Kotsch, R.W., D'Itri, F.M., and Upchurch, S.B., A direct method for measuring inorganic and organic carbon in recent sediments, *J. Sed. Petrol.*, 46, 1026, 1976.

Lambert, D. and Maher, W., An evaluation of the efficiency of the alkaline persulphate digestion method for the determination of total phosphorus in turbid waters, *Water Res.*, 29, 7, 1995.

Leventhal, J.S. and Shaw, V.E., Organic matter in Appalachian Devonian black shale: I. Comparison of techniques to measure organic carbon, II. Short range organic carbon content variations, *J. Sed. Petrol.*, 50, 77, 1980.

Lucotte, M. and D'Anglejan, B., A comparison of several methods for the determination of iron hydroxides and associated orthophosphates in estuarine particulate matter, *Chem. Geol.*, 48, 257, 1985.

Martinova, M.V., Nitrogen and phosphorus compounds in bottom sediments: mechanisms of accumulation, transformation and release, *Hydrobiologia*, 252, 1, 1993.

Mehta, N.C., Legg, J.O., Goring, C.A.I., and Black, C.A., Determination of organic phosphorus in soils: I. Extraction method, *Soil Sci. Soc. Proc.*, 18, 443, 1954.

Meybeck, M., Carbon, nitrogen and phosphorus transported by world rivers, *Am. J. Sci.*, 282, 401, 1982.

Mortimer, C.H., The exchange of dissolved substances between mud and water in lakes — I, *J. Ecol.*, 29, 280, 1941.

Mortimer, C.H., The exchange of dissolved substances between mud and water in lakes — II, *J. Ecol.*, 30, 147, 1942.

Mudroch, A. and Azcue, J.M., *Manual of Aquatic Sediment Sampling*, CRC/Lewis, Boca Raton, FL, 1995, 219.

Mudroch, A. and MacKnight, S.D., *Handbook of Techniques for Aquatic Sediments Sampling*, 2nd ed., CRC/Lewis, Boca Raton, FL, 1994, 236.

Mulholland, P.J. and Elwood J.W., The role of lake reservoir sediments as sinks in the perturbed global carbon cycle, *Tellus*, 34, 490, 1982.

Murphy, J. and Riley, J.P., A modified single solution method for the determination of phosphate in natural waters, *Anal. Chim. Acta.*, 27, 31, 1962.

Nakas, J.P. and Litchfield, C.D., Application of the diacetyl-monoxime thiosemicarbazide method to the analysis of urea in estuarine sediments, *Estuarine Coastal Mar. Sci.*, 5, 143, 1977.

Namiki, H., Spectrophotometric determination of phosphate in sewage by extraction of molybdenum blue, *Bunseki Kagaku*, 10, 945, 1961.

National Laboratory for Environmental Testing, Manual of Analytical Methods, Vol. 1, Major Ions and Nutrients, National Laboratory for Environmental Testing, Canada Centre for Inland Waters, 867 Lakeshore Rd., P.O. Box 5050, Burlington, Ontario, Canada, L7R 4A6, 1994.

Nelson, D.W. and Sommers, L.E., Total nitrogen analysis of soil and plant tissue, *J. Assoc. Off. Anal. Chem.*, 63, 770, 1980.

Nelson, D.W. and Sommers, L.E., Total carbon, organic carbon, and organic matter, in Methods of Soil Analysis — Part 2 — Chemical and Microbiological Properties, Page, A.L., Miller, R.H., and Keeney, D.R., Eds., 2nd ed., Agronomy Series No. 9 (Part 2), American Society of Agronomy, Inc. and Soil Science of America, Inc., Madison, WI, 1982, 539.

Nommik, H., A modified procedure for determination of organic carbon in soils by wet combustion, *Soil Sci.*, 111, 330, 1971.

Olsen, S.R. and Dean, L.A., Phosphorus, in Methods of Soil Analysis — Part 2 — Chemical and Microbiological Properties, 1st ed., Black, C.A., Ed., Agronomy Series No. 9 (Part 2), American Society of Agronomy, Inc., Madison, WI, 1965, 1035.

Olsen, S.R. and Sommers, L.E., Phosphorus, in Methods of Soil Analysis — Part 2 — Chemical and Microbiological Properties, Page, A.L., Miller, R.H., and Keeney, D.R., Eds., 2nd ed., Agronomy Series No. 9 (Part 2), American Society of Agronomy, Inc. and Soil Science of America, Inc., Madison, WI, 1982, 403.

Page, A.L., Miller, R.H., and Keeney, D.R., Methods of Soil Analysis — Part 2 — Chemical and Microbiological Properties, 2nd ed., Agronomy Series No. 9 (Part 2), American Society of Agronomy, Inc. and Soil Science of America, Inc., Madison, WI, 1982.

Parker, J.G., A comparison of methods used for the measurement of organic matter in marine sediment, *Chem. Ecol.*, 1, 201, 1983.

Patton, C.J. and Truitt, E.P., Methods of analysis by the U.S. Geological Survey National Water Quality Laboratory — Determination of Total Phosphorus by Kjeldahl Digestion Method and An Automated Colorimetric Finish that Includes Dialysis, Open-file report 92-146, U.S. Geological Survey, Box 25425, Mail Stop 517, Federal Centre, Denver, CO, 80225, 1992.

Pettersson, K., Bostrom., B., and Jacobsen, O., Phosphorus in sediments — speciation and analysis, *Hydrobiologia*, 170, 91, 1988.

Psenner, R., Pucsko, R., and Sager, M., Fraktionierung organischer und anorganischer phosphorverbindungen von sedimenten. Versuch einer Definition okologisch wichtiger Fraktionen, *Arch. Hydrobiol./Suppl.*, 70, 111, 1984.

Rao, C.R.M. and Reddi, G.S., Decomposition procedure with aqua regia and hydrofluoric acid at room temperature for the spectrophotometric determination of phosphorus in rocks and minerals, *Anal. Chim. Acta*, 237, 251, 1990.

Redfield, A C., Ketchum, R.H., and Richards, F.A., The influence of organisms on the composition of seawater, in *The Sea*, Vol. 2, Hill, M.N., Ed., Interscience, New York, 1963, 26.

Renmin, L., Daojie, L., Ailing, S., and Guihua, L., Chemiluminescence determination of nitrate with photochemical activation in a flow injection system, *Talanta*, 42, 437, 1995.

Rice, D.L., The detritus nitrogen problem: new observations and perspectives from organic geochemistry, *Mar. Ecol. Prog. Ser.*, 9, 153, 1982.

Riemann, B. and Sondergaard, M., *Carbon Dynamics in Eutrophic, Temperate Lakes*, Elsevier Science, New York, 1986, 284.

Riley, J.T., Burris, S.C., Stidam, J.M., Wu, A., and Zhang, D., Utilization of elemental analysers in university laboratories, *Am. Lab.*, Oct., 12, 1994.

Ruttenberg, K., Development of a sequential extraction method for different forms of phosphorus in marine sediments, *Limnol. Oceanogr.*, 37, 1460, 1992.

Sandstrom, M.W., Tirendi, F., and Nott, A., Direct determination of organic carbon in modern reef sediment and calcareous organisms after dissolution of carbonate, *Mar. Geol.*, 70, 321, 1986.

Saunders, J.F., III and Lewis, W.M., Transport of phosphorus, nitrogen and carbon by the Ampure River, Venezuela, *Biogeochemistry*, 5, 323, 1988.

Schuman, G.E., Stanley, M.A., and Knudsen, D., Automated total nitrogen analysis of soil and plant samples, *Soil Sci. Soc. Am. Proc.*, 37, 480, 1973.

Selmer-Olsen, A.R., Determination of ammonium in soil extracts by an automated indophenol method, *Analyst*, 96, 565, 1971.

Sommers, L.S., Harris, R.F., Williams, J.D.H., Armstrong, D.E., and Syers, J.K., Determination of total and organic phosphorus in lake sediments, *Limnol. Oceanogr.*, 15, 301, 1970.

Sommers, L.S. and Nelson, D.W., Determination of total phosphorus in soils: a rapid perchloric acid digestion procedure, *Soil Sci. Soc. Am. Proc.*, 36, 902, 1972.

Stevenson, F.J., Nitrogen-organic forms, in Methods of Soil Analysis — Part 2 — Chemical and Microbiological Properties, Page, A.L., Miller, R.H., and Keeney, D.R., Eds., 2nd ed., Agronomy Series No. 9 (Part 2), American Society of Agronomy, Inc. and Soil Science of America, Inc., Madison, WI, 1982, 625.

Stewart, B.A., Porter, L.K., and Beard, W.E., Determination of total nitrogen and carbon in soils by a commercial Dumas apparatus, *Soil Sci. Soc. Am. Proc.*, 28, 366, 1964.

Sugavara, K. and Kanamori, S., Spectrophotometric determination of submicrogram quantities of orthophosphate in natural waters, *Bull. Chem. Soc. Jpn.*, 34, 258, 1961.

Sutcliffe, D.W. and Jones, J.G., Eutrophication: research and application to water supply, *Freshwater Biol. Assoc.*, 2, 231, 1992.

Svensson, B.H. and Soderlund, R., Nitrogen, Phosphorus and Sulphur — Global Cycles, SCOPE Report 7, Ecol. Bull. 22, Stockholm, Sweden, 1976, 192.

Syers, J.K., Harris, R.F., and Armstrong, D.E., Phosphate chemistry in lake sediments, *J. Environ. Qual.*, 2, 1, 1973.

Syers, J.K., Williams, J.D.H., and Walker, T.W., The determination of total phosphorus in soils and parent materials, *N.Z.J. Agric. Res.*, 11, 757, 1968.

Trabalka, J.R. and Reichle, D.E., *The Changing Carbon Cycle — A Global Analysis*, Springer-Verlag, New York, 1986, 592.

Upchurch, J.B., Edzwald, J.K., and O'Melia, C.R., Phosphates in sediments of Pamlico estuary, *Environ. Sci. Technol.*, 8, 57, 1974.

U.S. Environmental Protection Agency, Methods for Chemical Analysis of Water and Wastes, Environmental Monitoring and Support Laboratory, Cincinnati, OH, 1983.

Van Eck, G.T.M., Forms of phosphorus in particulate matter from the Hollands Diep/Haringvliet, The Netherlands, *Hydrobiologia*, 92, 665, 1982.

Van Schouwenburg, J.C. and Walinga, I., The rapid determination of phosphorus in the presence of arsenic, silicon and germanium, *Anal. Chim. Acta*, 37, 271, 1967.

Verardo, D.J., Froelich, P.N., and McIntyre, A., Determination of organic carbon and nitrogen in marine sediments using the Carlo Erba NA-1500 Analyser, *Deep Sea Res.*, 37, 157, 1990.

Vollenweider, R.A., Scientific Fundamentals of the Eutrophication of Lakes and Flowing Waters with Particular Reference to Nitrogen and Phosphorus as Factors in Eutrophication, Organization Economic Cooperation and Development, Paris, DAS/CSI/68, 1968, 27.

Wada, E. and Hattori, A., *Nitrogen in the Sea: Forms, Abundance and Rate Processes*, CRC Press, Boca Raton, FL, 1991, 208.

Watanabe, F.S. and Olsen, S.R., Colorimetric determination of phosphorus in water extracts of soil, *Soil Sci.*, 93, 183, 1962.

Wildung, R.E., Schmidt, R.L., and Gahler, A.R., The phosphorus status of eutrophic lake sediments as related to changes in limnological conditions — total, inorganic, and organic phosphorus, *J. Environ. Qual.*, 3, 133, 1974.

Williams, J.D.H., Mayer, T., and Nriagu, J.O., Extractability of phosphorus from phosphate minerals common in soils and sediments, *Soil Sci. Soc. Am. J.*, 44, 462, 1980.

Williams, J.D.H., Syers, J.K., and Walker, T.W., Fractionation of soil inorganic phosphate by a modification of Chang and Jackson's procedure, *Soil Sci. Soc. Am. Proc.*, 31, 736, 1967.

Won, C.H., Sensitive spectrophotometric determination of orthophosphate in natural waters by a modified molybdenum (V) thiocyanate method, *Nippon Kagaku Zasshi*, 85, 859, 1964.

Wong, H.K.T. and Kemp, A.L.W., The determination of total nitrogen in sediments using an induction furnace, *Soil Sci.*, 124, 1, 1977.

chapter eight

Supercritical fluid extraction of organic contaminants in sediments

Hing-Biu Lee and Dennis R. Gere

8.1 Introduction

Supercritical fluid extraction (SFE) is a technique in which the extracting "liquid" is a supercritical fluid. The sample is usually in a solid or semisolid state and, in environmental applications, can be sludge, soil, sediment, or biological tissue. Even liquids may be extracted by SFE if the liquids are first passed through a solid phase extraction (SPE) material, such as an Empore filter, to sorb the analytes of interest onto a solid bed, with the solvent (not of analytical interest) passing to waste. The solid phase material containing the analytes of interest is then placed in the SFE extraction vessel for the last stage of sample preparation. Examples of this will be discussed in later sections of this chapter.

Supercritical fluid extraction appears, to the beginner in this field, to be a complicated technology. The authors would like to dispel some of the perceptions that lead to this concern by showing examples of robust, yet simple and efficient, SFE methods applied to environmental samples.

One of the major concerns in environmental protection is the need to reduce, in the analytical laboratory, the use of organic solvents that are restricted by the Montreal Protocol. This international treaty addresses the reduction of use of chemicals containing halogens such as chloro-fluoro hydrocarbons and other halocarbons, which cause ozone depletion in the atmosphere (Hileman, 1993; Zurer, 1993). Many of these solvents are commonly used in environmental sample preparation. Supercritical fluid extraction uses primarily supercritical CO_2 as the extraction fluid and thus can substantially reduce the solvent usage and disposal as expected by the Montreal Protocol.

Another challenge in today's laboratory is the need to improve productivity. This consideration involves the idea that the SFE may be considered as a stand-alone operation. That is, an operator can place weighed samples in the appartus, walk away, and come back when the sample is extracted. The analytes are in an appropriate solution for direct introduction (i.e., injection) into analytical equipment. It is expected that there will be continued improvements and developments of the method that will allow enhanced automation.

There are substantially reduced costs derived from the use of SFE vs. traditional extraction in the areas of solvent purchase costs, solvent disposal costs, and reduced labor charges. Switching to SFE also significantly reduces the human intervention experienced in classical manual extraction techniques, such as Soxhlet and liquid-solid extractions. The elimination of potential human error further reduces labor costs and improves the overall data quality in today's environmental analysis laboratory.

Supercritical fluid extraction has a good potential for selective extractions. It is therefore very useful as a sample cleanup technique prior to an analytical determination such as gas chromatography. David et al. (1992, 1993) described the SFE studies in which lipid-free fractions of polychlorinated biphenyls (PCBs) were extracted from sea gull eggs. Independent extraction and analysis of the lipid fraction of the eggs indicated a fat content of 35% by weight, while, after the selective SFE extraction with judicious choice of density/temperature conditions at each fractionation step, the fat or lipid content was less than 0.1%. These studies provided the source information later used to develop a robust method of extracting PCBs from environmental samples that may contain significant amounts of lipids.

It is appropriate to discuss briefly the topic of overall sample size for SFE. Most SFE methods in environmental analysis have been developed for general sample sizes of 1 to 5 g. Traditional Soxhlet extractions require sample sizes significantly larger (i.e., as much as 30 to 50 g) to achieve the same degree of precision. The smaller sample size of the SFE method requires serious consideration of the time-honored statistical sampling protocols (Diehl, 1974). If the material to be analyzed is homogeneous, subsampling is no problem. Any portion of the mass may be taken as the sample for the analysis. If the sample is inhomogeneous, however, the problem is not so simple, for a small portion taken at one point may not at all represent the composition of the total mass. Obviously, the problem is more difficult if the particles are large and if they vary greatly in composition from piece to piece. Under such conditions, the sample first taken, the so-called gross sample, is quite large, and this large preliminary sample is subjected to a careful process of alternate crushing and dividing until a suitable amount of material of much smaller particle size remains. In actual practice in industrial and clinical laboratories, consideration must always be given to the origin of the sample and the manner in which it was taken and subsequently handled. Often the best analysis is one that is not made at all because the sample was faulty and the labor of any analysis would only have been wasted.

8.2 Principles of supercritical fluids

In practice, there are many and varied definitions of the supercritical fluid state. Some discussions suggest that supercritical fluid is a fourth state of matter. That is simply not true, and the assertion leads to some confusion because a supercritical fluid may have properties of more than one of the common three states of matter, i.e., solid, liquid, and gas. It may be somewhat easier to explain this by simple phase diagrams for the appreciation of what is the physico-chemical significance of a supercritical system. Figure 8-1 is a phase diagram for water in which pressure is plotted against temperature. The actual form of the diagram roughly looks like a forked branch of a tree. For the sake of qualitative discussion, the X-axis is neither linear nor logarithmic. Some literary freedom has been taken in drawing this diagram for the sake of simplification of the introductory remarks. A typical physical chemistry text will provide more accuracy for the advanced reader, if desired.

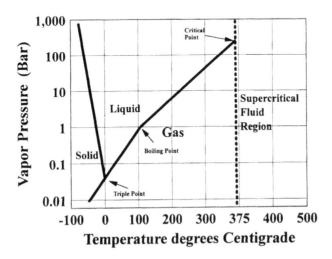

Figure 8-1 Phase diagram of water.

The three physical states of matter are depicted (i.e., zones from left to right): solid, liquid, and gas. A fourth zone but not a fourth state of matter at the far right is called the supercritical fluid region. The solid lines dividing the individual zones are the boundaries between the solid-liquid, solid-gas, and liquid-gas phases. The dotted line that sets off the supercritical fluid zone separates liquids from supercritical fluids and gases from supercritical fluids. This suggests that a supercritical fluid is neither a gas nor a liquid, but, in fact, that supercritical fluids possess some characteristics of both gases and liquids. There are also three points in this diagram that merit some discussion. The first is the point that is the juncture of the solid-gas, the solid-liquid, and the liquid-gas lines. This is known as the *triple point* and is the one set of invariant conditions in which solid, liquid, and gaseous states can coexist. For water, this point is at 0°C and 0.06 bar pressure. It is also commonly known as the freezing-melting point of water. Another invariant point on the diagram is the *critical point*. For water, this is at 374°C and 232 bar. A third point, T_b, is commonly known as the *boiling point* and is located at 100°C and 1.0 bar in Figure 8-1. The boiling point is a dependent-variable, which can be any point upon the line dividing liquid and gas phases; thus, it is not a true constant such as the triple point and the critical point.

In terms of importance or priority, the second point of consideration is the *critical temperature*. This invariant point, unique to each chemical, is defined by the temperature only. The critical pressure value appears to have a lesser or secondary significance since it is merely the vapor pressure for the chemical that exists at the critical temperature. The critical or supercritical fluid region exists at all pressures, at or above the critical temperature for a pure substance. At or above this critical temperature, there exists only one phase, completely independent of the pressure, that is, no matter how high or low the pressure is, the one phase will not condense to a liquid.

This becomes somewhat awkward when considering that there is no formal distinction between a "liquid" and a "gas". When two fluid phases of a pure substance coexist, the denser one may be called the liquid and the less dense one the gas. When there is only one fluid phase, any distinction is arbitrary. Nevertheless, the properties of a dilute gas, such as low density, high compressibility, and positive temperature coefficient of viscosity, are very different from those of a liquid at temperatures well below the critical temperature (i.e., high density, low compressibility, and negative temperature coefficient of viscosity).

An intriguing way of defining a critical (or supercritical) fluid is the consideration of kinetic energy and the potential energy of a simple closed system. The kinetic energy is a measure of the molecular motion activity or a manifestation of the degree of heat within a system. A higher kinetic energy implies a higher temperature and thus more motion of a given set of molecules within this closed system. Visualize a small cube in space containing one mole of a pure substance, 18 g of water or 44 g of CO_2, with the molecules vibrating, rolling, moving in three dimensions. The potential energy is the intermolecular glue holding similar molecules of a pure substance together in a condensed phase, such as a liquid. In Figure 8-1, visualize a point on the graph positioned on the X-axis at $-100°C$ and on the Y-axis at 100 bar. As one moves from left to right, traveling along an imaginary line parallel to the X-axis, the kinetic energy (i.e., the temperature) increases while the potential energy stays at a constant value, which is intrinsic to the type and number of atoms making up the molecules.

Approaching the solid-liquid line in Figure 8-1, the increase in kinetic energy now exceeds the potential energy. This represents enough energy to exceed the bonding strength of the three-dimensional physical connections, i.e., the transition from a solid to the less condensed liquid state. The kinetic energy continues to increase as our imaginary progression across the graph in Figure 8-1 moves over the liquid-gas line boundary. At that junction or line, sufficient kinetic energy is available in the system to allow molecules free three-dimensional travel over significantly longer distances (i.e., increased mean free path, often 100 to 1000 times the molecular diameter). This internal energy allows gases, as compared to liquids, at a constant pressure, significantly improved transport properties, such as lower viscosity and higher diffusivity.

In the gas region, before encountering the critical temperature or region, if two gas molecules of the same type collide, the collision is described as an inelastic collision. When gas molecules of the same kind collide, the probability is that the molecules will stick together for a short but finite period of time. This has an impact upon the transport properties of viscosity diffusivity (i.e., restricting the limiting values). One of the assumptions of our travel across the phase diagram on our route parallel to the X-axis is that the temperature is changing, but the pressure is constant. One consequence of this is that the density will decrease continuously. If we changed our assumption to include a constant density, then the transport properties would not change compared to the constant pressure scenario described above.

Continuing the imaginary trip from left to right in Figure 8-1, next encounted is the gas-supercritical region border. This transition is much more subtle than the previous transitions, as is denoted by a dotted line in Figure 8-1 rather than a solid line. It is instructive to remember that this border is the point at which the enthalpy of vaporization goes to a zero value or disappears from consideration. Thus, at and beyond the dotted transition line, there is no further energy needed to take a mole of pure compound from one "state" (i.e., gas) to the next "region" (i.e., supercritical fluid).

What then is the difference? The difference is that, in the supercritical region, molecules have the long mean free paths that they experienced in a gas, and now if one molecule collides with another similar molecule, the collision is described as an *elastic* collision. The molecules collide, and then they bounce off each other without a finite "sticking-together". This is another way of saying that at the critical temperature, the kinetic energy is equal and equivalent to the potential energy. Because this is independent of the pressure, the supercritical region goes from the bottom of the graph (i.e., the X-axis) all the way to an infinitely high pressure with no distinction at the pressure corresponding to the Y-axis position of the critical temperature. The only thing that is happening as one moves up in this diagram in the supercritical region is that the density increases with the pressure.

Consider what is more favorable in the supercritical region as the system conditions are changed (at constant pressure, but at ever-increasing temperature). The density continues to decrease. The transport properties of viscosity and diffusion or diffusivity become more favorable in the sense that the individual molecules of the same kind have less *drag* or retardation of their velocity as they pass close to each other or actually collide. These two properties have very positive effects upon extraction.

In extraction, more favorable diffusion implies a more efficient penetration of matrices with the supercritical fluid and thus a more favorable mass transfer, shorter extraction times as compared, for example, with liquid-solid Soxhlet-type extraction. More favorable viscosity implies the ability to use relatively high volume (or mass, or molar) flow rates while retaining small volume capillary tubing. Further, this low viscosity means that the fluid dynamics in the packed-matrix-extraction-thimble zone also will not contribute significantly to the overall pressure gradient from the pump through all of the apparatus to the restrictor-nozzle zone. This allows a well-controlled set of parameters for the extraction, such as flow rate of the fluid, density, pressure, diffusion, and repeatability of the method.

8.2.1 Supercritical fluid extraction hardware and the flowing fluid process

The SFE hardware used today in analytical laboratories, whether it is home-built or one of the commercially available systems, is virtually all derived from equipment, drawings, and patents of chemical engineering. The analytical equipment is usually just scaled down. This transformation is actually, in some cases, more difficult than it might appear at first consideration. This is especially true for the expansion nozzle (i.e., restrictor) needed to interface with the analytical SFE equipment (Figure 8-2).

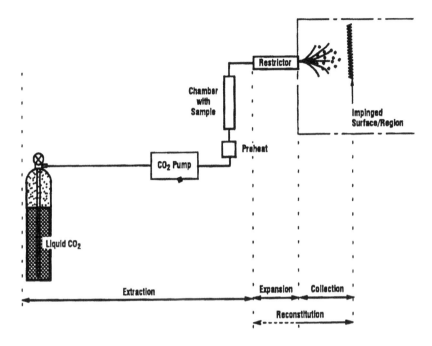

Figure 8-2 Generic supercritical fluid extraction hardware.

The equipment or hardware begins with the source of CO_2. In analytical SFE, liquid CO_2 of the highest purity (>99.9999%) that is contained in metal cylinders under relatively high pressure is used. The pressure in a given cylinder is the actual vapor pressure at ambient temperature. Liquid CO_2 has a vapor pressure of approximately 58 bar at 21°C. The triple point is low in temperature (–57°C), but high in pressure (5.1 bar).

A typical cylinder contains approximately 27 kg of CO_2. When it is full, the upper one-third of the cylinder volume is occupied by the gas phase and the lower two-thirds by the liquid phase. This phase ratio changes continually as the liquid is drawn out from the bottom of the cylinder by a dip-tube (siphon-tube or an eductor-tube), which comes down from the top on-off valve and extends down below the original liquid-gas interface. If the tank can be drawn down to the bottom of the dip-tube, the gas occupies four-fifths of the volume and the liquid remaining occupies only one-fifth of the volume. The dip-tube usually stops short of the bottom of the cylinder in order to avoid drawing-up into the tubing any particulate matter that may be at the bottom of the cylinder. When the liquid level drops below the open bottom end of the dip-tube, no more CO_2 can be drawn into the SFE apparatus.

The combination of remaining gaseous CO_2 and a small pool of liquid CO_2 below the bottom of the dip-tube usually make up approximately 4.6 kg of CO_2 that are left in the tank at the end of use. The actual fraction or weight of CO_2 left is a function of the actual dip-tube installed by the gas manufacturer. This can vary widely, and the reader is warned to check this carefully with the CO_2 supplier.

The substantial vapor pressure in the cylinder under ambient conditions is sufficient to feed enough liquid CO_2 for a normal reciprocating pump. Older design pumps, such as large syringe pumps, require an additional pressure charge to the tank (usually with helium) to artificially increase the vapor pressure to as much as 170 bar. Although this works in principle, there are many drawbacks, such as an inability to facilitate low density extractions, which favor selective extraction of volatile fractions (aroma constituents). Thus, this type of pump is no longer incorporated in the construction of new equipment. Pumps with pressure-measuring transducers provide an indication of the actual pressure, or in more sophisticated equipment, provide a feedback signal for pressure-temperature-density control. After the pump, the CO_2 (still in the liquid phase) passes through small diameter tubing until it approaches the extraction thimble. At this point or near to it, the temperature is adjusted to the choice desired for the supercritical extraction. That is, the CO_2 finally is adjusted to conditions (temperature-pressure-density) of the supercritical state. In Figure 8-2, this is indicated as a *preheat zone.*

As the CO_2 passes through the preheat zone and into the extraction thimble, the CO_2 becomes a supercritical fluid. The exact conditions are set by the temperature and pressure. The flow rate of the CO_2 does not affect the nature of the phase. After passing through the necessary capillary tubing, the CO_2, now containing any analytes or solutes that could possibly be dissolved from the matrix in the sample thimble, moves on to the *restrictor.*

The restrictor (nozzle) serves one important function and several secondary functions. The restrictor is the throttle point or the narrowing of the passageway for the moving fluid. It is the most restricted flow region in the entire flow path. The pump is the source of pressure as well as the flow rate determining device. The pump needs a control point downstream to limit the passage of molecules per unit time. This restriction then holds back the previously unlimited flow of molecules to a definite, predetermined level. Ideally, the restrictor serves to restrict the flow until the density of the fluid required for extraction is achieved.

In a simple, home-built apparatus, the restrictor often has dimensions of small diameter capillary tubing. For instance, a 15 cm length of 10-μm internal diameter tubing placed at the end of the apparatus may point into or within a sample collection thimble. This is one example of a so-called *fixed restrictor.* A fixed restrictor has a finite length and diameter. For a given volumetric (mass or molar) flow rate, it produces a pressure measurable some place between the pump and the final outlet point into the ambient. If the flow rate varies by itself or is deliberately varied by the experimenter, the pressure adjusts proportionately, as the restriction itself is fixed and constant.

The finite length of tubing has a distributed resistance, meaning that the resistance is additive from the top to the bottom of the capillary. As a result, the pressure drops continuously, but not necessarily linearly, from the operating extraction pressure of perhaps several hundred bar at the beginning of the restrictor down to one bar (ambient pressure) at the end of the restrictor. This decrease in pressure yields a decrease in density, which always leads to diminished solubility of analytes.

At the terminal end of the restrictor, the goal is the complete removal of solubility power of the flowing CO_2 with precipitation of the solutes. However, in the case of such a fixed restrictor with distributed resistance, the loss of solubility power will be gradual and continual throughout the length of capillary tubing. Thus, it is likely that precipitation will begin prematurely within the restrictor itself. If the concentration of solute is low and the resulting precipitate is formed in very small particle size, it may physically be blown through the open space in the restrictor and into the collection zone. More likely, however, it will adhere to the inner walls of the tubing and further restrict flow with an increase in pressure locally. In the worst case, the tubing closes itself completely to further flow, which causes a catastrophic system overpressure and cessation of extraction.

This situation is further compounded by the accompanying Joule-Thomson cooling effect, owing to the significant expansion of the CO_2 from a highly dense fluid, perhaps at several hundred bar pressure, down to a nondense gas at ambient pressure. This cooling can produce a temperature gradient over a very short distance of several hundred degrees centigrade. The cooling of the narrow opening of the tubing encourages freezing of high concentration bulk solutes such as water or lipids, for example. Localized heating of the capillary tube can help minimize this effect, but the actual temperature setting is problematic. If the tubing is fused silica, it often breaks under these conditions of stress. Stainless steel capillary is somewhat more tolerant, but most people choose to dry their sample to avoid this. That avoids the problem if water is the cause of the effect. If lipids or oils are causing the problem, it can be even more difficult.

This situation and other considerations, such as the distributed resistance across a finite distance of capillary tubing, lead to the design and use of fixed restrictors of many varieties. Conceptually this is easy: one simply takes a small piece of capillary-fused silica tubing, and with a small flame, the end of the tubing is drawn down to a very small diameter. In practice, this can be difficult to control. The first question might be what diameter one would desire. The second question might be how a given size opening at the terminus of the tubing might be obtained. If the first two questions can be answered, a third question might be how we precisely and accurately repeat the process whenever needed. The answers to all of these three questions are difficult to obtain for the average laboratory staff member.

The first question brings up the observation that a fixed restrictor links the flow and pressure to each other. That is, a fixed restrictor will provide a specific pressure for a defined flow rate. As the flow rate would vary, the pressure would vary, as a dependent variable. That means that as the flow would increase, the pressure will linearly increase, and, similarly, a decrease in flow would yield a linear decrease in pressure. If this were combined in an environment where the resistance changes as a solute flux passes, then the flow-pressure domain would be uncontrolled, unpredictable, and very difficult to reproduce between any two fixed restrictors.

Flow, pressure (or density), temperature, and time must be controlled precisely in order to yield the same molar quantity or mass of supercritical fluid needed to partition a given amount of analyte from a unit amount of matrix. Quoting extraction efficiency or percent recovery under such uncontrolled parameters is thus an empirical matter, where coincidence is random or indiscriminate.

There are many variations on fixed restrictors, including a carefully defined fritted zone at the end of a capillary tube. Most of these have led to an unending number of comments at SFE conferences, such as "the biggest hurdle in SFE is limited restrictor availability and technology." For more discussion on fixed restrictors, especially for supercritical fluid chromatography (SFC), the reader is referred to the work by Greibrokk et al. (1987, 1993).

Gere et al. (1982) described a variable back-pressure regulator as a postcolumn restrictor for SFC, although this was by no means the first of such devices to be made from an off-the-shelf apparatus. For at least 20 years, SFE and SFC have been practiced successfully, under certain favorable combinations of experimental parameters, with back-pressure regulators at the terminus of the apparatus. These favorable circumstances usually involve larger (than analytical scale) SFE. Larger scale work with larger flow pumps, flow rates, and appropriately-sized ancillary hardware can tolerate rather large void labyrinth volumes in the zone of expansion from the high pressure fluid to ambient fluid. If this large, poorly swept volume can be tolerated, back-pressure regulators have a fine characteristic: they can decouple flow and pressure. Thus, one gains independent control over two very important parameters involved in the integrity of SFE.

These large scale devices were not adequate for small scale volumes needed for analytical SFE with trace concentrations of analytes. Thus, it took some additional years of development before miniaturized, automated back-pressure regulator restrictors were commercially available.

Finally, in Figure 8-2, attention is focused on the zone just beyond the restrictor, where the expanding CO_2 and precipitating analytes impinge upon a solid surface or a retaining liquid. The zone discussed above in detail can be defined as the *expansion zone*, whereas the zone now discussed is defined as the *collection zone*. Collectively, these two zones make up the *reconstitution zone*, as shown in Figure 8-2.

Simplistically, the collection zone can be made up of an empty thimble, such as a test tube or small vial, with the exit of the restrictor placed into this zone, perhaps vertically, with the end of the restrictor close to but not touching the bottom or the walls of the thimble. Usually in a simple, home-built apparatus, the thimble or vial is glass. Thus, the bottom or the walls of the thimble are (hopefully) impinging surfaces or areas. In a first approximation, a solute would be expected to merely precipitate out of solution as the expanding CO_2 changes to a low density gas. Then the precipitating solutes hypothetically would drop to the bottom of the thimble, and a clean pure phase separation (i.e., gas-solid) would take place. Reality is often quite different; much adjustment needs to take place in this zone in order to achieve something close to 100% collection of the solute in concentration ranges from parts per billion (e.g., PCBs in sediment) up to 50% (e.g., total fat in a chocolate candy). Several physico-chemical parameters cause these deviations from the ideal. They include but are not limited to volatility of the solute, degree of coprecipitation of solid CO_2 (followed almost immediately with uncontrolled sublimation of the solid), aerosol formation, surface tension, occlusion in solid CO_2, rebound from impinging surface, and many other interacting phenomena.

Beyond simple empty thimbles, a time-honored approach has been to bubble the expanding CO_2 into another liquid. Although an improvement, this still suffers from many of the other complications and a few new ones, such as volatility and aerosol formation between the solutes and the chosen trapping liquid.

Most recently, some types of commercial equipment have included temperature-controlled solid trapping zones and also a liquid-reconstitution dispensing pump. This appears to have the maximum flexibility in offering the potential for full recovery over the widest range of analytes. For example, the use of a nonpolar solid, PoraPak Q, at perhaps $-10°C$ can trap volatile compounds, such as pentane, benzene, or toluene, quantitatively. Although it may be possible to accomplish the same collection-reconstitution with liquid trapping, it is more problematic to choose optimal liquids. For instance, an alcohol may indeed provide near optimal trapping for many organics but may be very inappropriate for introduction into a GC capillary column. This then would lead to an evaporative solvent exchange to the proper solvent for the GC, but the evaporation step introduces several new difficulties. Fortunately, if one has the apparatus for subambient as well as ambient and above ambient solid trapping, it is easy to convert to liquid trapping with very minor changes, if needed.

8.2.2   Advantages of the supercritical fluid extraction method

Supercritical fluid extraction works very much like conventional liquid extraction, but with these important differences: a) the extraction is usually carried out above the critical temperature of the fluid (This is not always necessary if the extraction pressure is above the critical pressure.), and b) the solvent power of CO_2 fluid depends upon density (a manifestation of temperature and pressure).

Traditional liquid extraction uses an assortment of solvents to extract different analytes. Hydrocarbon solvents such as hexane and isooctane are good for extracting nonpolar analytes. Benzene and toluene work for aromatic compounds. Methanol and others are useful for polar analytes. No single solvent is good for all possible samples.

The solvent power of a supercritical fluid increases as its density increases. Thus, a single supercritical fluid may extract a variety of components, depending on the pressure and temperature applied to the system. It is possible to fractionate by extracting the sample sequentially at two or more different densities.

Supercritical fluids are attractive candidates for sample extraction because they provide an unusual combination of properties:

1. Gas-like properties allow them to penetrate a sample much more easily than liquid solvents can.
2. Liquid-like properties allow them to dissolve analytes from the sample matrix, leaving behind unwanted interference, such as salts and inorganic compounds.
3. Variable solvent power allows a single supercritical fluid to substitute for a variety of conventional solvents.

8.2.3   Practical candidate fluids for supercritical fluid extraction

Many fluids have been used over the past few years for SFE. Table 8-1 lists the critical temperature and pressure for some of them.

Both the chemistry and the mechanics of a supercritical fluid system depend on temperature and pressure. Excessively high pressure or temperature requirements could make the apparatus too expensive

Table 8-1   Candidates for Supercritical Fluid Extraction Fluids

Fluid	Critical temperature (°C)	Critical pressure (bar)
Ethylene	9.3	50.4
Carbon Dioxide	31.1	73.8
Ethane	32.3	48.8
Nitrous Oxide	36.5	72.7
Propylene	91.9	46.2
Propane	96.7	42.5
Ammonia	132.5	112.8
Hexane	234.2	30.3
Water	374.2	220.5

to be practical. Commercial equipment economics follow severe price/volume curves. In today's laboratory, regulations of safety in the workplace place further constraints on the choices of fluids.

Hydrocarbons are tempting, but the flammability and explosion hazards cause one to pause before using them for widespread applications. Since ammonia is very unpleasant to work with, a fume hood or another venting precaution is needed to keep it out of the laboratory atmosphere as it is very toxic and is actually a weak reducing agent. Nitrous oxide has been used and cited in several early publications. It is polar and has reasonable values of critical temperature and pressure. However, there are several recent citations of violent explosive reactions between nitrous oxide, an oxidizing agent, and organic compounds such as ethanol (Sievers and Hansen, 1991) or oils and lipids (Raynie, 1993). For this reason, great caution should be used before this fluid is considered. Common opinion now considers nitrous oxide not suitable for a general purpose SFE fluid.

Water is included in the table and is also a tempting choice. Although home-built equipment has been used for experiments cited in the literature repeatedly over the past 20 years, engineering, safety, materials of construction, and ultimately the cost of robust commercial equipment have precluded widespread application of supercritical water. Perhaps this will change with time, but the economics currently do not point in a positive direction.

In summary, CO_2 is still the practical choice because of several coincidental properties. The critical temperature is near ambient and easily reached and controlled. The critical pressure is reasonably reached and controlled by slightly modified HPLC pumps. Carbon dioxide is nonflammable, nontoxic, odorless, and chemically inert. It is also widely available at a reasonable price in a variety of purity levels. It also does not present any unusual or expensive disposal problems.

In the near future, attention will be given to the so-called *alternate refrigerants*. They will soon be widely available at a price and purity comparable to CO_2. They are already screened for their potential acceptance with the Montreal Protocol requirements. A limited number of studies indicate several candidates may be shortly available.

8.3 Applications of supercritical fluid extraction in environmental analysis

In the following sections, a brief review on the applications of SFE in the determination of organic contaminants in environmental samples is given. It should be noted that a comprehensive review of this topic is beyond the scope of this article, and the reader is referred to several other excellent reviews on similar subjects that have been published in the past few years (King, 1989; Hawthorne, 1990; Pipkin, 1990; Knipe et al., 1992; Camel et al., 1993; Hawthorne et al., 1993a; Janda et al., 1993; Kane et al., 1993; Chesler et al., 1994; Gere and Derrico, 1994a,b; Gere et al., in press). In this chapter, emphasis is on the extraction conditions for the various organics in aquatic sediments and related samples. Since there are no standardized SFE methods available for these samples, a recommended procedure for the extraction of a class of compounds is given at the end of each section if most or all of the following conditions are met:

1. A detailed description of the method performance, i.e., precision, accuracy, method detection limit, etc., has been generated by one or more laboratories.
2. The procedure has produced satisfactory and reproducible recoveries when it was applied to standard or certified reference materials (SRMs/CRMs).
3. If SRMs are not available, the SFE recoveries have been similar to the Soxhlet results generated on some naturally contaminated samples.
4. The entire method has either been validated by an interlaboratory study or by two or more independent laboratories.

8.3.1 Polychlorinated biphenyls (PCBs) and chlorobenzenes

One of the first SFE procedures for the removal of trace organics such as PCBs from solid samples was reported by Schantz and Chesler (1986). In their work, extraction of PCBs from a sediment sample was described. Extraction was carried out at 40°C with a CO_2 density of 0.93 g/mL (345 bar) for 4 h. The recovery of Aroclor 1254 in that sample was quantitative, as it was similar to the Soxhlet value. No information for the recovery of individual PCB congeners was given. Later, quantitative recoveries of total PCBs from sediments were reported by other investigators in shorter extraction times using CO_2 modified by methanol or at higher extraction temperatures. For example, up to 100% recovery of total PCBs was achieved (Onuska and Terry, 1989a), in less than 10 min, from a certified sediment reference material (EC-1) developed by Environment Canada (Lee and Chau, 1987a,b). In that work, optimal results were obtained by extraction with 2% methanol-modified CO_2 at 203 bar and 40°C.

The presence of modifiers such as methanol or toluene was shown to improve the recovery of spiked PCBs from soils by 10 to 15% (Liu et al., 1991). Extractions were carried out at 40 or 50°C with CO_2 at 0.73 or 0.75 g/mL for 60 min. In another report, optimal results for six PCB congeners from spiked soil samples were obtained by a 10-min static extraction followed by a 20-min dynamic extraction using CO_2 at 50°C and 196 bar (van der Velde et al., 1992).

The efficiencies of supercritical chlorodifluoromethane (Freon-22), N_2O, and CO_2 for the extraction of PCBs from a river sediment (NIST SRM 1939) were compared (Hawthorne et al., 1992a). Extractions were performed for 40 min at either 405 bar and 50°C (for CO_2 and N_2O) or at 111 bar and 100°C (for Freon-22), so that the densities of the fluids were approximately 0.93 g/mL in all cases. Under such conditions, both CO_2 and N_2O generated an average recovery of 62% (vs. the certified values) for the nine PCB congeners. In contrast, average PCB recoveries of 90% or higher were obtained by extractions with Freon-22 and 5% methanol-modified CO_2 (at 70°C and 405 bar).

Using certified reference materials from U.S. EPA and NIST, the effects of temperature and pressure on the SFE efficiencies of PCBs were studied (Hawthorne et al., 1993b). Experiments were carried out with pure CO_2 at 50 and 200°C and pressures at 152, 355, and 658 bar. In that work, PCBs were more effectively extracted at any of the above pressures when the extraction was carried out at the higher temperature. The recoveries of 12 PCB congeners from a river sediment (SRM 1939) ranged from 67 to 163% of the certified values obtained by Soxhlet extraction. The same authors later reported the role of modifiers for the SFE of PCBs in SRM 1939 (Langenfeld et al., 1994). This river sediment was extracted at 80°C and 405 bar for 15 min (5-min static and 10-min dynamic) and a flow rate of 1 mL/min. Under such conditions, they found that the modifier identity was more important than its concentration for increasing extraction efficiencies. Modifiers such as methanol, acetic acid, and aniline were more effective than dichloromethane (DCM), toluene, hexane, and acetonitrile in the enhancement of PCB recovery.

The extraction of PCBs from a sewage sludge was successfully attempted by using pure CO_2 at 355 bar and 80°C (Porter et al., 1992).

A static extraction of 15 min and a dynamic extraction of 45 min with a gas flow rate of 300 ± 30 mL/min were performed. The SFE and Soxhlet recoveries of four PCB congeners (PCBs 101, 138, 153, and 180) were identical. Another method for the extraction of PCBs in sulfur-containing sediments under supercritical conditions was described by Bøwadt and Johansson (1994). This procedure involved a 20-min static extraction with pure CO_2 at a density of 0.75 g/mL (218 atm) at 60°C, followed by a 40-min dynamic extraction at the same density and temperature and a flow rate of 1 mL/min. Sulfur was conveniently removed during the extraction by mixing the sample with activated copper powder. The extracted PCBs were adsorbed on a Florisil trap and were eluted by n-heptane at the end of the extraction.

The optimization of supercritical CO_2 extraction for PCBs in sediment CRM EC-1 was studied (Lee and Peart, 1994). The extracts were analyzed by GC/MS for PCB homologs from the tri- to the octachlorobiphenyls. Extractions were carried out at 40, 60, 80, 100, and 120°C and 343 bar, with unmodified CO_2 for 21 min (1-min static and 20-min dynamic extractions). While the recovery of PCBs increased with increasing extraction temperature, there was a much bigger dependence of PCB recovery on temperature for the higher (e.g., hepta- and octachlorobiphenyl) homologs than the lower ones (e.g., tri- and tetra-chlorobiphenyls). Since the overall contribution to total PCBs by the hepta- and octachlorobiphenyls was small, there was practically no observable difference in total PCB concentrations for the 80, 100, and 120°C extractions.

The efficiency of several solid-phase traps in the SFE of PCBs from sewage sludge samples was evaluated (Bøwadt et al., 1994). The traps were filled with approximately 1 cc of stainless steel beads, ODS, silica gel, and Florisil. Extractions were carried out at 60°C and a fluid density of 0.75 g/mL for 30 min with pure CO_2 as well as CO_2 modified with 2 or 5% methanol, as well as 2% ethanol. They found that, if heptane was used as a trap eluent, only the ODS and Florisil traps were able to provide a PCB extract clean enough for direct GC/ECD analysis. These two sorbents were also suitable for the trapping of PCBs, even if 5% methanol-modified CO_2 was used for extraction, provided that the trap was maintained at 65°C (boiling point of methanol). A comparison of SFE and Soxhlet extraction was conducted as part of the certification of PCB congeners in an industrial soil (BCR CRM 481) (Bøwadt et al., 1995).

Although they were much less studied in comparison to PCBs, chlorobenzenes are another class of U.S. EPA priority pollutants that can be extracted simultaneously by the same procedure. The Soxhlet and SFE recoveries of several native chlorobenzenes, hexachloro-1,3-butadiene, and octachlorostyrene from a lake sediment sample were compared (Lee and Peart, 1994). After addition of 500 µL of water, the sample (1 g) was extracted by CO_2 at 100°C and 343 bar for 21 min (1-min static and 20-min dynamic) at a flow rate of 2 mL/min. While the results for octachlorostyrene obtained by the two techniques were quite similar, the SFE recovery of chlorobenzenes and hexachloro-1,3-buta-diene were 10 to 50% higher than the Soxhlet results. Lower evaporative losses of these volatile compounds by the SFE technique were presumably the cause of the discrepancy in recoveries.

8.3.1.1 *Recommended extraction procedure for PCBs in solid matrices (abstracted from the U.S. EPA draft method 3562 for solid wastes)*

1. Weigh 1.0 to 5.0 g of the dry and homogenized sample into a weighing dish. Mix the sample with 2 g of activated copper powder (electrolytic grade).
2. If necessary, spike the mixture with 150 µL of a surrogate (e.g., hexabromobenzene, 1,2,3,4-tetrachloronaphthalene, or octachloronaphthalene, etc., at 10 mg/mL).
3. Transfer the sample quantitatively to the extraction vessel on top of a 2-cm layer of Celite or anhydrous sodium sulfate. Place another 2-cm layer of Celite at the top of the sample. Fill the void volume, if any, of the extraction vessel with an inert, porous material, such as precleaned Pyrex glass wool, Celite, etc.
4. Extraction conditions:

CO_2 pressure (bar):	305
CO_2 density (g/mL):	0.75
Extraction fluid composition:	unmodified CO_2
Extraction chamber temperature (°C):	80
Static extraction time (min):	10
Dynamic extraction time (min):	40
Extraction fluid flow rate (mL/min):	2.5

5. Extract collection conditions:

Sorbent trap packing material:	Florisil
Trap temperature (°C):	15 to 20
Nozzle temperature (°C):	45 to 55 (variable restrictor)

6. Elution of the extract from the sorbent trap:

Rinse solvent:	n-heptane
Rinse volume (mL):	3.2 (in two rinses)
Rinse solvent flow rate (mL/min):	1
Trap temperature (°C):	38
Nozzle temperature (°C):	30

7. After each sample, the trap is cleaned and regenerated by rinsing it with 4.0 mL of a 1:1 (v/v) mixture of DCM and acetone followed by 3.0 mL of n-heptane.

8. The combined n-heptane extract is evaporated and adjusted to 1.0 mL for GC/ECD analysis. Further cleanup is normally not necessary.

8.3.1.2 Discussion

1. Activated copper was incorporated with the sample in order to eliminate the coextracted sulfur, which could cause plugging of the restrictor or inline filter and could stop extraction.
2. This method has been validated by three independent laboratories using sediment, soil, and sludge standard reference materials developed by Environment Canada, National Institute for Standards and Testing (NIST), and BCR European Union. Precision and accuracy data for the following 12 most commonly found PCB congeners (IUPAC numbers 28, 52, 101, 105, 118, 128, 138, 149, 153, 156, 170, and 180) from trichloro- to heptachlorobiphenyls were obtained from replicate extractions. The mean recovery (%) for all congeners varied from 87 to 105% and the mean relative standard deviation from 3.1 to 4.9%.
3. Based on a 1 g sample and a 1-mL final volume, the method detection limits for PCB congeners with GC-ECD analysis are in the order of 6.6 µg/kg.

8.3.2 Polynuclear aromatic hydrocarbons (PAHs)

Hawthorne and Miller (1986) demonstrated that SFE could be used for the rapid and quantitative recovery of several PAHs from solid samples such as diesel exhaust particulate (NIST SRM 1650) and from spiked Tenax-GC sorbent traps. Extractions were carried out at 304 bar for either 90 min (at 45°C with CO_2) or 30 min (at 65°C with 5% methanol-modified CO_2). Partial fractionation of *n*-alkanes from PAHs in SRM 1650 was achieved by varying the pressure, i.e., the solvating power, of the supercritical fluid. While over 84% of the aliphatic hydrocarbons were readily extracted at 76 bar, the majority of PAHs were only extracted when the pressure was raised to 300 atm. Meanwhile, a quantitative SFE procedure for five PAHs in an urban dust standard reference material (SRM 1649) was reported (Schantz and Chesler, 1986). Using supercritical CO_2 at 345 bar and a density of 0.93 g/mL, the extraction was carried out at room temperature for 4 h. While the SFE recoveries of fluoranthene, benzo[*a*]anthracene and benzo[*a*]pyrene were in good agreement with the certified values, those for indeno[123*cd*]pyrene and benzo[*ghi*]perylene were 30 and 18%, respectively, higher than the certified values.

Determination of PAHs in solid matrices by on-line SFE-capillary GC methods was described by various authors, including Wright et al.

(1987), Hawthorne and Miller (1987a), Hawthorne et al. (1989), and Levy et al. (1989). Supercritical fluid extraction of PAHs from environmental solids such as urban dust, fly ash, and river sediment, using ethane, CO_2, N_2O, and methanol-modified CO_2 and N_2O, was also described (Hawthorne and Miller, 1987b). A 60-min extraction with supercritical N_2O modified by 5% methanol was found to give quantitative results for the five PAHs in SRM 1649. Presumably due to its higher dipole moment, Freon-22 was also shown to yield high extraction efficiency for PAHs in solids (Hawthorne et al., 1992a).

Because of the undesirable properties of supercritical N_2O, most workers focused on the use of CO_2 as the extractant for PAHs. However, incomplete recovery of some native PAHs from soil or sediment, particularly those with a molecular mass of 252 and higher, was observed under some extraction conditions. The effects of temperature and pressure on the SFE recoveries of PAHs using unmodified CO_2 as the extractant were studied (Langenfeld et al., 1993). At 50°C, raising the extraction pressure (from 355 to 658 bar) had no effect on the extraction efficiencies from the SRM 1649 and a highly contaminated soil (U.S. EPA certified). Higher PAH recoveries were observed at 200°C for both samples; however, higher pressure has a more dramatic effect on the PAH recovery from the air particulate sample than the soil sample. It should be noted that the best recoveries for benzo[*ghi*]perylene and indeno[123*cd*]pyrene in SRM 1649 from a 40-min extraction were 60 and 45%, respectively. Levy et al. (1993) also reported that raising the extraction pressure and temperature increased the SFE recovery of PAHs from naturally contaminated soil and sediment samples using unmodified CO_2. These authors found that, for quantitative recovery of PAHs with a molecular weight 228 or higher, higher extraction pressure (456 vs. 355 or 253 bar) was needed for the soil, and higher temperature (150 vs. 100°C or lower) was needed for the sediments.

The effects of a solvent modifier, extraction temperature, and number of consecutive extractions on the recoveries of PAHs from a certified reference material EC-1 were evaluated (Lee and Chau, 1987a; Lee et al., 1993b). Although spiking 500 µL of water, methanol, or DCM to 1 g of sediment improved the recovery of all PAHs, the effect was much greater for those PAHs of molecular weight 228 or higher. Higher extraction temperature also had a positive effect on the extraction efficiencies of all PAHs, as the recoveries increased progressively from 60 to 120°C. With a static addition of modifier, a single extraction would produce low recovery (<65%) for PAHs with five or more rings (molecular weight 252 or higher). This situation was remedied by two more consecutive extractions of the same sample at 120°C and 338 bar with the addition of fresh modifier for each extraction or by the continuous introduction of a modifier by means of a second pump.

The benefits of using modified CO_2 to extract PAHs from sediments and soils were also reported by other workers. Using pure CO_2 (in a single-step extraction) and 10% methanol-modified CO_2 (in a multistep extraction), SFE of PAHs from a few soil standard reference materials, including HS-3 (National Science and Engineering Research Council of Canada) was reported (Lopez-Avila et al., 1990). The recoveries for most PAHs in HS-3 were lower than 50% (four were even lower than 20%) if pure CO_2 was used. The presence of methanol improved the recoveries considerably, as only four of the 16 PAHs were less than 50% recovered. The extraction of PAHs at ng/g level from a marine sediment (NIST SRM 1941) and a mussel tissue (NIST SRM 1974) was described by Porter et al. (1992). The extraction was performed by using 10% DCM-modified CO_2 at 507 bar, 125°C, and a flow rate of 600 ± 50 mL/min for 40 min. The SFE recoveries ranged from 87 to 148% (for the 11 PAHs in SRM 1941) and from 84 to 126% (for the 9 PAHs in SRM 1974) of the corresponding certified values. Quantitative recovery of spiked PAHs in soils was achieved by the addition of DCM (400 µL/g sample) as a static modifier (Dankers et al., 1993). The extractions were carried out with CO_2 of 0.76 g/mL density and a temperature of 90°C.

The role of various organic modifiers for the supercritical CO_2 extraction of PAHs in reference materials was examined (Langenfeld et al., 1994). Their results indicated that low molecular weight PAHs were best extracted with modifiers such as aniline, acetic acid, acetonitrile, methanol/toluene, hexane, and diethylamine. In contrast, modifiers capable of dipole-induced interactions and π–π interactions such as toluene, diethylamine, and DCM were the best modifiers for the SFE of high molecular weight PAHs. The relative extraction rates of spiked vs. native PAHs from environmental samples using SFE were compared (Burford et al., 1993). In this detailed extraction kinetics study, samples such as petroleum waste sludge, urban air particulates (NIST SRM 1649), and railroad bed soil were sequentially extracted with pure supercritical CO_2 and 10% methanol-modified CO_2 at 60°C and 405 bar. Regardless of the spiking method, the extraction rates for most of the spiked deuterated PAHs were up to tenfold higher than those of the same native PAHs. In most cases, a 30-min extraction with pure CO_2 recovered over 90% of the spiked deuterated PAHs, yet approximately 25 to 80% of the native PAHs were extracted. These results clearly demonstrate that the extraction conditions established from spiked recovery alone may not be valid for the quantitative extraction of incurred organics in aged environmental samples.

Using 8% (mol) modified CO_2 at 392 bar and an extraction temperature of 80°C, results similar to or better than Soxhlet extraction for the 16 U.S. EPA PAHs from a real world loam soil sample were obtained (Reindl and Höfler, 1994). Quantitative recovery of 1-nitropyrene from a diesel exhaust particulate standard reference material (NIST SRM

1650) as well as the extraction of some nitro-PAHs from bus soot by SFE procedures were demonstrated (Paschke et al., 1992). The highest results were obtained by extractions with Freon-22 or 10% toluene-modified CO_2 at 100°C and 405 bar for 45 min at a flow rate of 0.3 mL/min. The SFE recoveries for the PAHs in SRM 1650 using the above two extracting fluids were much higher than those reported by NIST based on Soxhlet extraction.

An unique example for the extraction of PAHs from environmental solids with sub- and supercritical water was described (Hawthorne et al., 1994). Water has a critical temperature of 374°C and a critical pressure of 221 bar. Extraction of the native PAHs from a soil sample by water was attempted at 50, 120, 200, 250, 300, and 400°C and various pressures. Since the solubilities of hydrocarbons in water are low due to its high polarity, the recoveries of PAHs were very low at 50 and 120°C. However, the dielectric constant for water drops drastically at higher temperatures and under moderate pressure. Hence, quantitative recoveries of all PAHs were obtained by a 15-min extraction at 250°C and 50 bar using subcritical water.

8.3.2.1 Recommended procedure for the extraction of PAHs in solid samples (abstracted from the U.S. EPA proposed method 3561 for solid wastes)

1. Determine the sample's moisture content from an aliquot of the homogenized sample.
2. Prepare the extraction thimble as described earlier for PCBs (steps 1 to 3) and replace the surrogate standard by a 10-mg/mL solution of *m*-quaterphenyl in a 1:1 acetonitrile/THF mixture.
3. For subsequent HPLC analysis, extract the sample with the following three-step procedure:

	Extraction 1	Extraction 2	Extraction 3
Extraction conditions:			
CO_2 pressure (bar):	120	338	338
CO_2 density (g/mL):	0.30	0.63	0.63
Extraction fluid composition:	CO_2	CO_2–MeOH–H_2O 95/1/4 (v/v/v)	CO_2
Extraction chamber temperature (°C):	80	120	120
Static extraction time (min):	10	10	5
Dynamic extraction time (min):	10	30	10
Extraction fluid flow rate (mL/min):	2.0	4.0	4.0

	Extraction 1	Extraction 2	Extraction 3
Extract collection conditions:			
Sorbent trap material:	ODS	ODS	ODS
Trap temperature (°C):	–5	80	80
Nozzle temperature (°C):	80	80	80
Reconstitution of extract:			
Rinse solvent:	THF–CH$_3$CN 1/1 (v/v)	none	THF–CH$_3$CN 1/1 (v/v)
Rinse volume (mL):	0.8	NA	0.8
Rinse solvent flow rate (mL/min):	1.0	NA	1.0
Trap temperature (°C):	60	NA	80
Nozzle temperature (°C):	45	NA	45

4. For subsequent GC analysis, substitute the extraction fluid with CO_2–MeOH–DCM (95/1/4, v/v/v) and rinse solvent with a mixture of 3:1 (v/v) DCM and isooctane.
5. Combine the extracts from extractions 1 and 3 and readjust the volume to 1.0 mL for HPLC (U.S. EPA Method 8310) or GC/MS (U.S. EPA Method 8270) analysis.

8.3.2.2 Discussion

1. This SFE method eliminated the use of a large quantity of organic solvent in the extraction of PAHs from solids. The cleaner SFE extract required no column or gel permeation cleanup before LC or GC analysis, thereby further reducing the amount of organic solvent used per sample.
2. To improve the extraction recovery of native PAHs from sediment samples, three consecutive extractions were performed for each sample. The first extraction was carried out at a lower temperature (80°C), with CO_2 at a low density (0.30 g/mL), to remove the lower molecular weight hydrocarbons. The remainder of the PAHs including the high molecular weight hydrocarbons in the sample were extracted at a higher temperature (120°C) and CO_2 density (0.63 g/mL) in the presence of a methanol/water mixture as a solvent modifier. The third extraction with pure CO_2 rid the system of modifiers and readied it for the subsequent extraction.
3. This procedure was validated with sediment reference materials certified for PAHs. In general, the SFE and Soxhlet results were comparable. However, the SFE recoveries for naphthalene and

methylated naphthalenes were approximately 150 and 125%, respectively, of the Soxhlet values. Higher SFE recoveries in these cases were presumably due to much lower evaporative loss of these volatile compounds since the SFE procedure required very little or no solvent evaporation.

4. Method 8310 is an HPLC method for PAHs with either UV/VIS or fluorescence detection with lower limits of detection between 0.010 and 1.00 mg/kg. Method 8270 is a GC/MS method with a detection limit of 0.70 mg/kg.

8.3.3 Total recoverable petroleum hydrocarbons (TPHs)

Over the last few years, application of SFE to the analysis of TPHs from soil and solid wastes has received significant attention. The extraction of hydrocarbons up to C_{35} with supercritical CO_2 has been reported (Monin et al., 1991). Emery et al. (1992) studied the SFE recovery of diesel fuel adsorbed on montmorillonite, kaolinite, and illite clay samples. These samples were extracted for 20 min with CO_2 (density 0.8 g/mL) either at 45 or 80°C. Quantitative (>95%) recoveries of $n\text{-}C_{14}$ to $n\text{-}C_{22}$ alkanes were obtained at the lower temperature and a flow rate of 1 mL/min from the illite and kaolinite clays either spiked or coated with the hydrocarbon mixture. In contrast, quantitative recovery was not achieved for many of the hydrocarbons, even at the higher extraction temperature and a flow rate of 2 mL/min from the calcium montmorillonite clay coated with hydrocarbons. The presence of water in the sample also reduced the extraction efficiency.

An offline SFE-IR method for the determination of hydrocarbons in soils was described (Lopez-Avila et al., 1992). The extraction was performed at 344 bar and 80°C using supercritical CO_2 for 30 min. The extracted hydrocarbons were collected in 3 mL of tetrachloroethene, which was found to be the better solvent than Freon-113 for the collection of hydrocarbons containing 30 or more carbons. Side-by-side comparison of SFE and Soxhlet extraction (with Freon-113) showed that both methods produced equivalent results. Using spiked and naturally contaminated samples of hydrocarbon levels from 1450 to 32,600 µg/g, the SFE method accuracy and precision ranged from 80 to 104% recovery and from 4 to 20% RSD, respectively. This SFE procedure was further evaluated in the field and in the laboratory, and the results were compared with those obtained by Soxhlet extraction (Hawthorne et al., 1993b). For those gasoline-, diesel-, motor-oil-, and crude-oil-contaminated soil samples tested, the field-extracted and the laboratory-extracted samples by the SFE technique produced virtually identical results, within 20% of the Soxhlet data.

A method for the determination of hydrocarbons in soils contaminated with diesel fuels using subcritical CO_2 extraction and offline SFC analysis with a flame ionization detector was described (Brooks and Uden, 1993). Aliquots (1 to 2 g) of the sample were extracted for 10 min at 294 bar and ambient temperature (23 to 25°C), and the extracts were collected in 10 mL of DCM. Extraction efficiency was >90% at a spiking level of 30 µg/g, regardless of the organic content of the soil, even after an aging period of 5 d. Selective extraction of hydrocarbons from C_{12} to C_{22} in contaminated soils was achieved by extraction with subcritical water at 50 bar and 250°C (Hawthorne et al., 1994). Under such conditions, the heavier hydrocarbons (C_{24} and higher) were largely unextracted.

An offline, coupled SFE and GC method for the automated analysis of petroleum hydrocarbons in soil was developed by Wylie et al. (1994). Soil samples spiked with the NIST SRM 1642b (sulfur in distillate fuel oil) were extracted by CO_2 at 80°C and 339 bar at a flow rate of 3 mL/min for 30 min. The hydrocarbons were adsorbed on an ODS trap that was cooled to –10°C during extraction for maximum recovery (102%). The trap was then eluted with 1.5 mL of isooctane into a vial, which was transported from the extractor to the GC autosampler by a robotic arm of the latter. To save time, the sample extract was subsequently analyzed by GC-FID while a second sample was being extracted.

8.3.3.1   Recommended procedure for the extraction of TPHs from solid samples (modified from the U.S. EPA proposed method 3560 for solid wastes)

1. Determine the sample's moisture content from an aliquot of the homogenized sample.
2. Weigh 3 g of the sample into a precleaned aluminum dish. A drying agent (e.g., anhydrous magnesium sulfate or diatomaceous earth) may be added to the sample that contains water in excess of 20%.
3. Transfer the weighed sample to an extraction vessel that has a volume slightly larger than the sample. Use two plugs of silanized glass wool to hold the sample and fill the void volume. Alternatively, a drying agent or clean sand can be used to fill the void volume.
4. Extract the sample with pure CO_2 at 344 bar and a temperature of 80°C for 30 min in the dynamic mode. Alternatively, extract the sample with CO_2 at a pressure greater than or equal to 344 bar at 150°C for 25 min and a gas flow of 3500 to 4000 mL/min. In the latter case, the use of a drying agent is not necessary.
5. Collect the extract in 3 mL of tetrachloroethylene. If a sorbent trap is used, set the trap temperature at –10°C during extraction

and at 60°C during rinsing. Rinse the trap with 1.5 mL of isooctane or tetrachloroethylene.
6. For samples known to contain elemental sulfur, use copper fil-
ings to remove the coextracted sulfur. The extract is ready for
analysis by U.S. EPA Method 8015, Non-halogenated Volatile
Organics by Gas Chromatography, or U.S. EPA Method 8440,
Total Recoverable Hydrocarbons by Infrared Spectrophotometry.

8.3.3.2 Discussion

1. Based on a 3-g sample, a final extract volume of 3 mL and
analysis with infrared detection, the method detection limit is
10 µg/g.
2. According to one estimate, a complete conversion from the Fre-
on extraction method to the U.S. EPA Draft Method 3560 would
eliminate the use of 30,000 L of liquid Freon-113 per year in the
United States (Lopez-Avila et al., 1992).

8.3.4 Pesticides and herbicides

Extraction conditions for organochlorine insecticides from spiked soil
and sand using supercritical CO_2 were evaluated (Lopez-Avila et al.,
1990). With the exception of methoxychlor, the SFE recovery for all
insecticides from spiked sand was quantitative (>75%) regardless of
the extraction temperature (50, 60, or 70°C) and pressure (152 or 253
bar). However, due to matrix effects, the recoveries for some insecticides were low when they were spiked onto dry soil samples. Supercritical fluid extraction of organochlorine insecticides from spiked soil
samples was attempted (van der Velde et al., 1992). The test compounds
included in this work were hexachlorobenzene, the BHC isomers, heptachlor epoxide, dieldrin, DDD, as well as *o,p*-, and *p,p*-DDTs. The use
of CO_2 at 50°C and 196 bar, 10-min static followed by 20-min dynamic
extraction with collection in isooctane, was found to be an optimal
extraction procedure. At a spiking level from 1 to 10 ng/g, the recoveries of organochlorines from peat soil ranged from 84 to 108%. The
SFE extracts were suitable for GC/ECD analysis without further
cleanup.

Improved SFE efficiency for the organochlorinated pesticides from
spiked soils was observed when a polar modifier, especially DMSO,
was added to CO_2 (Liu et al., 1991). The extraction efficiency of SFE for
six organochlorines and six organophosphates from soils was compared with sonication and Soxhlet extraction (Snyder et al., 1992).
Supercritical fluid extraction was carried out with 3% methanol-modified CO_2 at 355 bar and 50°C. The overall mean recoveries of all 12

pesticides for the sonication, Soxhlet, and SFE methods were 95, 93, and 92%, respectively. However, with an overall RSD of 2.9%, the SFE results were the most precise. Comparable SFE and sonication results were obtained from soils with incurred organochlorines at ng/g levels using this extraction method. Later, these authors also found that, while pesticide recoveries increased with CO_2 density, temperature over a range from 40 to 120°C had little effect on the extraction (Snyder et al., 1993). In fact, dichlorvos and endrin aldehyde showed a drop in recoveries at elevated temperatures due to thermal breakdown of these compounds.

A SFE method for the extraction of spiked and incurred pesticides from fatty and nonfatty food was also reported (Hopper and King, 1991). The sample was first mixed with a pelletized diatomaceous earth (Hydromatrix) as an enhancer prior to extraction. Supercritical fluid extraction was carried out at 80°C and 676 bar, with a total CO_2 flow of 100 L. Recoveries in excess of 85% for over 30 organochlorine and organophosphorus pesticides at incurred levels ranging from 0.005 to 2 µg/g were reported. The efficiency of supercritical CO_2 for the extraction of organochlorine, organophosphorus, and organonitrogen pesticides from spiked grain matrices was investigated by King et al. (1993). Supercritical fluid extraction was performed between 40 and 80°C with pressures from 135 to 676 bar. In most cases, pesticide recoveries exceeding 80% were observed over the above temperature and pressure ranges.

A procedure for the SFE of herbicides 2,4-dichlorophenoxyacetic acid (2,4-D) and dicamba using *in situ* chemical derivatization was developed (Hawthorne et al., 1992b). Quantitative recovery of the two herbicides from naturally contaminated agricultural soil at the low µg/g level was obtained by CO_2 extraction at 80°C and 405 bar in the presence of trimethylphenylammonium hydroxide in methanol as a methylating agent. However, three sequential derivatization/SFE steps, i.e., 15-min static followed by 15-min dynamic extractions, were required to achieve quantitative recovery of the two native herbicides from soil. It should be noted that the BF_3/methanol complex could also be used for the *in situ* SFE/methylation of 2,4-D but not dicamba since the reagent does not methylate benzoic acids.

Various procedures for the extraction of seven chlorophenoxy acid herbicides from soil were investigated (Lopez-Avila et al., 1993). Supercritical fluid extraction with CO_2 at 405 bar and 80°C without any derivatizing agent failed to recover 2,4-D from spiked sand and topsoil at low µg/g levels. Among the many reagents tested, the combination involving tetrabutylammonium hydroxide/methyl iodide (TBA/MI) as methylating agents was most effective; it was applicable to the qualitative determination of chlorophenoxy acid herbicides in soil spiked at 50 and 250 µg/g levels. The extraction was carried out at 405 bar and 80°C for 15-min static, followed by 15-min dynamic, at a flow rate of 1.5 mL/min and in the presence of 0.5 mL of 25% TBA in methanol and 0.5 mL of MI. The herbicides were converted, *in situ*, into their methyl esters during the extraction. Another example for the extraction of 2,4-D from a spiked Hoypus sandy loam soil using CO_2 was also reported (Rochette et al., 1993). No 2,4-D was recovered at 86 bar and 80°C using pure CO_2. Improved SFE recoveries of the herbicide were obtained by using a solvent modifier (methanol), an ion-pairing reagent (*m*-trifluoromethylphenyl trimethylammonium hydroxide), an ionic displacement reagent (0.2 *M* calcium chloride in methanol) and by *in situ* silylation (with a 2:1 mixture of hexamethyldisilazane and trimethylchlorosilane) as well as methylation (with BF_3/methanol) of the herbicide. Recovery of 2,4-D from the sandy loam sample was as high as 90% by the ionic displacement and the *in situ* methylation procedures.

Extraction of 47 organophosphorus (OP) pesticides spiked onto an inert matrix such as sand with pure and methanol-modified CO_2 has been reported (Lopez-Avila et al., 1990). Supercritical fluid extraction conditions were optimized for the isolation of eight OP pesticides from soil by Wuchner et al. (1993). Quantitative recoveries of OPs from an aged (7 to 9 d), slurry-spiked soil (4 µg/g) were achieved by an extraction with 3 to 5 mL of CO_2 premixed with methanol (2% m/m) at 50°C and 250 bar. Alternatively, similar results were also obtained by a static addition of 35 µL of methanol to the soil, followed by the extraction with 3 to 5 mL of pure CO_2.

The feasibility of an online SFE/SFC for the determination of sulfometuron methyl (Oust), a sulfonylurea herbicide, from agricultural products, has been demonstrated (McNally and Wheeler, 1988). Supercritical fluid extraction was carried out with 2% methanol-modified CO_2 at 40°C and 223 bar and a flow rate of 6 mL/min. Later, the same authors reported the SFE/SFC of diuron and linuron from spiked soil and grain samples (Wheeler and McNally, 1989). In order to achieve extraction recoveries of 95% or better for both compounds, the samples were extracted, under static conditions, at 120°C with CO_2 of densities 0.6 or 0.7 g/mL. The addition of 200 to 300 µL of methanol (for diuron) or ethanol (for linuron) to 1 g of soil or wheat was required in order to obtain these high recoveries. Extractions of sulfometuron methyl (Oust) and chlorsulfuron (Glean) by supercritical CO_2 and trifluoromethane have also been reported (Howard et al., 1993). Quantitative recoveries of the above two herbicides were obtained from spiked Celite samples with 2% methanol-modified CO_2. The extraction was performed at 50°C and 350 bar for 7 min (2-min static and 5-min dynamic) with a flow rate of 2 mL/min. The extracts were collected on a stainless-steel bead trap and were subsequently eluted by acetonitrile. While a 30% increase in extraction efficiency for chlorsulfuron and sulfometuron methyl was experienced when trifluoromethane was used as the solvent instead of pure CO_2, quantitative recoveries of the two herbicides from spiked Celite samples were only obtained by the use of 2% methanol-modified CO_2.

Factors affecting the SFE efficiency for the herbicide fluometuron and its metabolites from spiked and field soil samples were evaluated (Locke, 1993). The best recoveries were obtained if the extractions were carried out at 50°C with CO_2 of a density of 0.80 g/mL and a flow rate of 3 mL/min in the presence of water as a static modifier for 24 min (6-min static and 18-min dynamic). Under such optimal conditions, the SFE results of the herbicide were similar to those with a conventional Soxhlet extraction using methanol. The extraction of incurred pirimicarb from topsoil was shown to be ineffective with neat supercritical CO_2, N_2O, or $CHClF_2$ (Alzaga et al., 1995). The addition of a methanol modifier to CO_2 only led to a small increase in recovery. Quantitative (Soxhlet equivalent) recovery of pirimicarb in soil at ng/g levels was obtained by SFE at 100°C and 294 bar using 10% methanol in N_2O or 5% pyridine or triethylamine in CO_2 as the extracting fluid.

A comparison of SFE and liquid–liquid extraction for the isolation of eight selected pesticides stored in freeze-dried water samples was reported by Alzaga et al. (1994). These pesticides included triazine herbicides, organophosphorus, and other pesticides. Water samples (150 L) spiked with these pesticides were freeze-dried and homogenized. Aliquots of the freeze-dried samples were then extracted with 30 mL of CO_2 at 50°C and 196 bar and by hexane and DCM. Recoveries of the pesticides by the SFE method were consistently better than those obtained by solvent extraction.

The successful SFE of s-triazine herbicides such as atrazine, simazine, cyanazine, propazine, and terbutylazine at µg/g levels from spiked sediments was reported (Janda et al., 1989). Recoveries of >90% for all triazines were obtained by adding 20 µL of methanol directly to the sample (0.5 g) prior to the extraction with 18 mL of CO_2 at 48°C and 230 bar. The extract was collected in 1 mL of methanol. The extraction of s-triazine and phenylurea herbicides from spiked sediment by supercritical CO_2 was also described (Robertson and Lester, 1994). The highest recovery for s-triazines was obtained by a 30-min dynamic extraction with 20% acetone-modified CO_2 at 486 bar and 150°C with a flow rate of 2 mL/min. For the more labile phenylureas, an extraction temperature of 60°C was used.

8.3.4.1   Recommended procedure for the extraction of organochlorine insecticides (modified from the U.S. EPA draft method 3562 for solid wastes)

1. Follow steps 1 to 3 for the SFE of PCBs in solids described earlier.

2. Extract the sample using the following conditions:

CO_2 pressure (bar):	299
CO_2 density (g/mL):	0.87
Extraction chamber temperature (°C):	50
Extraction fluid composition:	unmodified CO_2
Static extraction time (min):	20
Dynamic extraction time (min):	30
Extraction fluid flow rate (mL/min):	1.0

3. Extract collection conditions:

Sorbent trap packing material:	ODS
Trap temperature (°C):	20
Nozzle temperature (°C):	50

4. Reconstitution conditions for the collected extracts:

Rinse solvent:	*n*-hexane
Rinse volume (mL):	1.3
Rinse solvent flow rate (mL/min):	2
Trap temperature (°C):	50
Nozzle temperature (°C):	30

8.3.4.2 Discussion

1. This procedure will also coextract, quantitatively or semiquantitatively, PCBs, chlorobenzenes and other chlorinated pesticides contained in the same sample. Further cleanup of the extract may be required to eliminate possible interference in the final GC-ECD analysis.
2. The following 14 pesticides have been tested in the development of this procedure: aldrin, β-, γ-, and δ-BHC, α-chlordane, *p,p'*-DDE, DDD, and DDT, dieldrin, endosulfan II, endrin, endrin aldehyde, heptachlor, and heptachlor epoxide. The mean recoveries (%) of these pesticides in three different soils and two different spiking levels (5 and 250 ng/g) varied from 74 to 108%. The mean relative standard variation and MDL for these pesticides varied from 4.0 to 7.4% and from 0.6 to 1.3 ng/g, respectively.

8.3.5 Phenols

The extraction of pentachlorophenol (PCP) from a soil reference material SRS 103-100 using 10% methanol-modified CO_2 was reported by Lopez-Avila et al. (1990). The recovery of PCP using this multistep extraction was 46% of its certified value. Meanwhile, two other reports described the successful SFE of chlorophenols from spiked soil samples (Janda and Sandra, 1990; Richards and Campbell, 1991). In all of the above cases, the extracts were either analyzed by GC/MS as free phenols or by GC after an offline derivatization into the acetyl derivatives. Supercritical fluid extraction of free chlorophenols from spiked soils (Liu et al., 1991) and wood chips (Kapila et al., 1992) was also attempted. The highest recovery of these phenols was obtained when the extraction was carried out at 50°C and 172 bar with methanol-modified CO_2.

Although not quantitative, the direct methylation of phenol and some methylphenols in a small volume (1 mL) of coal gasification wastewater using 20% TMPA was demonstrated (Hawthorne et al., 1992b). Supercritical fluid extraction was carried out with CO_2 at 80°C and 405 bar for 30 min (15-min static and 15-min dynamic). For wood soot leachates (50 mL), the samples were first acidified and filtered through Empore C_{18} disks. The methyl and methoxy phenols adsorbed on the disks were then methylated and extracted as described above. If a mixture of TMPA and 2,2,2-trifluoroethanol was used, these authors were able to convert phenol into its trifluoroethyl ether derivative (Hawthorne et al., 1992b).

A quantitative, *in situ* SFE and acetylation procedure was developed for the determination of native PCP and other chlorophenols in soil (Lee et al., 1992). Phenols were extracted from soil and acetylated during a static extraction with CO_2 at 365 bar and 80°C in the presence of acetic anhydride and triethylamine. Quantitative recovery of the di-, tri-, tetra-, and pentachlorophenols was obtained by a 10-min extraction (5-min static and 5-min dynamic) from soil samples fortified to 0.5 and 5 µg/g. In comparison, this SFE method and the steam distillation procedure produced very similar results for 2,3,5-trichlorophenol, the tetrachlorophenols, and PCP in a reference soil sample SRS 103-100. The same technique, with minor modifications, can also be applied to a wide variety of phenols. For example, chlorinated phenolics such as guaiacols, catechols, vanillins, and syringols in sediments collected downstream of chlorine-bleaching pulp mills were extracted and acety-lated by this procedure (Lee et al., 1993a). For optimal results, the extraction was carried out with CO_2 and acetic anhydride at 365 bar and 110°C in the presence of triethylamine. Presumably due to incom-plete derivatization in the static extraction step, lower extraction tem-peratures produced much lower yields of the acetyl derivatives, par-ticularly for the guaiacols. Recoveries of these phenolics at spiking levels of 500 and 50 ng/g ranged from 84 to 100%. A method for the determination of 4-nonylphenol, an environmental estrogen that was linked to the feminization of male fish, in sewage treatment plant sludge and aquatic sediment was developed by the application of this SFE/*in situ* acetylation procedure (Lee and Peart, 1995). This alkylphe-nol is a biorefractory metabolite of nonylphenol ethoxylates, a major class of nonionic surfactant used in many countries.

8.3.5.1 Recommended procedure for the extraction and acetylation of chlorinated phenolics

1. Cut a few circles of filter paper of the same diameter as the extraction thimble. Place one circle at the bottom of the thimble and then seal it with a cap.
2. Weigh 200 mg of Celite and then 1 g (range 0.1 to 2 g) of the homogenized and dry sediment/soil sample into the thimble. If needed, spike a surrogate solution containing a suitable phenol (e.g., a bromophenol or an alkyl phenol, depending on the parameters of interest) to the sample.
3. Add 30 µL of triethylamine to the sample and then cover it with another 200 mg of Celite.
4. Add 30 µL (for chlorophenols) or 120 µL (other phenols) of acetic anhydride to the top Celite layer. Place a glass rod or other inert, porous material to fill the void volume and seal the other end of the thimble with a cap.
5. Extraction conditions:

CO_2 pressure (bar):	365
CO_2 density (g/mL):	0.71
Extraction temperature (°C):	110
Static extraction time (min):	10
Dynamic extraction time (min):	5

6. Extract collection conditions:

Sorbent trap material:	ODS
Trap temperature (°C):	15
Nozzle temperature (°C):	45

7. Elution of extract from the trap:

Rinse solvent:	DCM
Rinse volume (mL):	1.0×2
Rinse solvent flow rate (mL/min):	2.0
Trap temperature (°C):	38
Nozzle temperature (°C):	38

8. Partition the combined organic extracts with 3 mL of 1% K_2CO_3 solution to remove the coextracted acetic acid and anhydride. Exchange solvent into isooctane and adjust the final volume of the organic extract to 1 mL for GC/ECD or GC/MS analysis.

8.3.5.2 Discussion

1. This procedure is highly time efficient since it combines extraction and derivatization in the same process. Typically, the SFE requires 45 min or less compared to hours of either Soxhlet extraction or steam distillation. The extraction method only con-

sumes a few milliliters of solvent as well as microliters of the
derivatization reagents.

2. The acetyl derivatives of phenols produce abundant and char-
    acteristic ions that are readily detected by GC/MS. Those deriv-
    atives for phenols with two or more chlorine atoms per molecule
    also have high electron capture detector sensitivity.
3. The present method has been validated by the Ontario Ministry
    of the Environment and Energy and is being used as one of the
    routine methods for the determination of chlorophenols in sed-
    iments.
4. Based on a 1 g sample and a final volume of 1 mL, the MDL of
    this method with GC-ECD detection limit is 10 μg/kg.
5. The same procedure has also been successfully applied to the
    determination of methylphenols, methylchlorophenols, and ni-
    trophenols in sediments and solid wastes.

8.3.6   Resin and fatty acids

A SFE method for the determination of phospholipid fatty acids in
whole lyophilized *E. coli* cells was developed (Hawthorne et al., 1992b).
The extraction was carried out with CO_2 at 405 bar and 100°C for 30
min (15-min static and 15-min dynamic). Methylation of the fatty acids
by trimethylphenylammonium hydroxide (either 1.5 or 15% in methanol) occurred during the static extraction stage.

A SFE method for the determination of native resin and fatty acids
from river sediments collected at pulp and paper mill locations was
also developed (Lee and Peart, 1992). Without any modifier, none of
the resin acids and only a small amount of palmitic acid could be
recovered by CO_2 at 80°C and 365 bar in a 15-min extraction (5-min
static and 10-min dynamic). Static addition of 300 μL of methanol to
500 mg of sediment improved the recovery of all acids; however, acetic
acid or formic acid was an even better modifier. With a 1:1 mixture of
methanol and formic acid, the best modifier found for these acids, the
recovery was over 85% (vs. Soxhlet) for all compounds. It should be
noted that the SFE recovery for palustric acid was 267% of the Soxhlet
result. In addition, neoabietic acid, a compound that was never recovered by the Soxhlet procedure (Lee and Peart, 1991), could also be
extracted by the SFE technique. Presumably, decomposition of these
two labile resin acids in the Soxhlet procedure was the major cause for
the different results.

8.3.7   Polychlorinated dibenzo-p-dioxins and furans

The SFE of 2,3,7,8-tetrachlorodibenzo-*p*-dioxin (TCDD) from spiked
sediment was reported (Onuska and Terry, 1989b). At a spiking level

of 200 µg/kg, the recovery of 2,3,7,8-TCDD was almost 100%. The extraction was carried out at 40°C and 314 bar for 30 min with supercritical CO_2 modified with 2% methanol. Their results also indicated that nitrous oxide and 2% methanol-modified nitrous oxide also extracted 2,3,7,8-TCDD efficiently, while pure CO_2 and sulfur hexafluoride produced recoveries of approximately 50%. Alexandrou and Pawliszyn (1989) investigated the SFE of native polychlorinated dibenzo-*p*-dioxins (PCDDs) and dibenzofurans (PCDFs) in a municipal incinerator fly ash using CO_2, 10% benzene-modified CO_2 and N_2O. In contrast to the results reported in the previous study for spiked sediment, no significant amounts of dioxins were extracted by pure CO_2 or even 10% methanol-modified CO_2 at 405 bar for 2 h from the fly ash sample. In contrast, the SFE recoveries of the dioxins and furan were quantitative in 2 h (compared to a 20-h Soxhlet extraction), when 10% benzene-modified CO_2 or pure nitrous oxide was used. Another approach to achieve high recoveries for these toxins was to destroy the fly ash matrix by exposing it to HCl (1 N).

In two other publications by Onuska and Terry (1991, 1992), the highest recoveries of PCDDs from fly ash were obtained by using 2% methanol-modified N_2O at 405 bar and 45°C on a hydrochloric- or formic acid-treated sample. The SFE of PCDDs and PCDFs from naturally contaminated soil samples was also described (von Holst et al., 1992). After the addition of 200 µL of methanol to 5 g of soil, the sample was extracted at 80°C with CO_2 of densities 0.25 and 0.8 g/mL for 18 or 25 min. The recoveries of 2,3,7,8-TCDD and TCDF at ng/kg levels were similar to or better than those obtained by Soxhlet extraction.

A three-step extraction and cleanup procedure for the fractionation of fly ash SFE extracts containing PCBs, chlorobenzenes, PCDDs, and PCDFs was described (Alexandrou et al., 1992a). The municipal fly ash (1 g) was extracted as previously described (Alexandrou and Pawliszyn, 1989) with nitrous oxide for 60 min at 414 bar. The SFE extract, after concentrating to 100 µL and spiking onto Florisil, was cleaned by a 15-min CO_2 extraction at 207 bar, which removed over 75% of the chlorobenzenes and PCBs. The PCDDs and PCDFs were then back-extracted from Florisil by N_2O at 414 bar for 90 min. All extractions were carried out at 40°C. Another similar SFE and cleanup procedure optimized for the GC/MS determination of PCDDs and PCDFs in fly ash and pulp samples was also described (Alexandrou et al., 1992b).

8.3.8   Organotins

Supercritical fluid extraction of six tetraalkyltin and seven organotin compounds from spiked topsoil samples with CO_2 or CO_2 modified with 5% methanol was investigated by Liu et al. (1993). Tetraalkyltin species were readily extracted (90 to 110% recovery) with pure CO_2 at

101 bar and a temperature of 40°C. For the ionic organotins, more drastic conditions using modified CO_2 at 456 bar and 80°C for 40-min (static) and 20-min (dynamic) extractions at a flow rate of 1.5 mL/min were needed. Moreover, the addition of sodium diethyldithiocarbamate (NaDDC) as a complexing agent to improve the solubility of tin complexes in the supercritical fluid was required to improve the recovery to 70% or above. For the SFE extracts of ionic organotins, an offline Grignard reaction was performed prior to GC/AED analysis.

Later, a SFE procedure for the determination of tributyltin in sediment was described (Dachs et al., 1994). Optimum recovery of tributyltin (82%) was obtained by a 30-min extraction at 343 bar and 60°C using 20% (v/v) methanol doped with HCl in CO_2. This procedure was applied to the extraction of tributyltin in two CRMs certified for organotins, namely, PACS-1 obtained from the National Research Council of Canada and CRM-462 obtained from the Community Bureau of Reference (BCR). The uncorrected recovery for tributyltin by this procedure was 69.4% from PACS-1 and approximately 75% from CRM-462. In another publication, a SFE procedure with an *in situ* derivatization step for the determination of butyl- and phenyltin compounds in sediment was developed (Cai et al., 1994). Derivatization was carried out under CO_2 at 355 bar and 40°C during the 10-min static extraction in the presence of hexylmagnesium bromide. Extraction of the hexyl derivatives was completed with 10 mL of CO_2 at the same temperature and pressure and at a flow rate of 1 to 1.5 mL/min. The recoveries of dibutyltin and tributyltin from PACS-1 were 38 and 78%, respectively, and those from CRM-462 were 63 and 91%, respectively. Monobutyltin was not recovered by this method.

More recently, an offline complexation/SFE and GC/AED procedure for the determination of organotin species in soils and sediments was reported (Liu et al., 1994). Instead of solid NaDDC, a DCM solution of diethylammonium diethyldithiocarbamate (DEA-DDC) was added to the sample and used as a complexing agent for the ionic tin compounds. DEA-DDC was preferred since it is more soluble than NaDDC in supercritical CO_2. The organotins were extracted by 5% methanol-modified CO_2 at 456 bar at 60°C for 50 min (20-min static and 30-min dynamic) with a flow rate of 1.5 mL/min. The recoveries of mono-, di-, and tributyltin in PACS-1 obtained by this procedure were 9, 95, and 108%, respectively.

Recoveries of 85 and 79% for tributyl- and dibutyltin, respectively, from PACS-1 were also reported by using supercritical CO_2 modified with 10% methanol (Chau et al., 1995). Optimal results were obtained by mixing 0.5 g of PACS-1 with 0.1 g of NaDDC, a 70°C extraction temperature, and a 30-min extraction with CO_2 pressures at 253 bar (2 min), 355 bar (2 min), and 507 bar (26 min). Although the SFE recovery for monobutyltin from spiked sediment was only 62%, it was the highest recovery so far reported by the SFE technique.

A method for the determination of alkyltins and alkylleads in solids by Freon-22 extraction under supercritical and subcritical conditions followed by micellar electrokinetic capillary chromatography was described (Li and Li, 1995). Much higher recoveries of trimethyllead, triethyllead, and tributyltin halides from spiked soil samples were obtained with subcritical Freon-22 at 50°C than supercritical Freon-22 at 100°C, both at an extraction pressure of 245 bar. However, the addition of NaDDC to the sample as a complexing agent was required in order to improve the recoveries of the trimethyltin and trimethyllead halides from <60% to >89%.

8.4 Auxiliary techniques in supercritical fluid extraction

8.4.1 Sulfur removal

Sulfur and sulfur compounds are present in nearly all bottom sediment and soil samples. Due to the nonpolar nature of elemental sulfur, it is readily extracted by organic solvents and is coeluted with other nonpolar and less polar analytes such as PCBs and organochlorines in column cleanup procedures. Because of its response to electron capture detectors, which are routinely used for the final analysis of many environmental samples containing polychlorinated analytes, an extra sulfur removal procedure is required. The more commonly used reagents to eliminate sulfur interference are metallic mercury, activated copper powder, activated Raney nickel, and tetrabutylammonium sulfite.

In supercritical fluid extractions, the presence of a large amount of sulfur in the sample also causes problems, such as the plugging of restrictors or inline filters, both of which would eventually stop the flow of CO_2. A technique for using granular copper to remove elemental sulfur in SFE extracts of soils was described (Pyle and Setty, 1991). It was found that a 2-g copper scavenger column placed after the extraction chamber was successfully applied to the extraction of the NIST SRM 1941, which contains approximately 25% sulfur by weight. No interruption in flow was experienced during the extraction. Sulfur removal in the SFE of sludge and SRM 1941 was also achieved by placing 2 g of copper granules (10-40 mesh) at the outlet end of the extraction cell (Porter et al., 1992). No evidence of restrictor plugging by sulfur was observed during those extractions. Another approach reported for sulfur removal under SFE conditions involved mixing the sediment (2 g) with prerinsed copper powder (1.5 g, electrolytic grade) (Bøwadt and Johansson, 1994). This method works well for the extraction of PCBs in sediments with sulfur contents from 0.8 to 2.5%. In this case, copper powder was activated by rinsing it sequentially with deionized water, acetone, and hexane. The residual solvent was evaporated on a Rotovap, and the copper was stored under argon.

8.4.2   Lipid removal

A supercritical fluid cleanup technique for the separation of organochlorine insecticides from fats has been developed by France et al. (1991). The lipid matrices examined in this study were either chicken fat with incurred residues or commercial lard spiked with the insecticides at low $\mu g/g$ levels. The separation was performed at 40°C and a pressure from 190 to 270 bar using a modified supercritical fluid chromatograph equipped with a stainless steel column (7 cm × 4.6 mm i.d.) packed either with 5% deactivated neutral alumina (1.4 g) or silica gel (0.5 g). The mobile phase was either CO_2 (alumina column) or 2% methanol-modified CO_2 (silica gel column). It was noted that the amount of methanol in the CO_2 was critical when the silica gel column was used. While the addition of methanol was necessary to improve the recoveries of endrin and dieldrin, large amounts of lipids would be coeluted if 3% of methanol in CO_2 was used. The precision and accuracy of the SFE cleanup method for lindane, heptachlor, heptachlor epoxide, dieldrin, endrin, and o,p'-DDT compared favorably with those obtained by the conventional column cleanup method.

Based on a similar concept, a quantitative SFE method for the determination of PCBs in fish with high (approximately 30% by weight) lipid content was developed (Lee et al., 1995). Fish tissues mixed with granular anhydrous sodium sulfate were extracted in the presence of activated basic alumina at 100°C and 345 bar with pure CO_2. The incorporation of alumina in the extraction thimble successfully reduced the amount of lipid in the extract to 2% or less, thereby eliminating the need of the time-consuming and solvent-wasting gel permeation cleanup. If necessary, the remaining lipid could be removed easily by a drop of concentrated sulfuric acid. This procedure was also applicable to the simultaneous determination of organochlorine insecticides such as hexachlorobenzene, p,p'-DDE, and mirex in fish tissues.

8.4.3   In situ derivatization

The most widely used supercritical fluid, CO_2, is a nonpolar solvent because of its lack of a dipole moment. As a result, polar organics have relatively low solubility in CO_2, leading to the less than quantitative SFE recovery of such compounds relative to Soxhlet extractions using polar solvents. In many cases, this problem can be alleviated by the addition of a modifier to CO_2 to increase the polarity of the extracting fluid, or by switching to a more polar supercritical fluid, such as N_2O. Another popular approach is the so-called *in situ* derivatization/SFE technique. It involves the conversion of a polar, active group (usually an active hydrogen) into a less polar functional group, which is more readily extracted. Because of its lower polarity, the derivative is also more amenable to column cleanup if necessary. This technique is also very time efficient since the derivatization is carried out *in situ* during the static extraction stage. With a judicious choice of reagents, derivatives that are more sensitively and selectively detected by electron-capture and mass spectrometric detectors than their parent compounds can be formed.

Hill et al. (1991) demonstrated the feasibility of simultaneous derivatization and extraction of polar compounds with supercritical CO_2 as a rapid alternative to solvent extraction followed by derivatization. After an initial SFE with CO_2 at 304 bar and 60°C on a marine sediment, additional reactive analytes were extracted in a second SFE in the presence of a mixture of hexamethyldisilane and trimethylchlorosilane.

In order for the *in situ* derivatization to be successfully incorporated into the extraction scheme, conditions for the two procedures must be compatible with each other. For example, the temperature for extraction and derivatization should be similar, and the reagents and derivatives must be stable under SFE conditions. Thus, if a modifier is used for the extraction, it must not react with the derivatizing agent. Examples for the application of the *in situ* derivatization technique in the determination of organic contaminants in environmental samples are summarized in Table 8-2.

8.4.4 *Supercritical fluid extraction of organics in water samples via solid phase extraction*

Despite the numerous applications of SFE of organic contaminants from solid matrices, there were few examples reported for water samples. Using a setup similar to the conventional purge-and-trap system for semivolatile analytes, direct aqueous extraction of phenols, phosphonates, caffeine, and triprolidine with CO_2 has been demonstrated (Hedrick and Taylor, 1990, 1992). Since their approach was limited to small sample sizes (3 mL), it was not applicable to trace analysis in environmental samples. The major difficulties of direct SFE of water samples as cited by these authors were the mechanical mobility of the matrix as well as the relatively high solubility of water in CO_2. More recently, another example for the direct SFE of organochlorines in aqueous samples was given by Barnabas et al. (1994a). In addition to less than quantitative recoveries and a long extraction time, this method was again limited by a small sample size (45 mL) and thus could not be applied to trace analysis.

Table 8-2 Examples of *In Situ* Derivatization of Contaminants in Soils and Aquatic Sediments Under Supercritical Fluid Extraction Conditions

Analytes	Reagents	Derivatives	Recoveries (%)	References
Acids, resin and fatty	PFBBr/TEA	PFB esters	35–45 (native resin acids)	Lee and Peart, 1992a
Herbicides, 2,4-D, dicamba	TMPA	Methyl esters	116–137 (native herbicides)	Hawthorne et al., 1992b
Herbicides, 2,4-D	HMDS/TMCS	Silyl ester	31 (spiked sandy loam)	Rochette et al., 1993
Herbicides, 2,4-D	BF$_3$/methanol	Methyl ester	90 (spiked sandy loam)	Rochette et al., 1993
Herbicides, phenoxy acid	TMPA	Methyl esters	13–96 (spiked sand)	Lopez-Avila et al., 1993
Herbicides, phenoxy acid	TBA/methyl iodide	Methyl esters	69–141 (spiked clay)	Lopez-Avila et al., 1993
Herbicides, phenoxy acid	PFBBr/TEA	PFB esters	24–83 (spiked topsoil)	Lopez-Avila et al., 1993
Phenols, chloro-	Acetic anhydride/TEA	Acetates	99 (native PCP in soil)	Lee et al., 1992b
Phenols, catechols/guaiacols	Acetic anhydride/TEA	Acetates	87–98 (spiked sediments)	Lee et al., 1993a
Phenol, alkyl-	Acetic anhydride/TEA	Acetates	96–98 (spiked sediment)	Lee and Peart, 1995
Organotin, tri-/dibutyltin	Hexylmagnesium bromide	Hexyl derivatives	78 (native TBT in PACS-1)	Cai et al., 1994

Note: PFBBr = pentafluorobenzyl bromide, TEA = triethylamine, TMPA = trimethylphenylammonium hydroxide, HMDS = hexamethyldisilazane, TMCS = trimethylchlorosilane, TBA = tetrabutylammonium hydroxide, PCP = pentachlorophenol, TBT = tributyltin chloride.

Supercritical fluid extraction of liquid samples is more conveniently carried out in a two-step process: enrichment of the organics with a solid phase adsorbent followed by the SFE of the adsorbent. The indirect SFE of sulfometuron methyl and chlorsulfuron in water (0.1 or 1 L) was reported (Howard and Taylor, 1992). Water samples were first filtered through Empore C_{18} disks, and the adsorbed urea herbicides were extracted from the disks with 2% methanol-modified CO_2 at 350 bar and 50°C. The total extraction time was 26 min (2-min static and 24-min dynamic at a flow rate of 2 mL/min). Recoveries of the analytes at spiking levels of 50 and 500 µg/L were over 90% using a stainless steel trap. The SFE of phenols from an acidified wood soot leachate (50 mL) adsorbed on an Empore disk has been described earlier (Hawthorne et al., 1992a,b).

The application of combined solid phase extraction using disks or cartridges made of a bonded C_{18} phase and SFE to extract semivolatile and nonvolatile organic compounds from water was demonstrated (Tang et al., 1993). The samples consisted of a selected group of PAHs, PCBs, organochlorine insecticides, and phthalate esters spiked into particulate-free, reagent-grade water. The adsorbed organics on the cartridge or disk were then extracted by 30 mL of CO_2 at 304 bar and 60°C. The SFE recoveries were similar to those obtained by cartridge extraction followed by solvent elution. However, they also observed that, for large sample sizes, the disks were more efficient than the cartridges in terms of total analysis time and recovery of the organics. In a related study, Ho and Tang (1992) optimized the SFE conditions for environmental pollutants such as PAHs and organochlorinated insecticides from spiked C_{18} solid phase extraction cartridges. Most of the above pollutants were quantitatively recovered by CO_2 at ≥40°C and 355 or 405 bar for an extraction time of 20 to 35 min. For those analytes that were not extracted efficiently with pure CO_2, the addition of 300 to 500 µL of methanol to the cartridge greatly improved the recoveries.

Examples for the extraction of organochlorines and herbicides (Barnabas et al., 1994a,b) as well as PAHs (Messer and Taylor, 1995) in water using solid phase extraction disks prior to SFE were also documented.

References

Alexandrou, N., Lawrence, M.J., and Pawliszyn, J., Cleanup of complex organic mixtures using supercritical fluids and selective adsorbents, *Anal. Chem.*, 64, 301, 1992a.

Alexandrou, N., Miao, Z., Colquhoun, M., Pawliszyn, J., and Jennison, C., Supercritical fluid extraction and cleanup with capillary GC-ion trap mass spectrometry for the determination of polychlorinated dibenzo-*p*-dioxins and dibenzofurans in environmental samples, *J. Chromatogr. Sci.*, 30, 351, 1992b.

Alexandrou, N. and Pawliszyn, J., Supercritical fluid extraction for the rapid determination of polychlorinated dibenzo-*p*-dioxins and dibenzofurans in municipal incinerator fly ash, *Anal. Chem.*, 61, 2770, 1989.

Alzaga, R., Bayona, J.M., and Barceló, D., Use of supercritical fluid extraction for pirimicarb determination in soil, *J. Agric. Food Chem.*, 43, 395, 1995.

Alzaga, R., Durand, G., Barceló, D., and Bayona, J.M., Comparison of supercritical fluid extraction and liquid–liquid extraction for isolation of selected pesticides stored in freeze-dried water samples, *Chromatographia*, 38, 502, 1994.

Barnabas, I.J., Dean, J.R., Hitchen, S.M., and Owen, S.P., Supercritical fluid extraction of organochlorine pesticides from an aqueous matrix, *J. Chromatogr.*, 665, 307, 1994a.

Barnabas, I.J., Dean, J.R., Hitchen, S.M., and Owen, S.P., Selective supercritical fluid extraction of organochlorine pesticides and herbicides from aqueous samples, *J. Chromatogr. Sci.*, 32, 547, 1994b.

Bøwadt, S. and Johansson, B., Analysis of PCBs in sulfur-containing sediments by off-line supercritical fluid extraction and HRGC-ECD, *Anal. Chem.*, 66, 667, 1994.

Bøwadt, S., Johansson, B., Pelusio, F., Larsen, B.R., and Rovida, C., Solid-phase trapping of polychlorinated biphenyls in supercritical fluid extraction, *J. Chromatogr.*, 662, 424, 1994.

Bøwadt, S., Johansson, B., Wunderli, S., Zennegg, M., de Alencastro, L.F., and Grandjean, D., Independent comparison of Soxhlet and supercritical fluid extraction for the determination of PCBs in an industrial soil, *Anal. Chem.*, 67, 2424, 1995.

Brooks, M.W. and Uden, P.C., Extraction and analysis of diesel fuel by supercritical fluid extraction and microbore supercritical fluid chromatography, *J. Chromatogr.*, 637, 175, 1993.

Burford, M.D, Hawthorne, S.B., and Miller, D.J., Extraction rates of spiked vs. native PAHs from heterogeneous environmental samples using supercritical fluid extraction and sonication in methylene chloride, *Anal. Chem.*, 65, 1497, 1993.

Cai, Y., Alzaga, R., and Bayona, J.M., *In situ* derivatization and supercritical fluid extraction for the simultaneous determination of butytin and phenyltin compounds in sediment, *Anal. Chem.*, 66, 1161, 1994.

Camel, V., Tambuté, A., and Caude, M., Analytical-scale supercritical fluid extraction: a promising technique for the determination of pollutants in environmental matrices, *J. Chromatogr.*, 642, 263, 1993.

Chau, Y.K., Yang, F., and Brown, M., Supercritical fluid extraction of butyltin compounds from sediment, *Anal. Chim. Acta*, 304, 85, 1995.

Chesler, T.L., Pinkston, J.D., and Raynie, D.E., Supercritical fluid chromatography and extraction, *Anal. Chem.*, 66, 106R, 1994.

Dachs, J., Alzaga, R., Bayona, J.M., and Quevauviller, P., Development of a supercritical fluid extraction procedure for tributyltin determination in sediments, *Anal. Chim. Acta*, 286, 319, 1994.

Dankers, J., Groeneboom, M., Scholtis, L.H.A., and van der Heiden, C., High-speed supercritical fluid extraction method for routine measurement of polycyclic aromatic hydrocarbons in environmental soils with dichloromethane as a static modifier, *J. Chromatogr.*, 641, 357, 1993.

David, F., Kot, A., Vanluchene, E., Sippola, E., and Sandra, P., Optimizing selectivity in SFE and chromatography for the analysis of pollutants in different matrices, Proc. of 2nd Eur. Symp. on Analytical SFC and SFE, Riva del Garda, Italy, 10-14, 1993.

David, F., Verschuere, M., and Sandra, P., Off-line supercritical fluid extraction — capillary GC applications in environmental analysis, *Fresenius J. Anal. Chem.*, 344, 479, 1992.

Diehl, H., *Quantitative Analysis Elementary Principles and Practice*, 2nd ed., Oakland Street Press, Ames, IA, 1974, 71.

Emery, A.P., Chesler, S.N., and MacCrehan, W.A., Recovery of diesel fuel from clays by supercritical fluid extraction-gas chromatography, *J. Chromatogr.*, 606, 221, 1992.

France, J.E., King, J.W., and Snyder, J.M., Supercritical fluid-based cleanup technique for the separation of organochlorine pesticides from fats, *J. Agric. Food Chem.*, 39, 1871, 1991.

Gere, D.R., Board, R.L., and McManigill, D., Supercritical fluid chromatography with small particle diameter packed columns, *Anal. Chem.*, 54, 736, 1982.

Gere, D.R. and Derrico, E.M., SFE theory to practice — first principles and method development, Part I, *LC-GC*, 12, 352, 1994a.

Gere, D.R. and Derrico, E.M., SFE theory to practice — first principles and method development, Part II, *LC-GC*, 12, 432, 1994b.

Gere, D.R., Randall, L.G., and Callahan, D., Supercritical fluid extraction: principles and applications, Chapter 11, *Instrumental Methods in Food Analysis*, Pare, J.R.J. and Belanger, J.M.R., Eds., Elsevier Science, Amsterdam, in press.

Greibrokk, T., Berg, B.E., Blilie, A.L., Doehl, J., Farbrot, A., and Lundanes, E., Techniques and applications in supercritical fluid chromatography, *J. Chromatogr.*, 394, 429, 1987.

Greibrokk, T., Berg, B.E., Hansen, E.M., and Gjorven, S., On-line enzymatic reaction, extraction, and chromatography of fatty acids and triglycerides with supercritical carbon dioxide, *High Resolution Chromatogr. Chromatogr. Commun.*, 16, 358, 1993.

Hawthorne, S.B., Analytical-scale supercritical fluid extraction, *Anal. Chem.*, 62, 633A, 1990.

Hawthorne, S.B. and Miller, D.J., Extraction and recovery of organic pollutants from environmental solids and Tenax-GC using supercritical CO_2, *J. Chromatogr. Sci.*, 24, 258, 1986.

Hawthorne, S.B. and Miller, D.J., Directly coupled supercritical fluid extraction-gas chromatographic analysis of polycyclic aromatic hydrocarbons and polychlorinated biphenyls from environmental solids, *J. Chromatogr.*, 403, 63, 1987a.

Hawthorne, S.B. and Miller, D.J., Extraction and recovery of polycyclic aromatic hydrocarbons from environmental solids using supercritical fluids, *Anal. Chem.*, 59, 1705, 1987b.

Hawthorne, S.B., Miller, D.J., and Krieger, M.S., Coupled SFE-GC: a rapid and simple technique for extracting, identifying, and quantitating organic analytes from solids and sorbent resins, *J. Chromatogr. Sci.*, 27, 347, 1989.

Hawthorne, S.B., Langenfeld, J.L., Miller, D.J., and Burford, M.K., Comparison of supercritical $CHClF_2$, N_2O, and CO_2 for the extraction of polychlorinated biphenyls and polycyclic aromatic hydrocarbons, *Anal. Chem.*, 64, 1614, 1992a.

Hawthorne, S.B., Miller, D.J., Nivens, D.E., and White, D.C., Supercritical fluid extraction of polar analytes using *in situ* chemical derivatization, *Anal. Chem.*, 64, 405, 1992b.

Hawthorne, S.B., Miller, D.J., Burford, M.D., Langenfeld, J.J., Eckert-Tilotta, S., and Louie, P.K., Factors controlling quantitative supercritical fluid extraction of environmental samples, *J. Chromatogr.*, 642, 301, 1993a.

Hawthorne, S.B., Miller, D.J., and Hegvik, K.M., Field evaluation of the SFE-infrared method for total petroleum hydrocarbon (TPH) determinations, *J. Chromatogr. Sci.*, 31, 26, 1993b.

Hawthorne, S.B., Yang, Y., and Miller, D.J., Extraction of organic pollutants from environmental solids with sub- and supercritical water, *Anal. Chem.*, 66, 2912, 1994.

Hedrick, J.L. and Taylor, L.T., Supercritical fluid extraction strategies of aqueous based matrices, *High Resolution Chromatogr. Chromatogr. Commun.*, 13, 312, 1990.

Hedrick, J.L. and Taylor, L.T., Direct aqueous extraction of nitrogenous bases from aqueous solution, *High Resolution Chromatogr. Chromatogr. Commun.*, 15, 151, 1992.

Hileman, B., Concerns broaden over chlorine and chlorinated hydrocarbons, *Chem. Eng. News*, 71(16), 11, 1993.

Hill, J.W., Hill, H.H., Jr., and Maeda, T., Simultaneous supercritical fluid derivatization and extraction, *Anal. Chem.*, 63, 2152, 1991.

Ho, J.S. and Tang, P.H., Optimization of supercritical fluid extraction of environmental pollutants from a liquid-solid extraction cartridge, *J. Chromatogr. Sci.*, 30, 344, 1992.

Hopper, M.L. and King, J.W., Enhanced supercritical fluid carbon dioxide extraction of pesticides from foods using pelletized diatomaceous earth, *J. Assoc. Off. Anal. Chem.*, 74, 661, 1991.

Howard, A.L. and Taylor, L.T., Quantitative supercritical fluid extraction of sulfonyl urea herbicides from aqueous matrices via solid phase extraction disks, *J. Chromatogr. Sci.*, 30, 374, 1992.

Howard, A.L., Yoo, W.J., Taylor, L.T., Schweighardt, F.K., Emery, A.P., Chesler, S.N., and MacCrehan, W.A., Supercritical fluid extraction of environmental analytes using trifluoromethane, *J. Chromatogr. Sci.*, 31, 401, 1993.

Janda, V. and Sandra, P., The use of supercritical-fluid extraction in analysis, *Chem. Listy*, 84, 366, 1990.

Janda, V., Bartle, K.D., and Clifford, A.A., Supercritical fluid extraction in environmental analysis, *J. Chromatogr.*, 642, 283, 1993.

Janda, V., Steenbeke, G., and Sandra, P., Supercritical fluid extraction of s-triazine herbicides from sediment, *J. Chromatogr.*, 479, 200, 1989.

Kane, M., Dean, J.R., Hitchen, S.M., Dowle, C.J., and Tranter, R.L., Experimental design approach for supercritical fluid extraction, *Anal. Chim. Acta*, 271, 83, 1993.

Kapila, S., Nam, K.S., Liu, M.H., Puri, R.K., and Yanders, A.F., Promises and pitfalls of supercritical fluid extraction in polychlorinated compound analyses, *Chemosphere*, 25, 11, 1992.

King, J.W., Fundamentals and applications of supercritical fluid extraction in chromatographic science, *J. Chromatogr. Sci.*, 27, 355, 1989.

King, J.W., Hopper, M.L., Luchtefeld, R.G., Taylor, S.L., and Orton, W.L., Optimization of experimental conditions for the supercritical carbon dioxide extraction of pesticide residues from grains, *J. Assoc. Off. Anal. Chem. Int.*, 76, 857, 1993.

Knipe, C.R., Gere, D.R., and McNally, M.E. Supercritical fluid extraction, developing a turnkey method, ACS Symposium Series No. 488, *Supercritical Fluid Technology*, Bright, F. and McNally, M.E., Eds., American Chemical Society, Washington, D.C., 1992.

Langenfeld, J.J., Hawthorne, S.B., Miller, D.J., and Pawliszyn, J., Effects of temperature and pressure on supercritical fluid extraction efficiencies of polycyclic aromatic hydrocarbons and polychlorinated biphenyls, *Anal. Chem.*, 65, 338, 1993.

Langenfeld, J.J., Hawthorne, S.B., Miller, D.J., and Pawliszyn, J., Role of modifiers for analytical-scale supercritical fluid extraction of environmental solids, *Anal. Chem.*, 66, 909, 1994.

Lee, H.B. and Chau, A.S.Y., Analytical reference materials, Part VI, Development and certification of a sediment reference material for selected polynuclear aromatic hydrocarbons, *Analyst*, 112, 31, 1987a.

Lee, H.B. and Chau, A.S.Y., Analytical reference materials, Part VII, Development and certification of a sediment reference material for total polychlorinated biphenyls, *Analyst*, 112, 37, 1987b.

Lee, H.B. and Peart, T.E., Determination of resin and fatty acids in sediments near pulp mill locations, *J. Chromatogr.*, 547, 315, 1991.

Lee, H.B. and Peart, T.E., Supercritical carbon dioxide extraction of resin and fatty acids from sediments at pulp mill sites, *J. Chromatogr.*, 594, 309, 1992.

Lee, H.B. and Peart, T.E., Optimization of supercritical carbon dioxide extraction for polychlorinated biphenyls and chlorinated benzenes from sediments, *J. Chromatogr.*, 663, 87, 1994.

Lee, H.B. and Peart, T.E., Determination of 4-nonylphenol in effluent and sludge from sewage treatment plants, *Anal. Chem.*, 67, 1976, 1995.

Lee, H.B., Peart, T.E., and Hong-You, R.L., *In situ* extraction and derivatization of pentachlorophenol and related compounds from soils using a supercritical fluid extraction system, *J. Chromatogr.*, 605, 109, 1992.

Lee, H.B., Peart, T.E., and Hong-You, R.L., Determination of phenolics from sediments of pulp mill origin by *in situ* supercritical carbon dioxide extraction and derivatization, *J. Chromatogr.*, 636, 263, 1993a.

Lee, H.B., Peart, T.E., Hong-You, R.L., and Gere, D.R., Supercritical carbon dioxide extraction of polycyclic aromatic hydrocarbons from sediments, *J. Chromatogr.*, 653, 83, 1993b.

Lee, H.B., Peart, T.E., Niimi, A.J., and Knipe, C.R., Rapid supercritical carbon dioxide extraction method for determination of polychlorinated biphenyls in fish, *J. Assoc. Off. Anal. Chem. Int.*, 78, 437, 1995.

Levy, J.M., Cavalier, R.A., Bosch, T.N., Rynaski, A.F., and Huhak, W.E., Multidimensional supercritical fluid chromatography and supercritical fluid extraction, *J. Chromatogr. Sci.*, 27, 341, 1989.

Levy, J.M., Dolata, L.A., and Ravey, R.M., Considerations of SFE for GC/MS determination of polynuclear aromatic hydrocarbons in soils and sediments, *J. Chromatogr. Sci.*, 31, 349, 1993.

Li, K. and Li, S.F.Y., Determination of alkyllead and alkyltin compounds in solid samples by supercritical and subcritical fluid extraction and MEKC, *J. Chromatogr. Sci.*, 33, 309, 1995.

Liu, M.H., Kapila, S., Yanders, A.F., Clevenger, T.E., and Elseewi, A.A., Role of entrainers in supercritical fluid extraction of chlorinated aromatics from soils, *Chemosphere*, 23, 1085, 1991.

Liu, Y., Lopez-Avila, V., Alcaraz, M., and Beckert, W.F., Determination of organotin compounds in environmental samples by supercritical fluid extraction and gas chromatography with atomic emission detection, *High Resolution Chromatogr. Chromatogr. Commun.*, 16, 106, 1993.

Liu, Y., Lopez-Avila, V., Alcaraz, M., and Beckert, W.F., Off-line complexation/supercritical fluid extraction and gas chromatography with atomic emission detection for the determination and speciation of organotin compounds in soils and sediments, *Anal. Chem.*, 66, 3788, 1994.

Locke, M.A., Supercritical CO_2 fluid extraction of fluometuron herbicide from soil, *J. Agric. Food Chem.*, 41, 1081, 1993.

Lopez-Avila, V., Benedicto, J., Dodhiwala, N.S., Young, R., and Beckert, W., Development of an off-line SFE-IR method for petroleum hydrocarbons in soils, *J. Chromtaogr. Sci.*, 30, 335, 1992.

Lopez-Avila, V., Dodhiwala, N.S., and Beckert, W.F., Supercritical fluid extraction and its application to environmental analysis, *J. Chromatogr. Sci.*, 28, 468, 1990.

Lopez-Avila, V., Dodhiwala, N.S., and Beckert, W.F., Development in the supercritical fluid extraction of chlorophenoxy acid herbicides from soil samples, *J. Agric. Food Chem.*, 41, 2038, 1993.

McNally, M.E. and Wheeler, J.R., Supercritical fluid extraction coupled with supercritical fluid chromatography for the separation of sulfonylurea herbicides and their metabolites from complex matrices, *J. Chromatogr.*, 435, 63, 1988.

Messer, D.C. and Taylor, L.T., Method development for the quantitation of trace polyaromatic hydrocarbons from water via solid-phase extraction with supercritical fluid elution, *J. Chromatogr. Sci.*, 33, 290, 1995.

Monin, J.C., Barth, D., Perrut, M., Espitalie, M., and Durand, B., Extraction of hydrocarbons from sedimentary rocks by supercritical carbon dioxide, *Org. Geochem.*, 13, 1079, 1991.

Onuska, F.I. and Terry, K.A., Supercritical fluid extraction of PCBs in tandem with high resolution gas chromatography in environmental analysis, *High Resolution Chromatogr. Chromatogr. Commun.*, 12, 527, 1989a.

Onuska, F.I. and Terry, K.A., Supercritical fluid extraction of 2,3,7,8-tetrachlorodibenzo-*p*-dioxin from sediment samples, *High Resolution Chromatogr. Chromatogr. Commun.*, 12, 357, 1989b.

Onuska, F.I. and Terry, K.A., Supercritical fluid extraction of polychlorinated dibenzo-*p*-dioxins from municipal incinerator fly ash, *High Resolution Chromatogr. Chromatogr. Commun.*, 14, 829, 1991.

Onuska, F.I. and Terry, K.A., Supercritical fluid extraction of polychlorinated dibenzo-*p*-dioxins from municipal incinerator fly ash, *Chemosphere*, 25, 17, 1992.

Paschke, T., Hawthorne, S.B., Miller, D.J., and Wenclawiak, B., Supercritical fluid extraction of nitrated polycyclic aromatic hydrocarbons and polycyclic aromatic hydrocarbons from diesel exhaust particulate matter, *J. Chromatogr.*, 609, 333, 1992.

Pipkin, W., Practical considerations for supercritical fluid extraction, *Am. Lab.*, Nov., 40D, 1990.

Porter, N.L., Rynaski, A.F., Campbell, E.R., Saunders, M., Richter, B.E., Swanson, J.T., Nielsen, R.B., and Murphy, B.J., Studies of linear restrictors and analyte collection via solvent trapping after supercritical fluid extraction, *J. Chromatogr. Sci.*, 30, 367, 1992.

Pyle, S.M. and Setty, M.M., Supercritical fluid extraction of high sulfur soils, with use of a copper scavenger, *Talanta*, 38, 1125, 1991.

Raynie, D.E., Warning concerning the use of nitrous oxide in supercritical fluid extraction, *Anal. Chem.*, 65, 3127, 1993.

Reindl, S. and Höfler, F., Optimization of the parameters in supercritical fluid extraction of polynuclear aromatic hydrocarbons from soil samples, *Anal. Chem.*, 66, 1806, 1994.

Richards, M. and Campbell, R.M., Comparison of supercritical fluid extraction, Soxhlet, and sonication methods for the determination of priority pollutants in soil, *LC-GC Int.*, 4, 33, 1991.

Robertson, A.M. and Lester, J.N., Supercritical fluid extraction of *s*-triazines and phenylurea herbicides from sediment, *Environ. Sci. Technol.*, 28, 346, 1994.

Rochette, E.A., Harsh, J.B., and Hill, H.H., Jr., Supercritical fluid extraction of 2,4-D from soils using derivatization and ionic modifiers, *Talanta*, 40, 147, 1993.

Schantz, M.M. and Chesler, S.N., Supercritical fluid extraction procedure for the removal of trace organic species from solid samples, *J. Chromatogr.*, 363, 397, 1986.

Sievers, R.E. and Hansen, B., Supercritical fluid nitrous oxide explosion, *Chem. Eng. News*, 69(29), 2, 1991.

Snyder, J.L., Grob, R.L., McNally, M.E., and Oostdyk, T.S., Comparison of supercritical fluid extraction with classical sonication and Soxhlet extractions for selected pesticides, *Anal. Chem.*, 64, 1940, 1992.

Snyder, J.L., Grob, R.L., McNally, M.E., and Oostdyk, T.S., The effect of instrumental parameters and soil matrix on the recovery of organochlorine and organophosphate pesticides from soils using supercritical fluid extraction, *J. Chromatogr. Sci.*, 31, 183, 1993.

Tang, P.H., So, J.S., and Eichelberger, J.W., Determination of organic pollutants in reagent water by liquid-solid extraction followed by supercritical fluid elution, *J. Assoc. Off. Anal. Chem. Int.*, 76, 72, 1993.

van der Velde, E.G., de Haan, W., and Liem, A.K.D., Supercritical fluid extraction of polychlorinated biphenyls and pesticides from soil, *J. Chromatogr.*, 626, 135, 1992.

von Holst, C., Schlesing, H., and Liese, C., Supercritical fluid extraction of polychlorinated dibenzodioxins and dibenzofurans from soil samples, *Chemosphere*, 25, 1367, 1992.

Wheeler, J.R. and McNally, M.E., Supercritical fluid extraction and chromatography of representative agricultural products with capillary and microbore columns, *J. Chromatogr. Sci.*, 27, 534, 1989.

Wright, B.W., Frye, S.R., McMinn, D.G., and Smith, R.D., On-line supercritical fluid extraction-capillary gas chromatography, *Anal. Chem.*, 59, 640, 1987.

Wuchner, K., Ghijsen, R.T., Th. Brinkman, U.A., Grob, R., and Mathieu, J., Extraction of organophosphorus pesticides from soil by off-line supercritical fluid extraction, *Analyst*, 118, 11, 1993.

Wylie, P.L., Knipe, C.R., and Gere, D.R., Coupling off-line SFE to GC for the analysis of petroleum hydrocarbons in soil, *Am. Environ. Lab.*, 18, 3/94, 1994.

Zurer, P.S., Looming ban on production of CFCs, halons spurs switch to substitutes, *Chem. Eng. News*, 71(46), 12, 1993.

Index

A

Absolute method, of instrumental
     neutron activation analysis,
     154–156
Acid digestion
  acid properties, 92–94
  advantages, 92
  fusion decomposition and,
     comparison, 100
  $HClO_4$, HNO_3, 92–94
  hydrofluoric acid, 93–94
  of mercury, 105
  nitric acid, 92–93
  sample preparation, 95
  testing parameters
    closed systems, 96–100
    open systems, 94–95
American Society for Testing and
     Materials method, for bulk
     density measurements, 12
Aminization, 180
Ammonia determinations, using Kjeldahl
     method
  apparatus, 203–205
  parameters, 203
  procedure, 207
  reagents, 205
Ammonification, 180
Analytical method, 85
Analytical technique, 85
Aqua regia, *see also* Acid digestion
  description, 92
  of mercury, 105
  open system performance of, 94
  oxidizing strength of, 93
Argon, 113–114
ASTM method, *see* American Society for
     Testing and Materials method

Atomic absorption spectrometry
  advantages, 102
  background correction, 103
  cold-vapor, 104–105
  development, 102
  flame
    applications, 103–104
    historical uses, 87
    inductively coupled plasma
     emission spectrometer and,
     comparison, 87
    solvent extraction, 104
  graphite furnace
    analyte separation techniques,
     109–110
    applications, 107
    automation, 108–109
    description, 107–108
    development, 107
    historical uses, 87
    limitations, 108–109
    Pd-Mg modifier, 109–111
    simultaneous multielement,
     110–111
  hybride generation, 106–107
  illustration, 103
  instrumentation, 102–103
  interferences, 102–103
  principles, 102–103
  quartz tube
    illustration, 105
    interference, 106
    mercury determinations, 104–107
  slurry nebulization
    advantages, 111
    procedure, 111
    variability in, 111–112
Atterberg Limits, 11
Autoanalyzer techniques, 194–195

B

BET equation, 35–36
Boiling point, 232
Bomb digestion, 214
Bottom sediments, *see also* Sediments
  analysis of
    purpose, 1–2
    use of soil science techniques, 2
  definition, 1
  deposition patterns, 1
  nutrients, *see* Nutrients
  sources, 1–2
  trace elements, *see* Trace elements
Brinkman Particle Size Analyzer, for
      particle size analysis, 29–30
Bulk density
  definition, 11
  measurements
    description, 10
    gravimetric methods
      ASTM method, 12
      clod method, 13
      core method, 12
      excavation method, 13–14
      Syringe technique, 12–13
    purpose, 10
    radiation method, 14–15

C

Carbon, in sediments
  chemical forms, 179–180
  detection techniques, 177
  determination methods
    considerations, 189–190
    dry combustion, *see* Dry combustion
    sample preparation, 189
    selection of, 181–182
  distribution, 175–177
  organic forms
    description, 179–180
    determination of, 182, 196
  particulate matter, 178
  resuspension, 176
Carbon dioxide, use for supercritical fluid
      extraction, 235–236, 241,
      247–248
Carbon/hydrogen/nitrogen analyzer, 177
Cation exchange capacity, correlation
      with surface area, 35
Charge coupled devices, 114–115
Charge injection devices, 115–116
Chemical fractionation, 186, 192

apparatus, 217
flow chart, 216
principle, 215
procedure, 217–218
reagents, 217
Chlorobenzenes, supercritical fluid
      extraction of, 244–245
Chlorsulfuron, supercritical fluid
      extraction of, 255
Citrate-dithionite-bicarbonate method
  description, 185–186
  uses of, 192
Clay
  methods to determine surface size
    ethylene glycol monoethyl ether,
      36
    methylene blue, 37
    particle size classification, 23
Clod method, for bulk density
      measurements, 13
Closed decomposition systems
  advantages and disadvantages, 96
  definition, 96
  energy sources for, 96–97
  microwave radiation
    advantages and disadvantages,
      97–98
    applications, 97–98
    automated flow-through systems,
      98–99
    description, 96
    systems for, 96–97
    technological advances, 98–99
    testing equipment sensitivities, 96
Colorimetric techniques, 184, 191
Coulter Counter, for particle size analysis,
      26–27
Critical fluid, methods of defining,
      232–233
Critical point, 232
Critical temperature, 232

D

Data quality control, in testing
      laboratories, 219–220
Decomposition techniques
  acid digestion, *see* Acid digestion
  fusion, *see* Fusion decomposition
Dicamba, supercritical fluid extraction of,
      254
2-4,Dichlorophenoxyacetic acid,
      supercritical fluid extraction of,
      254

Digestive procedures
  acid
    acid properties, 92–94
    advantages, 92
    fusion decomposition and,
      comparison, 100
    HClO$_4$.HNO$_3$, 92–94
    hydrofluoric acid, 93–94
    of mercury, 105
    nitric acid, 92–93
    sample preparation, 95
    testing parameters
      closed systems, 96–100
      open systems, 94–95
  bomb, 214
  dry combustion
    for carbon determinations
      inorganic forms, 196
      organic forms, 187–188
      total forms, 188
    considerations, 188–189
    description, 181–182
    with digestion by HCl
      analysis, 212
      apparatus, 208, 210
      principle, 208
      procedure, 211–212
      reagents, 210–211
    instrumentation
      apparatus, 197
      commercial systems, 182
      reagents, 197
      sources, 195–196
    principle, 195–197
    sample preparation, 196
    species types, 188
  hot block, 214
  H$_2$SO$_4$.K$_2$S$_2$O$_8$
    analysis, 214
    apparatus, 213
    principle, 213
    procedures, 214
    reagents, 213–214
  wet
    advantages and disadvantages,
      182
    description, 182
Direct injection nebulizer, 117
Direct sample insertion
  inductively coupled plasma atomic
    emission spectrometry, 118
  inductively coupled plasma mass
    spectrometry, 129
Dissolved organic carbon, 178

Dry combustion technique
  for carbon determinations
    inorganic forms, 196
    organic forms, 187–188
    total forms, 188
  considerations, 188–189
  description, 181–182
  with digestion by HCl
    analysis, 212
    apparatus, 208, 210
    principle, 208
    procedure, 211–212
    reagents, 210–211
  instrumentation
    apparatus, 197
    commercial systems, 182
    reagents, 197
    sources, 195–196
  principle, 195–197
  sample preparation, 196
  species types, 188
Dry sieve analysis, of particle size, 23–25
Drying techniques, 23
Dumas combustion/oxidation method,
  182

E

EDS spectrometers, *see* Energy dispersive
  spectrometers
EGME, *see* Ethylene glycol monoethyl
  ether
Eh measurements
  biases, 49
  description, 48–51
  difficulties associated with, 48–49
  electrode use, 49–50, 52–55
  equilibrium potential, 49–50
  general procedure for
    equipment, 59–61
    field measurements, 51–55, 61–62
    solutions, 59–60
  mixed potentials, 49–50
  reference electrodes, *see* Reference
    electrodes
  sediment core, 54
Elastic condition, 234
Elastic scattering, of neutrons, 148
Electrical sensing zone method, for
  particle size analysis, 26
Electrodes, *see* Reference electrodes
Electrothermal vaporization
  inductively coupled plasma atomic
    emission spectrometry, 117–118

inductively coupled plasma mass
        spectrometry, 128–129
Elements in sediment, *see* Trace elements
        in sediment
Energy dispersive spectrometers, 74
Epithermal neutron activation analysis,
        162
Epithermal neutrons, 148
Ethylene glycol monoethyl ether, for
        determining surface area of
        sediments, 36
ETV, *see* Electrothermal vaporization
Eutrophication
        definition, 2
        nutrient accumulation and, 175
Excavation method, for bulk density
        measurements, 13–14

F

FANES, *see* Furnace atomic nonthermal
        excitation spectrometry
Fatty acids, supercritical fluid extraction
        of, 260
Fine-grained sediments
        effect on contaminant mobilization,
                20–21
        reactivity of, 20
Fixed restrictors, 236
Flame atomic absorption spectrometry
        applications, 103–104
        comparison with inductively coupled
                plasma emission spectrometer,
                87
        historical uses, 87
        solvent extraction, 104
Flow injection techniques
        inductively coupled plasma atomic
                emission spectrometry,
                123–124
        inductively coupled plasma mass
                spectrometry, 128
Fluometuron, supercritical fluid
        extraction of, 256
Fluorescence spectrometry, *see* X-ray
        fluorescence spectrometry
Flux reagents, for fusion decomposition,
        100
Fractionation, *see* Chemical fractionation
Furnace atomic nonthermal excitation
        spectrometry, 110–111
Fusion decomposition
        acid decomposition and, comparison,
                100

automation, 101
description, 100
disadvantages, 100
flux reagents, 100
lead fire assay, 101
Nis fusion, 101
oxidative reagents, 100
reductive fusion with litharge, 100
trace elements commonly used, 100

G

Gamma ray attenuation technique,
                sediment properties testing
                using
        bulk density measurements, 14–15
        water content measurements, 18
Gas, comparison with liquid, 232
Geochemical prospecting, 2
Geological sample analysis, 76–77
Germanium detectors, high-purity
        counting efficiency, 160
        development, 156
        energy resolution, 158
        illustration of, 159
        irradiation conditions, 161–162
        photopeak efficiency of, 158, 160
        summing effect, 160
Graphite furnace atomic absorption
                spectrometry
        analyte separation techniques,
                109–110
        applications, 107
        automation of, 108–109
        description, 107–108
        development, 107
        historical uses, 87
        limitations, 108–109
        Pd-Mg modifier, 109–111
        simultaneous multielement,
                110–111
Gravel, particle size classification of,
        22
Gravimetry
        for bulk density measurements
                ASTM method, 12
                clod method, 13
                core method, 12
                excavation method, 13–14
                Syringe technique, 12–13
        instrumental methods and,
                comparison, 19–20
        with oven-drying method, for water
                content measurements, 15–16

H

HClO$_4$.HNO$_3$ procedure
  description, 92
  oxidizing power, 92–93
  trace element dissolution using, 92–93
Herbicides
  in situ derivatization, 266
  supercritical fluid extraction of,
      254–256
HF, *see* Hydrofluoric acid
HF-HCLO$_4$.HNO$_3$.HCl procedure, *see*
      HClO$_4$.HNO$_3$ procedure
High-purity germanium detectors
  counting efficiency, 160
  development, 156
  energy resolution, 158
  illustration of, 159
  irradiation conditions, 161–162
  photopeak efficiency of, 158, 160
  summing effect, 160
Hot block digestion, 214
HPGe detectors, *see* High-purity
      germanium detectors
H$_2$SO$_4$.K$_2$S$_2$O$_8$ digestion
  analysis, 214
  apparatus, 213
  principle, 213
  procedures, 214
  reagents, 213–214
Hybride generation
  atomic absorption spectrometry,
      106–107
  illustration of, 119
  inductively coupled plasma atomic
      emission spectrometry
    advantages, 118
    applications, 121
    illustration of, 119
  inductively coupled plasma mass
      spectrometry, 131
  nebulization and, comparison, 118
Hydrocarbons, 241
Hydrofluoric acid
  open system, 94–95
  oxidizing strength of, 93
  parameters for use of, 94

I

Ignition method, *see* Dry combustion
      technique, with digestion by
      HCl
Image analysis for particle size, 31–33

In situ derivatization, 264–266
Inductively coupled plasma atomic
      emission spectrometry
  advantages, 112
  applications
    detection limits, 118–119
    determination of trace elements in
        sediments, 120–122
  argon use, 113–114
  detection
    preconcentration techniques
      ion exchange, 122–124
      solvent extraction, 122–123
    ultrasensitive, 122–123
  flame atomic absorption spectrometry
      and, comparison, 87
  flow injection, 123–124
  function, 113
  historical uses, 87
  hybride generator systems
    advantages, 118
    applications, 121
    illustration of, 119
    interference, 122
  instrumentation, 112–115
  interferences, 116
  nebulization
    alternative approaches, 117–118
    description, 117
    pneumatic, 135
  optimization in, 115–116
  principles of, 112–115
  recent improvements, 114–115
  sample introduction, 113–114
  schematic of, 113
  simultaneous and sequential
      measurements, 114
  treatises, 112
  uses of, 87
Inductively coupled plasma mass
      spectrometry
  advantages and disadvantages,
      124–125
  applications
    determination of trace elements in
        aquatic sediments, 131–132,
        135–136
    geochemical explorations, 132–133
    Pb contamination, 134
    rare earth element determinations,
        133–134
  calibration, 127
  description, 87
  detection limits, 127

development, 124
hybride generator systems, 131
instrumentation, 125–127
interferences, 125–127
nebulization alternatives
　direct sample insertion devices, 129
　electrothermal vaporization,
　　128–129
　flow injection, 128
　laser sampling, 129–130
　slurry, 130–131
principles of, 125–127
sample homogeneity, 130
schematic of, 125
Inelastic scattering, of neutrons, 148
Insecticides, supercritical fluid extraction of
description, 253
recommended procedure, 256–257
Instrumental methods
gravimetric methods and, comparison,
　19–20
for particle size distribution analysis
　electrical sensing zone, 26
　image analysis, 31–33
　laser diffraction spectroscopy, 30–31
　laser-time of transition theory, 29–30
　X-ray sedimentation, 26, 28
Instrumental neutron activation analysis
accuracy, 165, 167
analytical sensitivity
　definition, 163
　effect of neutron flux, 163
　parameters to optimize, 163–164
applications
　other types, 168
　sediment samples, 167–168
detection limits, 156, 164–166
high-purity germanium detectors
　counting efficiency, 160
　development, 156
　energy resolution, 158
　illustration of, 159
　irradiation conditions, 161–162
　photopeak efficiency of, 158, 160
　summing effect, 160
introduction, 147
methods
　absolute, 154–156
　overview, 154
　relative, 154
neutron sources, 149–150
radionuclides, 156–157
reference sources, 147–148
scheme, 150–153

sediment sample preparation, 162–163
theory, 147–153
uses, 147
Iron hydroxides, extraction methods for,
　187
Iron–phosphorus interaction, in
　sediments, 180–181

J

Joule-Thomson cooling effect, 237

K

KCl (*2N*) extraction
apparatus, 203–205
principle, 203
procedure
　automated analysis, 207
　extraction, 207
　manual analysis, 207
reagents
　ammonia, 205
　nitrates and nitrites, 205–206
　urea analysis, 206
Kjeldahl method
advantages and disadvantages, 190
apparatus, 199
applications, 190–191
description, 183
pretreatment, 190
principle, 198–199
procedure
　automated analysis, 200, 202
　description, 190
　manual analysis, 202
　sediment digestion, 202
reagents
　for ammonia analysis, 202
　for sediment digestion, 199

L

Laser diffraction spectroscopy, 30–31
Laser-time of transition theory, for
　　particle size analysis, 29–30
Lead fire assay, 101
Lipids, supercritical fluid extraction of,
　264–265
Liquid, comparison with gas, 232
Liquid extraction, comparison with
　　supercritical fluid extraction,
　　240
Lithium metaborate, 100–101

M

Mass spectrometer, *see* Inductively
     coupled plasma mass
     spectrometry
MB method, *see* Methylene blue method
Mercury detection in sediments
  acid digestion, 105
  cold-vapor atomic absorption
     spectrometry, 104–105
  hybrides, 106
  inductively coupled plasma mass
     spectrometry, 134
Mercury porosimetry, 18
Methylene blue method, 37
Microwave radiation, for closed
     decomposition systems
  advantages and disadvantages, 9
     7–98
  applications, 97–98
  automated flow-through systems,
     98–99
  description, 96
  systems for, 96–97
  technological advances, 98–99
Mineralization, 180
Mixed potentials, 49
Molybdenum blue method
  description, 187–188
  procedure, 193
Mud, particle size classification of, 22
Multimolecular adsorption theory, 35–36

N

Nebulization
  conventional types, alternatives to
    direct sample insertion, 118, 129
    electrothermal vaporization,
     117–118, 128–129
    gaseous introduction, 118
    ultrasonic, 117–118
  inductively coupled plasma atomic
     emission spectrometry
    alternative methods, 117–118
    description, 117
  inductively coupled plasma mass
     spectrometry, 130–131
  slurry, atomic absorption spectrometry
    advantages, 111
    procedure, 111
    variability in, 111–112
Neutron
  classifications, 148

description, 148
epithermal, 148
rapid, 148
reaction types, 148–149
thermal, 148
Neutron activation analysis
  epithermal, 162
  instrumental
    accuracy, 165, 167
    analytical sensitivity
     definition, 163
     effect of neutron flux, 163
     parameters to optimize, 163–164
    applications
     other types, 168
     sediment samples, 167–168
    detection limits, 156, 164–166
    high-purity germanium detectors
     counting efficiency, 160
     development, 156
     energy resolution, 158
     illustration of, 159
     irradiation conditions, 161–162
     photopeak efficiency of, 158, 160
     summing effect, 160
    introduction, 147
    methods
     absolute, 154–156
     overview, 154
     relative, 154
    neutron sources, 149–150
    radionuclides, 156–157
    reference sources, 147–148
    scheme of, 150–153
    sediment sample preparation,
     162–163
    theory of, 147–153
    uses of, 147
Neutron capture, 149
Neutron flux, 163–164
NiS fusion, 101
Nitrate determinations, using Kjeldahl
     method
  apparatus, 203–205
  parameters, 203
  procedure, 207
  reagents, 205–206
Nitric acid
  properties of, 92–93
  trace element dissolution using, 92–93
Nitrification, 180
Nitrite determinations, using Kjeldahl
     method
  apparatus, 203–205

parameters, 203
procedure, 207
reagents, 205–206
Nitrogen, in sediments
  ammonia, *see* Ammonia
  chemical forms, 180
  determination methods, 182–184
    colorimetric techniques, 184, 191
    considerations, 190–191
    dry combustion, *see* Dry combustion
    Dumas technique, 182
    inorganic species, 183–184
    Kjeldahl method, *see* Kjeldahl
      method
    total species, 182–183
  distribution, 2, 175
  nitrates, *see* Nitrates
  nitrites, *see* Nitrites
  organic forms
    description, 179
    types, 184
  particulate organic matter, 178
  sources, 2
  species types, 177
  transported forms, 178
  urea, *see* Urea
Nitrous oxide, use for supercritical fluid
    extraction, 241
Nuclear fission, 149
Nuclear reaction, 148–149
Nutrients, in sediments
  chemical forms
    carbon, *see* Carbon, in sediments
    nitrogen, *see* Nitrogen, in sediments
    phosphorus, *see* Phosphorus, in
      sediments
    suspended particulate matter,
      178–179
  detection techniques, 177
  determination methods
    chemical fractionation, *see* Chemical
      fractionation
    considerations, 188–189
    description, 193–195
    dry combustion, *see* Dry combustion
      techniques
    $H_2SO_4.K_2S_2O_8$ digestion, *see* $H_2SO_4.$
      $K_2S_2O_8$ digestion
    KCl (2N) extraction, *see* KCl (2N)
      extraction
    Kjeldahl method, *see* Kjeldahl
      method
    sample preparation techniques,
      188–189

distribution, 175–177
types of, 175

O

Organophosphorus pesticides (OP),
    255
Organotins, supercritical fluid extraction
    of, 261–263
Orthophosphate, *see also* Phosphorus, in
    sediments
  definition, 181
  determination techniques
    automated flow manifold, 209
    dry combustion, with digestion by
      HCl, 208
    molybdenum blue method, 193
    overview, 187
  extraction from sediments, 212
Oust, *see* Sulfometuron methyl
Oxidation techniques, *see* Wet oxidation
    techniques
Oxidation-reduction potential, *see* Redox
    potential
Oxidative reagents, for fusion
    decomposition, 100

P

Pallman-Wiegner effect, 58
Particle size distribution
  analysis
    accuracy of, 21
    classical methods
      pipette, 25–26
      sieve analysis, 23–25
    description, 21
    instrumental methods
      electrical sensing zone, 26
      image analysis, 31–33
      laser diffraction spectroscopy,
        30–31
      laser-time of transition theory,
        29–30
      X-ray sedimentation, 26, 28
    removal of organic material, 23
    sample preparation, 21, 23
  classification of, 22
  size differences
    clay-sized, 20
    description, 20–21
    purpose of, 20–21
    silt-sized, 20
Particulate organic carbon, 178

PCDD, *see* Polychlorinated dibenzo-*p*-dioxins
PCDF, *see* Polychlorinated dibenzofurans
PCP, *see* Pentachlorophenol
Pd-Mg modifiers, 109–111
Pentachlorophenol, supercritical fluid extraction of
    description, 257–258
    discussion, 259–260
    recommended procedure, 259
Pesticides, organophosphorus, *see* Organophosphorus pesticides
pH measurements
    description, 55–56
    difficulties associated with, 55–56
    electrode use, 47–48, 54–56
    general procedure for
        equipment, 59–61
        field measurements, 55–56, 61–62
        solutions, 59–60
    reference electrodes, *see* Reference electrodes
    suspension effect, 56–59
Phenols, *see* Pentachlorophenol
Phosphoric acid, 181
Phosphorus, in sediments
    aluminum-bound, 186
    calcium-bound, 186
    carbonate-bound, 186
    chemical forms, 180–181
    complexing agents, 186–187
    determination methods
        chemical fractionation, *see* Chemical fractionation
        comparison of, 191–192
        considerations, 191–193
        digestion techniques, 184–185
        dry combustion, with digestion by HCl, *see* Dry combustion technique, with digestion by HCl
        $H_2SO_4.K_2S_2O_8$ digestion, *see* $H_2SO_4.K_2S_2O_8$ digestion
        molybdenum blue method, 193
        organic complexing agents, 186–187
        partial extraction, 185–186
        total species, 184–185
    distribution, 2, 175
    environmental types, 177
    exchangeable, 186
    hydrolyzable, 186
    inorganic forms, 179, 180–181
    iron-bound, 186
    mineralization of, 180–181

organic forms, 180–181
orthophosphate, *see* Orthophosphate
particulate matter, 178
sources, 2
Photomultiplier tubes, 114
Phytoplankton, 179
Pipette analysis method, of particle size, 25–26
p-Nitrophenol, 38
pNP, *see* p-Nitrophenol
Polychlorinated biphenyls, supercritical fluid extraction of
    description, 242
    discussion, 246
    effect of temperature and pressure, 243
    recommended extraction procedure, 245–246
    solid-phase traps, 244
    use of modifiers, 243
Polychlorinated dibenzofurans, 261
Polychlorinated dibenzo-*p*-dioxins, 260–261
Polynuclear aromatic hydrocarbons, supercritical fluid extraction of
    carbon dioxide use, 247–248
    discussion, 250
    organic modifiers, 248
    overview, 246–247
    recommended procedure, 249–250
Porosity
    definition, 18
    measurements of, 18–19
Potassium chloride (2*N*) extraction, *see* KCl (2*N*) extraction

Q

Quartz tube atomic absorption spectrometry
    illustration of, 105
    interference, 106
    mercury determinations, 104–107

R

Radiation
    for closed decomposition systems, *see* Microwave radiation, for closed decomposition systems
    sediment properties testing using
        bulk density measurements, 14–15
        water content measurements, 18
Rapid neurons, 148

Rare earth element determinations, 133–134

Redox potential, Eh measurements to determine
biases, 49
description, 48–51
difficulties associated with, 48–49
electrode use, 49–50, 52–55
equilibrium potential, 49–50
general procedure for
equipment, 59–61
field measurements, 51–55, 61–62
solutions, 59–60
mixed potentials, 49–50
reference electrodes, *see* Reference electrodes
sediment core, 54

Reduction-oxidation potential, *see* Redox potential

Reductive fusion with litharge, 101

Reference electrodes
bias sources, 64
calibration of, 64
construction of, 62–63
function, 62
illustration of, 63
junctions, 63
storage and care of, 63–64

Relative method, of instrumental neutron activation analysis
accuracy, 165
description, 154

Resin, supercritical fluid extraction of, 260

Restrictor
description, 236
fixed, 236

S

Sample preparation
for analysis of physical properties of sediments, 9–10
methods of drying, 2
nutrient analysis, 188–189
particle size distribution analysis, 21, 23
sample collection, 4
water content analysis, 15–16
X-ray fluorescence spectrometry, 75–77

Sand, particle size classification of, 22

Sedigraph, 26, 28

Sediment sampling program
elements of, 4
preparation of, 4

Sedimentation rates, 10

Sediments, *see also specific sediment*
atomic absorption spectrometry analysis of, *see* Atomic absorption spectrometry
bioassessment of, 3–4
contamination of, 3
decomposition techniques
acid digestion, *see* Acid digestion
fusion, *see* Fusion decomposition
definition, 7
Eh measurements, *see* Eh measurements
inductively coupled plasma atomic emission spectrometry analysis of, *see* Inductively coupled plasma atomic emission spectrometry
neutron activation analysis, *see* Instrumental neutron activation analysis
nutrients, *see* Nutrients
particle size, *see* Particle size
pH measurements, *see* pH measurements
physical properties
bulk density, *see* Bulk density
collection methods, 4, 9–10
description, 7–8
porosity, *see* Porosity
sample preservation, 9–10
surface area, *see* Surface area
techniques for, 2–3
water content, *see* Water content
resuspension of, 176
standard reference materials, 89–91
trace elements, *see* Trace elements
X-ray fluorescence spectrometry analysis, *see* X-ray fluorescence spectrometry

Sensitivity, of instrumental neutron activation analysis
definition, 163
effect of neutron flux, 163
parameters to optimize, 163–164

Sieve analysis, of particle size
advantages, 34
description, 23–25

Simplex optimization method, 115

Slurry nebulization
atomic absorption spectrometry
advantages, 111
procedure, 111
variability in, 111–112

inductively coupled plasma mass
    spectrometry, 130–131
Solid phase extraction of water samples,
    267
Spectrometers, *see specific spectrometer*
Standard reference materials, 89–91
Sulfometuron methyl, supercritical fluid
    extraction of, 255
Sulfur removal, 263–264
Summing effect, 160
Supercritical fluid extraction
    advantages, 230, 240
    applications of, in environmental
        analysis
        chlorobenzenes, 242–246
        fatty acids, 260
        herbicides, 254–256
        insecticides, 253–255
        organotins, 261–263
        overview, 242
        pentachlorophenol
            description, 257–258
            discussion, 259–260
            recommended procedure, 259
        pesticides, 255
        polychlorinated biphenyls
            description, 242
            discussion, 246
            effect of temperature and
                pressure, 243
            recommended extraction
                procedure, 245–246
            solid-phase traps, 244
            use of modifiers, 243
        polychlorinated dibenzofurans,
            261
        polychlorinated dibenzo-*p*-dioxins,
            260–261
        polynuclear aromatic hydrocarbons
            carbon dioxide use, 247–248
            discussion, 250
            organic modifiers, 248
            overview, 246–247
            recommended procedure,
                249–250
        resin, 260
        total recoverable petroleum
            hydrocarbons, 251–253
    auxiliary techniques
        lipid removal, 264
        in situ derivatization, 264–265
        sulfur removal, 263–264
        water samples, 265–267
    candidate fluids for, 240–242

carbon dioxide sources, 235–236
cost reductions using, 230
description, 229
flowing fluid process, 235–240
hardware, 234–240
liquid extraction and, comparison,
    240
misconceptions regarding, 229–230
sample size, 230–231
zones, 239
Supercritical fluids
    advantages, 240
    comparison with gas and liquid, 234
    description, 231–232
    methods of defining, 232–233
    solvent power, 240
Surface area, of sediments
    cation exchange capacity and, 35
    definition, 35
    description, 34–35
    differences in, 34–35
    methods for determining
        BET method, 35, 37
        description, 35–36
        ethylene glycol monoethyl ether, 36
        methylene blue, 37
        p-Nitrophenol, 38
Suspended particles
    carbon species in, 176–177
    description, 176
    nutrients in, 178
Suspension effect
    bias sources, 58
    description, 56–57
    methods to prevent, 59
Syringe technique, for bulk density
    measurements, 12–13

T

TCDD, *see* 2,3,7,8-Tetrachlorodibenzo-*p*-
    dioxin
2,3,7,8-Tetrachlorodibenzo-*p*-dioxin,
    supercritical fluid extraction of,
    260–261
Thermal neutrons, 148
Time domain reflectometry
    description, 16
    instrumentation, 16–17
    soil applications, 17–18
Total dissolved salt content, 87
Total Kjeldahl nitrogen method, *see*
    Kjeldahl method
Total organic nitrogen, 180

Total recoverable petroleum
    hydrocarbons, supercritical
    fluid extraction of, 251–253
Total-reflection X-ray fluorescence
    spectrometry, 167
TPH, *see* Total recoverable petroleum
    hydrocarbons
Trace elements in sediments
  atomic absorption spectrometry
      detection and analysis
    advantages, 102
    background correction, 103
    cold-vapor, 104–105
    development, 102
    flame
      applications of, 103–104
      historical uses, 87
      inductively coupled plasma
          emission spectrometer and,
          comparison, 87
      solvent extraction, 104
    graphite furnace
      analyte separation techniques,
          109–110
      applications, 107
      automation of, 108–109
      description, 107–108
      development, 107
      historical uses, 87
      limitations, 108–109
      Pd-Mg modifier, 109–111
      simultaneous multielement,
          110–111
    hybride generation, 106–107
    illustration of, 103
    instrumentation of, 102–103
    interferences, 102–103
    principles of, 102–103
    quartz tube
      illustration of, 105
      mercury determinations, 104–107
    slurry nebulization, 111–112
  decomposition techniques
    acid digestion
      acid properties, 92–94
      advantages, 92
      fusion decomposition and,
          comparison, 100
      HClO₄.HNO₃, 92–94
      hydrofluoric acid, 93–94
      of mercury, 105
      nitric acid, 92–93
      sample preparation, 95
      testing parameters

    closed systems, 96–100
    open systems, 94–95
  criteria for selecting, 91
  overview, 91
  hybride evolution, 106–107
  inductively coupled plasma atomic
      emission spectrometry
    advantages, 112
    applications
      detection limits, 118–119
      determination of trace elements in
          sediments, 120–122
    argon use, 113–114
    detection
      preconcentration techniques, 122
      ion exchange, 122–124
      solvent extraction, 122–123
      ultrasensitive, 122–123
    flame atomic absorption
        spectrometry and, comparison,
        87
    flow injection, 123–124
    function, 113
    historical uses, 87
    hybride generator systems
      advantages, 118
      applications, 121
      illustration of, 119
      interference, 122
    instrumentation, 112–115
    interferences, 116
    nebulization
      alternative approaches, 117–118
      conventional types, 117
      description, 117
      pneumatic, 135
    optimization in, 115–116
    parameters to optimize, 115–116
    principles of, 112–115
    recent improvements, 114–115
    sample introduction, 113–114
    schematic of, 113
    simultaneous and sequential
        measurements, 114
    treatises, 112
    uses, 87
  inductively coupled plasma mass
      spectrometry
    advantages and disadvantages,
        124–125
    applications
      determination of trace elements in
          aquatic sediments, 131–132,
          135–136

geochemical explorations,
132–133
Pb contamination, 134
rare earth element
determinations, 133–134
calibration, 127
description, 87
detection limits, 127
development, 124
hybride generator systems, 131
instrumentation, 125–127
interferences, 125–127
nebulization alternatives
direct sample insertion devices,
129
electrothermal vaporization,
128–129
flow injection, 128
laser sampling, 129–130
slurry, 130–131
principles of, 125–127
schematic of, 125
standard reference materials, 89–91
Tributyltin, supercritical fluid extraction
of, 262
Triple point, 232
TXRF, *see* Total-reflection X-ray
fluorescence spectrometry

purpose of, 11, 15
sample preparation, 15–16
time domain reflectometry,
16–18
Wavelength dispersive systems, 73–74
WDS, *see* Wavelength dispersive systems
Wet digestion procedures
advantages and disadvantages, 182
description, 182
Wet oxidation techniques
advantages and disadvantages, 189
description, 181–182
interferences, 189
Kjeldahl
advantages and disadvantages,
190
apparatus, 199
description, 183
pretreatment, 190
principle, 198–199
procedure
automated analysis, 200, 202
description, 190
manual analysis, 202
sediment digestion, 202
reagents
for ammonia analysis, 202
for sediment digestion, 199

U

Ultrasonic nebulization, 117–118
Urea determinations, using Kjeldahl
method
apparatus, 203–205
parameters, 203
procedure, 207
reagents, 206
USN, see Ultrasonic nebulization

W

Water, use for supercritical fluid
extraction, 241
Water content
definition, 15
measurement methods
gravimetry with oven-drying
method, 15–16
indirect types, 16

X

X-ray fluorescence spectrometry
advantages, 69, 80–81
calibration of, 77–80
dispersion and detection of X-rays,
73–74
environmental sample analysis using
calibration of, 77–80
detection and counting, 73–74
sample preparation, 75–77
limitations, 80–81
theory of
absorption, 71–73
description, 70
excitation, 70
lines and notation, 71
scatter, 73
uses for, 69
X-ray sedimentation method, for particle
size analysis, 26, 28

.

Milton Keynes UK
Ingram Content Group UK Ltd.
UKHW031128141024
449569UK00006B/352